Y0-CBY-096
s-On TCP/IP

The McGraw-Hill Series on Computer Communications (Selected Titles)

ISBN	AUTHOR	TITLE
0-07-005147-X	Bates	*Voice and Data Communications Handbook*
0-07-005560-2	Black	*TCP/IP and Related Protocols*, 2/e
0-07-005590-4	Black	*Frame Relay Networks: Specifications and Implementation*, 2/e
0-07-011769-1	Charles	*LAN Blueprints: Engineering It Right*
0-07-011486-2	Chiong	*SNA Interconnections: Bridging and Routing SNA in Heirarchical, Peer, and High-Speed Networks*
0-07-016769-9	Dhawan	*Mobile Computing: A Systems Integrator's Handbook*
0-07-018546-8	Dziong	*ATM Network Resource Management*
0-07-020359-8	Feit	*SNMP: A Guide to Network Management*
0-07-021389-5	Feit	*TCP/IP: Architecture, Protocols and Implementation with IPv6 and IP Security*, 2/e
0-07-024563-0	Goralski	*SONET: A Guide to Synchronous Optical Networks*
0-07-024043-4	Goralski	*Introduction to ATM Networking*
0-07-031382-2	Huntington-Lee/Terplan/Gibson	*HP's OpenView: A Manager's Guide*
0-07-034249-0	Kessler	*ISDN: Concepts, Facilities, and Services*, 3/e
0-07-035968-7	Kumar	*Broadband Communications*
0-07-041051-8	Matusow	*SNA, APPN, HPR and TCP/IP Integration*
0-07-060362-6	McDysan/Spohn	*ATM: Theory and Applications*
0-07-044435-8	Mukherjee	*Optical Communication Networks*
0-07-044362-9	Muller	*Network Planning Procurement and Management*
0-07-046380-8	Nemzow	*The Ethernet Management Guide*, 3/e
0-07-051506-9	Ranade/Sackett	*Introduction to SNA Networking*, 2/e
0-07-054991-5	Russell	*Signaling System #7*
0-07-057724-2	Sackett/Metz	*ATM and Multiprotocol Networking*
0-07-057199-6	Saunders	*The McGraw-Hill High Speed LANs Handbook*
0-07-057639-4	Simonds	*Network Security: Data and Voice Communications*
0-07-060363-4	Spohn	*Data Network Design*, 2/e
0-07-069416-8	Summers	*ISDN Implementor's Guide*
0-07-063263-4	Taylor	*The McGraw-Hill Internetworking Handbook*
0-07-063301-0	Taylor	*McGraw-Hill Internetworking Command Reference*
0-07-063639-7	Terplan	*Effective Management of Local Area Networks*, 2/e

Hands-On TCP/IP

Paul Simoneau

McGraw-Hill

New York San Francisco Washington, D.C. Auckland Bogotá
Caracas Lisbon London Madrid Mexico City Milan
Montreal New Delhi San Juan Singapore
Sydney Tokyo Toronto

Library of Congress Cataloging-in-Publication Data

Simoneau, Paul.
Hands-on TCP/IP / Paul Simoneau.
p. cm.—(Computer communications series)
Includes index.
ISBN 0-07-912640-5 (pbk.)
1. TCP/IP (Computer network protocol) I. Title. II. Series.
TK5105.585.S587 1997
004.6′2—dc21 97-20199
CIP

McGraw-Hill

A Division of The ***McGraw-Hill*** *Companies*

3 4 5 6 7 8 9 0 DOC/DOC 9 0 2 1 0 9 8

P/N 057774-9
PART OF
ISBN 0-07-912640-5

The sponsoring editor for this book was Steven Elliot, the editing supervisor was Stephen M. Smith, and the production supervisor was Suzanne W. B. Rapcavage. It was set in Century Schoolbook by Priscilla Beer of McGraw-Hill's Professional Book Group composition unit.

Printed and bound by R. R. Donnelley & Sons Company.

This book is printed on recycled, acid-free paper containing a minimum of 50% recycled de-inked fiber.

To the wind beneath my wings
and
my friend Brady Hammond

Contents

Preface xv
Acknowledgments xix

Chapter 1. Stacked Protocols and Standards 1

Stacked Protocols 1
Xerox Network System 1
Systems Network Architecture 2
Open Systems Interconnect 3
TCP/IP Standards Control 9
Requests for Comments 11
Assigned Numbers RFC 13
For Your Information RFC 15

Chapter 2. TCP/IP Stack Overview 19

Layer Responsibilities 19
Process Layer 20
Host-to-Host Layer 20
Internet Layer 22
Network Interface Layer 23
Some Assembly Required 24

Chapter 3. Network Interface Layer 27

Ethernet II Addresses 27
Ethernet Header Fields 29
802.3 Is Not Ethernet 29
Packet Information 30
IEEE 802.3 32
Token Ring or IEEE 802.5 32
Serial Line Interface Protocol 34
Point-to-Point Protocol 35
Appendix A Ethernet Vendor Address Components 35
Appendix B Ether Types (Protocol Types) by Code 46
Appendix C Ether Types by Name 51

Chapter 4. IP Networks, Subnets, and Hosts 57

Overview 57
Class A 58
Class B 59
Class C 59
Class D 60
IP Communications Logic Process 60
Subnet 61
When to Subnet 62
Subnetting IP Networks 63
Subnet Calculations 63
Building the Mask 65
Applying the Mask 67
Interfaces Lost 68
IP Communications Logic Process (Step Two) 68
IP Communications Logic Process (Step Three) 70

Chapter 5. Address Matching 73

Overview 73
Domain Name System 74
Domain Name System Tree 74
Organizational Domain Grouping 76
Country Domains 77
Name Server 77
Resource Records 78
Down the Stack 79
IP Communications Logic Process (Conclusion) 79
Address Resolution Protocol Cache 80
Layout Format and Decoding Tips 81
Address Resolution Protocol 82
Hardware Type 83
Protocol Type 84
Hardware Length 84
Protocol Length 85
Operation 85
Source Hardware Address 85
Source Protocol Address 86
Target Hardware Address 87
Target Protocol Address 87
LAN Fill 87
ARP Implementation 88
Appendix A ISO Country Codes 91
Appendix B Protocol and Service Names for DNS WKS Records 97
Appendix C Address Resolution Protocol Parameters 102

Chapter 6. Internet Protocol 105

Self-Healing Networks 105
IP Header 105
IP Header Decode 106
Version 106
Header Length 107

Type-of-Service Byte 107
Precedence of the Data 107
Total IP Length Field 109
Datagram ID Number 109
Fragment Area 110
Reserved Bit 111
Don't Fragment Bit 111
Fragment Status 111
Fragment Offset 111
Fragment Reassembly 112
Fragmenting Fragments 112
Time to Live 113
A Time-to-Live Example 114
Protocol Field 115
IP Header Checksum 115
IP Addresses 115
Source IP Address 115
Target IP Address 120
IP Option Fields 120
Copy-Through-Gate Bit 121
IP Option Values 121
Route-Based Options 122
Internet Timestamp Collection Option 124
IP Sample Data Exchanges 125

Chapter 7. IP's Next Generation 127

Overview 127
The Contenders or Alphabet Soup 128
TUBA 129
CATNIP 130
SIPP 130
Proposal Reviews 131
A Revised Proposal 132
IPv6 132
Header Format 133
Extension Headers 137
IPv6 Addressing 141
Address Formats 142
Address Hierarchy 142
Provider-Based Unicast Address 144
IPv4 to IPv6 Transition 146

Chapter 8. User Datagram Protocol 149

Speed versus Reliability 149
Header Fields 150
Port Basics 151
Ports and Sockets 152
Applications 153
Sample Exchanges 153
Appendix A Well-Known Port Numbers from RFC 1700 155
Appendix B Registered Port Numbers from RFC 1700 164

Chapter 9. IP Routing 173

Bridging or Routing 173
Routing or Bridging 174
Routers and Gateways 174
Direct versus Indirect Routing 174
Table-Driven Routing 176
Local 176
Extended 177
Default 178
Automatic versus Manual Routing 178
Exterior or Interior Protocol 178
Routing Information Protocol 180
Open Shortest Path First 185
OSPF versus RIP 185
Variable Subnet Masking 186
An Example of OSPF Subnetting 187
Border Gateway Protocol 188
Routing Table Size and CIDR 189
Path Attributes 190
Routing Information Base 190
Messages 191

Chapter 10. Internet Control Message Protocol 193

Overview 193
Message Destinations 193
Messages 194
Echo Request (Type 8) and Response (Type 0) 195
Destination Unreachable Example 197
Source Quench Message 198
Redirecting Traffic with an ICMP Type 5 199
Time-Exceeded Message 201
Parameter Problem Message 202
Timestamp Request and Response Message 202
Subnet Mask Request or Response Message 204
ICMP Samples 204

Chapter 11. Transmission Control Protocol 207

Reliable Transport Services 207
TCP Header 207
Source Sequence Number 208
Acknowledgment Sequence Number 209
TCP Header Length 209
Session Bit Flags 210
Sender Window Size 210
TCP Checksum 211
Urgent Pointer 211
Option Fields (Type, Length, and Option) 211
The Three-Step Handshake 214
Congestion and TCP 215
Ending a TCP Virtual Connection Normally 216
Reset Session 216

TCP Sample Session 217

Chapter 12. Telnet 219

Overview 219
Client, Server, and Network Virtual Terminal 219
Telnet Option Negotiations 220
Telnet Sample Session 222
Telnet 3270 232

Chapter 13. Network Security 235

Security Threats 235
Availability Attacks 235
Confidentiality Threats 236
Integrity Risks 236
Authenticity Attacks 236
Other Types of Computer Crime 237
Security Procedures 237
Access Control 238
Passwords 239
Authentication 239
Data Integrity 240
Routing Control 241
Traffic Padding 242
Security Servers 242
Computer Security Organizations 243
FIRST 243
CERT Coordination Center 244
COAST 244
Network Scanning 245

Chapter 14. File Transfer Protocol 247

Multiple Sessions 247
FTP Commands 248
FTP Response Codes 248
FTP Sample Session 249
Appendix FTP Commands from RFC 959 260

Chapter 15. Simple Mail Transfer Protocol 267

Overview 267
E-mail Names 267
Mail Server 268
SMTP Commands 268
SMTP Response Codes 269
SMTP Service Start 271
Mail Origin 271
Mail Recipient 272
Carrying the Mail 272
SMTP Message End 273
Ending the SMTP Session 274
Ending the TCP Session 274

Multipurpose Internet Mail Extensions 274
MIME Content-Type Headers 275
Encoding 277

Chapter 16. Centrally Managed TCP/IP Addressing 279

Overview 279
RARP Overview 279
RARP Request 280
RARP Reply 281
No RARP Server 281
BootP Overview 282
BootP Request 282
BootP Response 283
DHCP Overview 284
DHCP Address Acquisition 285
Client Address Release 285
Lease Renewal/Rebinding 286
DHCP Message Format 287
DHCP Options 287
Virtual IP Networking Issues 288

Chapter 17. Trivial File Transfer Protocol 293

Overview 293
Client and Server Ports 293
Operation Codes 294
Read Request/Write Request Layout 294
Opcode 3—Data 295
Opcode 4—Ack 295
Error Handling 296
TFTP Challenges 296
A Sample Read Session 297
A Sample Write Session 298
A Sample Session Error Message 299

Chapter 18. Simple Network Management Protocol 301

Basics 301
A Model 301
The Agent 302
The Structure of Management Information 303
Abstract Syntax Notation One 304
Management Information Base 305
Network Management Applications 307
Private MIBs 307
SNMP Version 1 and 2 Protocol Data Units 308
The Desktop Management Task Force 308
Appendix SMI Network Management Private Enterprise Numbers 309

Chapter 19. Internet Services 355

Internet Access Methods 355
PPP and SLIP Access 356
Archie 356
Gopher 357
World Wide Web 357
Web Browsers 358
URL 359
The Hypertext Transfer Protocol 360
Java 361
Common Gateway Interface 362

Suggested Reading List 363
About the CD 365
Index 367

Preface

This book is arranged in a logical progression from the bottom to the top of the TCP/IP stack of protocols, as displayed below. We cover each of these protocols and build on the knowledge thus accumulated. We assume that you will read the chapters in order. Should you decide to jump around from one chapter to another, you may find that you missed an explanation in a chapter you did not read.

In the following paragraphs, we identify the aspects of TCP/IP that are covered in each chapter. For example, if you already know everything about the network interface layer, you can probably skip that chapter. Over the years, thousands of adult students have found this procedure to work better than a top-down approach for retaining knowledge.

Chapter 1 is an orientation. There you can learn why protocols work together in a stack by looking at the best-known stacked protocols. We review Xerox Network System (XNS), Systems Network Architecture (SNA), and Open Systems Interconnect (OSI). Using the OSI stack as a model, we look at the functions each layer performs and how the layers work together.

Since this chapter is also about standards, we step back to look at the history of TCP/IP and when the protocols became standards. We also review the organizations that participate in the standards process. Equally important is the documentation of the protocols, especially those that are standards. We

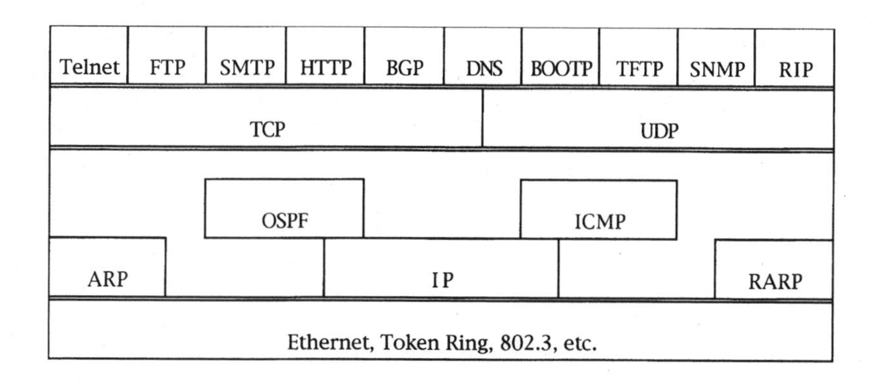

investigate the Requests for Comments (RFCs) and particularly the Standards RFCs, the Assigned Numbers RFC, and the For Your Information RFCs.

In Chap. 2 we complete the discussion of stacked protocols by examining the TCP/IP stack layer by layer. In this chapter you can learn the responsibilities of each layer of the TCP/IP stack and how they come together into the data that travels on local and wide area networks.

Chapter 3 begins our journey up the TCP/IP stack. This stop explores the network interface layer or bottom rung of the stack. You can learn about the structure of Ethernet II addresses and header fields. We see how and why Ethernet II is not the same as IEEE 802.3. You can discover the parameters that restric both IEEE 802.3 and Ethernet II.

We then show you the structure of an IEEE 802.3 header and what happens when it carries TCP/IP data. You can learn about the fields in a token ring or IEEE 802.5 header and what functions they perform. You will also learn how SLIP and PPP vary from each other by comparing their different headers.

By studying Chap. 4 you can learn the different classes of IP Version 4 networks and how they apply to organizations. You will discover the steps of the Internet Protocol communications logic process. One of the most important pieces of TCP/IP also comes up in this chapter: subnetting. Here you will learn when, why, and how to subnet. You can also find out what challenge occurs when we subnet an IP network.

In Chap. 5 you can learn the different aspects involved in matching different addresses. We begin by looking into matching names to IP addresses in the Domain Name System (DNS). Here you will learn the DNS tree structure and what the characters at the end of a domain name identify. You can also learn how DNS servers work together and what records they use.

We conclude the IP communications logic process in this chapter by digging in to Address Resolution Protocol (ARP). Here you can learn the fields in the ARP header and what functions they perform. We also bring up hexadecimal decoding, which you will need to read the ARP and other headers we discuss in this book. We finish the chapter by showing what may happen when you implement ARP.

Chapter 6 gets into the bits and bytes of the Internet Protocol (IP). By examining IP and its header you can learn how it complies with the functionality specified in the OSI model and how it performs those functions. You can also learn how, why, and what happens when routers fragment IP datagrams. We discuss how IP prevents loops, identifies the next protocol, and sets both the source and target addresses. By studying this chapter, you will learn what options you can add to IP to control or discover the path a datagram travels through a network.

In Chap. 7 you can learn where we stand with IP's next generation. We explain the contenders and the selection. You can learn the fields of the proposed header and why some fields moved from the header. Here you can study the ways that IPv6 will expand the flexibility of IP while making the TCP/IP process better and faster overall. To help you plan for the transition to Version 6, you can learn IPv6 addressing and compatibility with IPv4 addresses.

Chapter 8 examines the User Datagram Protocol (UDP). In this chapter you can learn why some applications use UDP while others use TCP. You will see how the the UDP header completes its task. You will also come to understand the reason for using ports in UDP and TCP. Finally, we provide a resource list of sample UDP ports along with lists of assigned and registered ports.

Next, instead of proceeding to TCP, we go back to the Internet layer to discuss routing, in Chap. 9. Here you can learn the differences between bridges and routers and why you would choose one over the other. From there we go into routing to explore gateways, direct versus indirect routing, routing tables, and automatic versus manual table maintenance. We then help you understand the open interior and exterior protocols by examining RIP, OSPF, and BGP. You can learn how they each handle the routing duties assigned to them. That learning process will include header exploration and discussions on CIDR, supernetting, and variable subnet masking.

Chapter 10 explores the Internet Control Message Protocol (ICMP) as it relates to IPv4. By studying this chapter you can learn the various messages that report variations from the IP specifications or perform troubleshooting diagnostics. We look into each message type and discuss the codes that provide more information to explain what happened. Here you will also learn why protocol analysis is such an excellent tool for troubleshooting.

We examine Transmission Control Protocol (TCP) in Chap. 11. Here you can learn how TCP handles a session by coming to understand its header fields and the function each plays. By studying this chapter you will see how TCP keeps track of every byte and how each end of the session controls the flow of data to itself. You will also learn how to recognize the beginning of a TCP session and whether the session ended normally or abnormally. We show you how to recognize congestion and what could cause problems. We conclude the chapter by giving you a sample TCP session to review.

In Chap. 12 we begin to delve into TCP-based applications by examining Telnet. After looking at the way Telnet operates by studying the option negotiations, you will learn the kinds of capabilities that the two systems present to each other. We follow that with a step-by-step examination of a Telnet session. This lets you learn what happens in a session as well as the potential problems. You can also learn how well the TCP works with its applications by seeing it work with Telnet.

After touching on some of the security gaps that are possible with a protocol analyzer, we direct our attention to network security in Chap. 13. Here you will learn the categories of security threats and what actions you can take to protect your network. Studying this and the rest of the chapter will help you understand your options, as well as find out organizations that offer help with security problems. By using their solutions and advice as well as looking into a method of scanning your network for gaps in security, you can avoid many common problems.

Chapter 14 takes us into File Transfer Protocol (FTP). In this chapter you can learn the commands and responses that FTP uses to establish and maintain sessions. By studying a sample FTP session you can learn to recognize FTP actions in preparing for a file, requesting a file, and transferring a file. The session will also teach you how to recognize each of the multiple files along with the accompanying TCP adjustments.

Simple Mail Transfer Protocol (SMTP) is explored in Chap. 15. Here you can learn what made e-mail the primary reason for Internet access during years preceding the creation of the World Wide Web. Studying this chapter will help you learn how similar SMTP is to FTP in its use of commands and response codes. Stepping through a sample SMTP session helps you understand how the protocol works as well as refreshing your understanding of TCP's functions in suport of SMTP.

Chapter 16 focuses on centrally managing IP addresses. We begin with RARP and work through BootP and DHCP. In this chapter you will learn how each one operates and why many managers are moving toward DHCP. We end the chapter with a brief discussion of the issues surrounding virtual IP networks. By studying this chapter you can learn the benefits and challenges that each protocol brings to the centralized control of IP addresses and what they could mean to your network.

We use Chap. 17 to cover Trivial File Transfer Protocol (TFTP). Here you can learn how a UDP-handled protocol can support unattended file transfers and what potholes await the manager who makes use of TFTP. We also teach you how regimented TFTP can be by looking at its structured messages. You will learn the reasons for the structure and how the structure serves to disseminate files and updates.

Simple Network Management Protocol (SNMP) is the topic of Chap. 18. We begin with the basic manager-agent model and each part's separate structure. You can learn how simple SNMP really is and what support pieces like SMI, ASN.1, and MIBs do to make it platform-independent. You can learn the differences between private management information bases and the current open MIB. We also teach you the variations proposed in Version 2 and the work done by the Desktop Management Task Force.

Chapter 19 discusses Internet services. Here you can learn many of the available methods for accessing the Internet, which are most popular, and why. We also touch on some of the Internet functions like Archie, Gopher, and the World Wide Web. By studying this chapter you will learn about Web browsers, URLs, and HTTP. We expand on HTTP by delving into Java and CGI.

Paul Simoneau

Acknowledgments

Just as no one experience makes up a lifetime, no one person creates a book. While it is impossible for me to give credit to all those who contributed to this collection of knowledge and experiences that now bears my name as author, these Acknowledgments are an attempt to get close. To those who feel that I have omitted mention of their participation, my sincerest apology and a request: Please contact me and give me the opportunity to correct the situation.

At the top of the appreciation list for me is the Creator, who provided all the basic tools, abilities, gifts, blessings, and life in the first (and second) place. Next in deep gratitude are my dear parents, Bill and Jane, who (like God) have loved me through my fumblings to the success I have. Their belief in me and encouragement have been nothing short of amazing. Their work on helping me fine-tune my tools and abilities is integral to where I am, where I am going, and what I am about.

Ginny, Patrick, Maggie, Amy, and Clint—my family—have done much more than encourage and support me. They have also done without me many times so that I could work on the books that led to and include this one. Ginny kept the house together, kept the family life going, and supported me in the face of a very hectic schedule of her own. She was the parent to attend events while I taught in another city during the day and wrote at night. Patrick and Maggie have not just given up time with Dad but had to get homework help by long-distance calls. Amy and Clint, while married and building their own life, have gone long periods without hearing from me as writing and rewriting consumed my nonworking hours. I love all of my family and hope that this begins to let them know how very much I appreciate them and their sacrifices.

Outside my family there are many who have been there when I needed to learn. Working with technical staff in many capacities helped me a lot. Jeff Rowlett, Carter Smythe, and Bob Brezany kept me technically correct. When I supervised others, they had lessons to teach me—particularly Scott, Bruce, Catherine, Marjorie, Phil, Regan, and Kim at Federal Systems Group.

My mentors Bob Binder, Kurt Wright, Dan Retter, Dwight Custer, Steve Knier, Russ Carleton, and others have guided me along the way and helped me prepare in different areas. In the role of my teacher they were not alone. I learned quite a bit from most of the thousands of people who each thought they were just a student in one of my classes.

Getting this book to where it is and where it can go is largely due to the generous efforts of the excellent instructors who have contracted with American Research Group to teach classes. I have had the pleasure of sharing teaching duties and ideas with John Allen, Bob Berbrick, Teresa Bisaillon, Richard Bruyere, Jack Buhse, Russ Carleton, Rick Chapin, George Churchwell, Carolyn Cutler, Barry Dilgard, David Ford, Rick Gallaher, Roger Herr, Mark Jones, Paul Pival, Ted Rohling, Gwen Snear, George Stiefelmeyer, Glenn Tapley, Don Vitz, J. D. Wegner, and Brad Werner.

The staff at American Research Group should also receive credit. Although there are too many to name individually, some of these people should be, anyway. Steve, Greg, Scott, Patrick, and Bill keep all the instructors and course directors on their toes as we strive for our best in all areas. Bruce and his staff go well beyond the call to make sure we have what we need to get to that best. Sarah and Sue always make sure we are welcomed wherever we go. Frank and his folks keep us going so many places. In particular, Dave Knier and Dave Mantica stand out in their work. The unsung heroes who are the Customer Service staff are another key to success.

Much help in improving the writing and artwork in this book came from Terri at ARG, Michelle at Just Your Type, Jody Zolli, Russ Carleton at InterLink, John Allen at Computer and Communication Services, and Terry Slattery at Chesapeake Computer Consultants.

The encouragement and support of friends like Glenn and Debbie, Barbara and Jim, Rick and Susan, Steve and Valerie, Enrique and Shirley, John and Peggy, Terry and Peggy, J. D. and Laurie, Bruce, Carl, John, Shelba, Patsy, and Yvette made the nights and weekends special when we got together during breaks in the writing.

Of course, this book would not be a book without the patient diligence of the staff at McGraw-Hill. Steven Elliot and Stephen Smith coordinated all the activity to make this come out as it should. And for all of those who worked behind the scenes like Joe Rivellese, who kept me graphic-correct, a huge thank you for all that I know you must have done to make what I did better.

Chapter

1

Stacked Protocols and Standards

Stacked Protocols

Once upon a time (in the early days of networking) stacked protocols or protocol suites began appearing in the local area network (LAN). A protocol, in this case, is a set of rules that tells each computer how, what, and when to send data between two computer interfaces. These rules work just like the rules for personal communications that tell us when to speak and what language to use so others can understand us.

In the same way that our personal communication rules differ to match the situation, protocols change to match the need. When we combine protocols into groups, with each protocol doing a specialized function, it is usually in a layered, modular format. By stacking protocols together, each can do its task. We can replace a protocol as long as the replacement works with its neighbors in the same manner. This offers great flexibility.

The authors of the Transmission Control Protocol/Internet Protocol (TCP/IP) stack designed their protocols in the same format. Another way to identify this stack of cooperating and interconnected protocols is by calling it a *suite,* like a suite of interconnected rooms. The primary goal of the TCP/IP suite is flexibility.

By working in a stack, each protocol at a layer supports the layers above it and provides services beyond the ability of the layers below it. This interlayer cooperation lets each protocol do what it does best and looks to the other protocols to perform their own functions. This way the suite can work well over multiple physical media types (network interface layer in Fig. 1.1) and support many different application processes (process layer in Fig. 1.1).

Xerox Network System

Bob Metcalfe and David Boggs at Xerox developed Ethernet in 1981. Ethernet became the standard it is today through Xerox's collaboration with

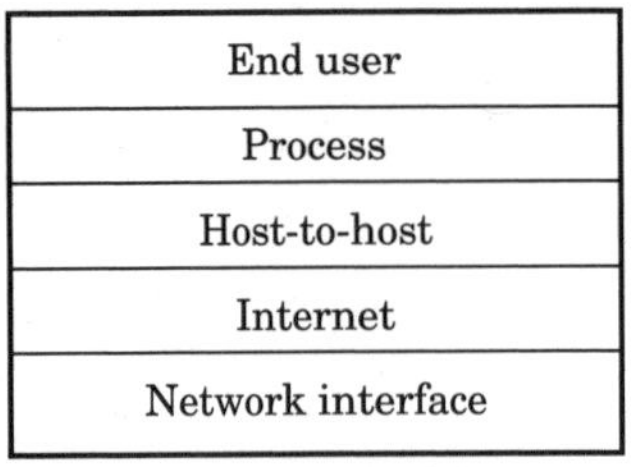

End user
Process
Host-to-host
Internet
Network interface

Figure 1.1 The TCP/IP stack.

Digital Equipment Corporation and Intel, and the term *D.I.X. Ethernet* identified this standard. While Systems Network Architecture (SNA) was the first stacked protocol, Xerox Network Systems (XNS) was the first protocol suite to run commercially over Ethernet (Fig. 1.2). Even though Xerox released XNS as an open architecture, it left too many possibilities for vendor-specific implementation at the higher layers to become a true standard.

While many vendors based their network implementations on XNS, each developed proprietary extensions and modifications. This meant that one vendor's implementation of XNS (some vendors even called their products XNS) would not work well or at all with another vendor's equipment or software.

The challenge is to design protocols with enough structure for interoperability, while allowing enough flexibility to encourage different vendors to implement it on their various systems. That challenge continues in some protocols today. By remaining flexible to encourage vendor creativity, these protocols can allow for too much vendor-dependent variation.

Systems Network Architecture

Another stacked protocol suite that made a major difference in networking was IBM's SNA (Fig. 1.3). Before SNA, the IBM solution to networking was the complex result of a marketing challenge. IBM had a few hundred communications products. With more than a dozen data link protocols, more than 30 bisynchronous teleprocessing access methods, and no one way to meet all customer needs, something had to give way.

IBM engineers typically had to customize bisynchronous communication solutions for each customer site. Significant growth often required a complete new system of hardware and customized software. Customers were under-

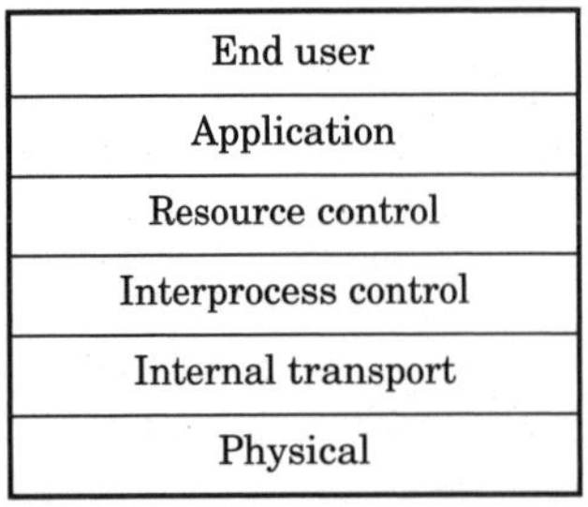

End user
Application
Resource control
Interprocess control
Internal transport
Physical

Figure 1.2 The XNS stack.

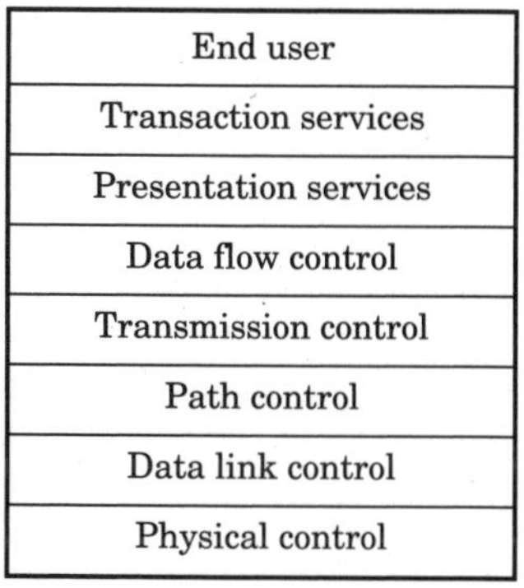

End user
Transaction services
Presentation services
Data flow control
Transmission control
Path control
Data link control
Physical control

Figure 1.3 The SNA stack.

standably reluctant to toss out the older (legacy) solutions and buy or lease the new system(s).

Customers obviously wanted to maintain compatibility with the existing solution. This would help them preserve their financial and system or process investment. This meant SNA had to be both more flexible and more complex than any other single IBM system to date. On the positive side, the International Standards Organization (ISO) patterned much of the Open Systems Interconnect (OSI) model after SNA, including stacked modularity, the number of layers, and the functionality of each of the layers.

Open Systems Interconnect

We include the OSI stack (Fig. 1.4) here because it has become an industry-accepted model for layered networking protocols even though much of that came from SNA. The Advanced Research Projects Agency network (ARPAnet), which is an ancestor of the Internet and the first major TCP/IP network, predates the OSI model by approximately 10 years. TCP/IP's protocols have no direct relation to OSI.

In essence, each layer in the model can perform within the limitations and capabilities provided by the layer below it in the stack while offering its ser-

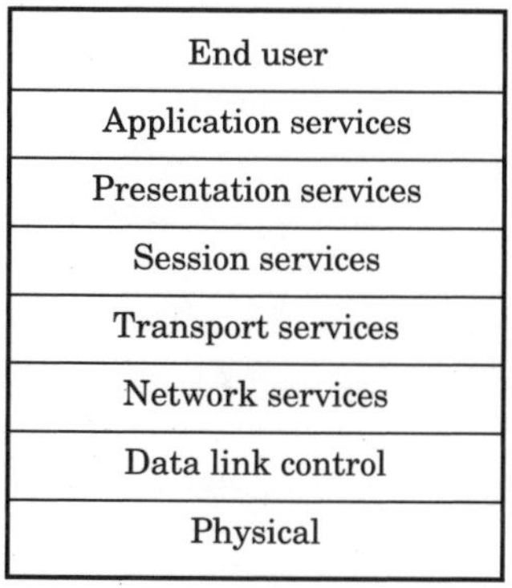

End user
Application services
Presentation services
Session services
Transport services
Network services
Data link control
Physical

Figure 1.4 The OSI stack.

vices to the layer above it. This way a specification could change a layer (or a protocol within a layer) without problems as long as the new protocol worked with the layers above and below in the same way as the old protocol. Each layer's protocols included defining the relationship to the layers and protocols around it.

Each layer in the model has assigned functions. From the bottom of the stack they are:

The *Physical Layer* (Layer 1) defines the physical design and properties of the network connector(s) that are acceptable for use in the identified network. That may be RJ-45 or RJ-11 for use with twisted-pair (shielded or unshielded) cabling or a BNC connector for use with coaxial cabling. This layer also specifies the electrical signaling that will represent the binary 1s and 0s of the data.

The *Data Link Layer* (Layer 2) provides flow control to handle the placement of the data on the network without causing abnormal functions such as collisions. When it works with IP, it passes the data to the Physical Layer in the proper order through a process called *sequencing*.

By offering a frame check sequence (FCS) or a cyclic redundancy check (CRC), the different protocols at the Link Layer can each support error control in their own way. For most it is simply error detection or telling the receiving Link Layer that the data in the packet is clean or that it contains one or more errors.

Most network technicians recognize Link Layer station addressing as the network interface card (NIC) or media access control (MAC) address. The Link Layer addresses frequently show in protocol analyzers on a LAN. Wide area network (WAN) Link Layer addresses may vary from a telephone number to no address at all. Each case will depend on the Link Layer protocol that the network manager chose.

The Link Layer should also provide a way to identify the protocol that it is carrying as its payload. This way the Link Layer can support multiple Network Layer protocols and identify each on a packet-by-packet basis. See the Content ID in Fig. 1.5.

OSI layer	Functions
Application services	Application uniformity and compatibility
Presentation services	Context and syntax for clear communications between platforms
Session services	Session control, synchronization, and support beyond connectivity
Transport services	End-to-end connectivity, data integrity, segmenting, and flow control
Network services	Logical addressing, fragmentation, and routing
Data link control	Logical link control, media access control, physical addressing, and content ID
Physical	Signal encoding, timing, and physical connection

Figure 1.5 OSI model layers and functions.

By supporting diagnostics and alarms, the Link Layer can know when Ethernet collisions occur. It can also tell when a token ring interface is beaconing and identify other network abnormalities that interfere with proper network functions. The alarms portion of this capability notifies the network (in the case of a collision) or the network interfaces (in the case of beaconing).

The *Network Layer* (Layer 3) has multiple responsibilities as well. It supports the upper layers by handling the data transmission and system connectivity. This includes using logical addressing (such as IP addresses) to identify the network and interface that is the source of the data as well as the logical address that is the target of the transmission. To match the logical network address to a particular physical Link Layer address, the Network Layer offers address resolution.

When the network layout does not have the target and source systems in the same location, it is possible that the network manager would choose to use routers to support the interconnectivity. The Network Layer is responsible for that routing function. It is frequently the layer in which the routing protocols themselves live. Routers use these protocols to share routing information with each other.

Since there are so many different Link Layer protocols available, there will be times when the size of the packet from one side of a router is too large for the Link Layer protocol serving the receiving side of the router. The Network Layer steps in here to provide fragmentation capabilities. The fragmentation process breaks the larger packet into smaller packets (fragments) that will fit in the receiving Link Layer protocol's parameters. The end or receiving system will reassemble these fragments.

Layer 3's Network Layer protocols also provide information to the target system. This includes:

- Identifying the datagram payload or the next protocol in the packet
- Giving instructions to routers in how to handle routing each datagram
- Providing the precedence or quality of service the router is to provide the datagram

The Network Layer is also the primary location of user-accessible diagnostics. In TCP/IP this is the location of Internet Control Message Protocol (ICMP), which provides most of the TCP/IP diagnostic capability. Processes, such as ping, send ICMP messages to test connectivity. We will discuss these in Chap. 10.

The *Transport Layer* (Layer 4) handles end-to-end (i.e., source-to-target) communications almost as if there were no layers below it. It is typically the layer that provides the data integrity that the lower layers sacrificed for flexibility to get the data through to the receiving end.

To manage a reliable transfer, the Transport Layer reports each end's capability parameters to the opposite end. If one of those parameters is a limit on the individual packet data size, the sending end will segment the data to accommodate the receiver's limitation. If the parameter is a limited inbound buffer, the sending Transport Layer will apply flow control to prevent buffer overflow.

The *Session Layer* (Layer 5) provides a control structure that establishes, manages, and properly concludes sessions, even though lower layers may fail to maintain their connection between applications. This gives an interrupted session, such as a file transfer, recovery support.

Recovery support is usually in the form of markers that identify the major completion points in the data transfer. The recovery can pick up from the last major point and resume the data transfer after the lower layers reestablish the conversation between the two systems.

Layer 5 also identifies which application the stack is carrying. This offers the ability to support multiple applications over the same protocol suite.

The *Presentation Layer* (Layer 6) transforms data to offer a standardized application interface, and common system commands and communication services like reformatting, compression, and data encryption. One standardized application programming interface, used by TCP/IP and OSI protocols, is Abstract Syntax Notation One (ASN.1). We will look at ASN.1 again in Chap. 18's discussion of SNMP.

The top layer (Layer 7) is the *Application Layer.* This layer provides common application services to users and separates the applications from network functionality.

The irony is that the OSI protocol suite, while receiving the endorsement of many international governments, seems to be one of the least popular methods for interconnecting computers. At the same time, TCP/IP (a far less than model protocol) is the most popular. It hasn't always been that way: TCP/IP has an interesting history. (See Fig. 1.6.)

1969. ARPAnet begins linking scientists and researchers using three computers in California and one in Utah. Those systems were at the University of California at Santa Barbara (UCSB) and Los Angeles (UCLA), Stanford Research International (SRI), and the University of Utah in Salt Lake City. Steve Crocker starts documenting the network by writing the first request for comments (RFC). RFCs provide information and specifications for the Internet, TCP/IP, and related protocols.

1972. The Internetworking Working Group (INWG) begins work toward setting protocol standards. Jon Postel publishes the Telnet specifications in RFC 318.

1973. The ARPAnet begins international service with connections to England (University College of London) and the Royal Radar Establishment in Norway. Bob Metcalfe outlines Ethernet in his doctoral (Ph.D.) thesis. Vinton Cerf and Bob Kahn present the beginning Internet ideas at the INWG meeting. File Transfer Protocol (FTP) specifications are published in RFC 454.

1969—ARPAnet begins
1972—INWG, Telnet
1973—ARPAnet to Europe, Ethernet thesis, Internet paper, FTP
1974—Telenet, TCP paper
1975—DCA manages ARPAnet
1976—UUCP
1977—UUCP with Unix, e-mail
1979—Usenet, DARPA sets ICCB
1982—TCP/IP standard, EGP, DDN
1983—Name server, Internet and MILnet, IAB, BSD Unix
1986—NSFnet
1987—Merit manages NSFnet, UUNET
1988—NSFnet to DS-1, CERT, GOSIP
1989—IETF, IRTF, commercial access
1990—Archie, ISP
1991—CIX, WAIS, PGP, NREN
1992—ISOC, WWW, NSFnet to DS-3
1993—InterNIC, Mosaic, UN, White House
1994—Senate and House, Internet "spam," First Virtual
1995—ANSnet, $50 DNS, Vatican, Java
1996—To OC-3, Internet expo

Figure 1.6 TCP/IP history time line.

1974. Bolt, Beranek and Newman (BBN) launches Telenet, the first commercial (public) packet delivery service. Cerf and Kahn publish a paper specifying the detailed design of Transmission Control Protocol (TCP).

1975. The Defense Communications Agency (DCA), now renamed the Defense Information Systems Agency (DISA), assumes operational management of the ARPAnet.

1976. AT&T's Bell Labs develops Unix-to-Unix Copy Protocol (UUCP).

1977. UUCP distributed with AT&T Unix. RFC 733 sets electronic mail specifications.

1979. Usenet, the source of over 10,000 special interest newsgroups, begins by linking Duke University and the University of North Carolina with UUCP. The Defense Advanced Research Projects Agency (DARPA) sets up the Internet Configuration Control Board (ICCB).

1982. Based on an INWG recommendation, DCA and DARPA set TCP/IP as the ARPAnet protocol. The Department of Defense (DoD) designates TCP/IP as its standard for networking. RFC 827 presents Exterior Gateway Protocol (EGP) specifications. DoD creates the Defense Data Network (DDN) for military communications.

1983. The University of Wisconsin develops the name server, relieving the need to know the exact address of a remote system. ARPAnet switches from Network Control Protocol to TCP/IP as its standard. ARPAnet divides into Internet and MILnet. MILnet then integrates with DDN. The Internet

Activities Board replaces the ICCB and begins to guide the evolution of TCP/IP. The University of California at Berkeley releases 4.2 BSD (Berkeley Software Distribution) Unix that includes TCP/IP.

1986. The National Science Foundation Network (NSFnet) begins service at 56 kbps to connect five supercomputer centers at Cornell University, the University of California at San Diego, the University of Illinois at Urbana-Champaign, the Pittsburgh Supercomputing Center, and Princeton University.

1987. NSF turns over NSFnet management to Merit Network, Inc. Unix-to-Unix Network (UUNET) begins providing Internet access.

1988. The NSFnet grows to 1.544 Mbps (DS-1). The Internet worm (a self-replicating software program that consumes system resources) crawls around the Internet and through approximately 6000 of the 60,000 attached hosts. The Computer Emergency Response Team (CERT) begins service in response to the worm. DoD specifies OSI as its long-term standard and TCP/IP as the interim solution. The federal government specifies that all U.S. government-purchased computers will support the Government Open Systems Interconnect Profile (GOSIP). Internationals joining the NSFnet include Canada, Denmark, Finland, France, Iceland, and Sweden.

1989. The Internet Architecture Board (IAB) reorganizes to include the Internet Engineering Task Force (IETF) and the Internet Research Task Force (IRTF). The IETF continues to guide the evolution of the TCP/IP protocol suite and other Internet operations. The IRTF organizes and explores advanced networking. Compuserve and MCI Mail begin the first commercial e-mail access to the Internet. Those joining the NSFnet include Australia, Germany, Holland (the Netherlands), Israel, Italy, Japan, Mexico, New Zealand, and Puerto Rico.

1990. McGill University releases the Internet search tool Archie. ARPAnet officially ends operation as the Internet and MILnet continues to provide replacement service. The first commercial Internet service provider (ISP), The World, begins offering dial-up Internet access. Internationals joining the NSFnet include Argentina, Austria, Belgium, Brazil, Chile, Greece, India, Ireland, South Korea, Spain, and Switzerland.

1991. The Commercial Internet Exchange (CIX) forms to offer Internet functions to commercial organizations after the NSF waives the Acceptable Use Policy (AUP). The University of Minnesota releases Gopher. Thinking Machines releases Wide Area Information Systems (WAIS). Philip Zimmerman releases the encryption software Pretty Good Privacy (PGP). The High Performance Computing Act clears the path for the National Research and Education Network (NREN). Internationals joining the NSFnet include Czech Republic, Hong Kong, Hungary, Poland, Portugal, Singapore, South Africa, Taiwan, and Tunisia.

1992. The Internet Society begins with Vinton Cerf as president. The European Laboratory for Particle Physics Research in Switzerland (CERN) begins the World Wide Web (WWW). The NSFnet contracts with Advanced Networks and Services (ANS) for the ANSnet to replace the NSFnet and upgrades the backbone of the Internet to 44.736 Mbps (DS-3). The Internet Hunt begins its monthly scavenger hunt across the Internet. Internationals joining the NSFnet include Cameroon, Croatia, Cyprus, Ecuador, Kuwait, Latvia, Luxembourg, Malaysia, Thailand, and Venezuela.

1993. NSF contracts create network information centers for civilian government agencies and commercial organizations, naming the common function the InterNIC. The National Center for Supercomputing Applications (NCSA) releases Mosaic. The White House joins the Internet. The United Nations comes on-line. Internet Talk Radio begins broadcasting. Those joining the NSFnet include Bulgaria, Costa Rica, Egypt, Guam, Indonesia, Kenya, Liechtenstein, Peru, Romania, the Russian Federation, Turkey, Ukraine, United Arab Emirates, and the Virgin Islands.

1994. The Internet celebrates 25 years. The United States Senate and the House of Representatives install Internet-accessible information servers. The National Institute for Standards and Technology (NIST) drops the OSI-only requirement by suggesting that the GOSIP specification include TCP/IP. The Arizona law firm of Canter and Siegel "spams" news groups with junk mail advertising their green card lottery services. The response to spamming was (and continues to be) very negative (flame or expletives-not-deleted) e-mail. First Virtual, the first "cyberbank," opens for business. Internationals joining the NSFnet include Algeria, Armenia, Bermuda, China, Colombia, Jamaica, Lebanon, Lithuania, Morocco, Nicaragua, Niger, Panama, Philippines, Sri Lanka, and Uruguay.

1995. The NSFnet returns to its roots as a research network. ANSnet continues as the backbone, along with interconnections to other ISP networks. WWW becomes the number-one use of the Internet. America Online (AOL), CompuServe and Prodigy begin to offer Internet access. Registration of Domain Names requires a $50 annual fee. The Vatican joins the Web. Netscape agrees to include Sun Microsystems' Java capabilities in its WWW browsers and servers.

1996. MCI begins offering its government-contracted very-high-speed Backbone Network Service (vBNS) at 155 Mbps (OC-3) to replace ANSnet. The Internet World Exposition, the first Internet-based world's fair, takes place.

TCP/IP Standards Control

In the TCP/IP history time line, we touched lightly on the structure of the Internet Architecture Board (IAB) and some of the organizations that support

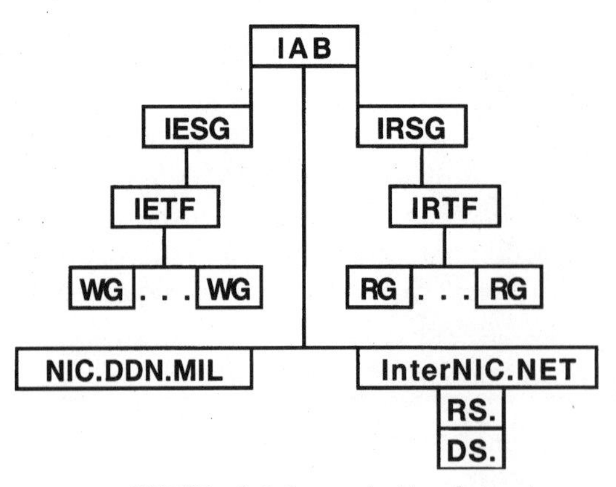

Figure 1.7 TCP/IP-related organizational structure.

the evolution of TCP/IP and the Internet's functionality. (See Fig. 1.7.) A 1989 reorganization set the Internet Engineering Task Force (IETF) and the Internet Research Task Force (IRTF) under the IAB. The IAB has the responsibility to:

- Review Internet standards
- Oversee the IETF and IRTF and ratify their major changes
- Act as an international policy liaison for the Internet community
- Manage the RFC publication process
- Identify strategic, long-term challenges and opportunities
- Resolve technical issues outside the IETF or IRTF purview

The IETF has a manager over each of eight working areas. These managers and an IETF chair make up the Internet Engineering Steering Group (IESG). The IETF coordinates the management, operation, and evolution of the Internet. It maintains existing protocols and promotes the development of new protocols. It also identifies operational and technical problems and provides a forum for exchanging technical information in the Internet community toward solution of those problems.

Each member of the Internet Research Steering Group (IRSG) chairs a volunteer Internet research group that functions much like an IETF working group but under the IRTF charter. The IRTF studies long-term (5- to 10-year) research issues and those currently affecting the Internet. For example, these issues include how to handle a billion Internet users or what to do as new technologies increase the user access speeds.

For years the DoD and the NSF funded a Network Information Center (NIC.DDN.MIL) for all Internet and MILnet administration and documentation. That group, previously under contract to Stanford Research International (SRI) and now contracted to Government Systems Inc. (GSI), is called the "Nick" (NIC).

An April 1993 cooperative agreement with the NSF set up three Internet NICs (InterNIC) to let the renamed D-NIC (Defense Network Information Center) focus on MILnet users. This was to relieve the strain that the extraordinary and unexpected growth in the number of networks and users had placed on GSI. The InterNIC's charter is to support the rest of us, commercial organizations and civilian government agencies.

RS.InterNIC.Net, awarded to Network Solutions, Inc. (NSI), provides registration services for organizations and individuals adding or changing registered information. NSI coordinates that registered information and assigns Internet domains and certain IP network addresses.

AT&T offers a database and directory services, under the contract, as DS.InterNIC.Net. Those services include

- A directory of directories
- Listings of FTP sites
- White and yellow pages
- Library catalogs
- Data archives such as Archie and the RFCs

General Atomics received the Information Services (IS.InterNIC.Net) contract to provide information for many areas of Internet activities. The government chose not to renew that contract, and DS.InterNIC.Net is providing those services now.

Requests for Comments

As we saw in the historical review, since 1969 we have been documenting the Internet and the open standards that make up TCP/IP and related Internet protocols in Requests for Comments. Internet-interested authors also write RFCs on many networking topics. One of the main reasons that the Internet has been able to develop so smoothly and so quickly is its documentation of the standards.

Anyone may write an RFC. You must comply with the instructions to RFC authors contained in RFC 1543 and submit your RFC to the IETF User Services Working Group for review. Once approved, the RFC Editor will assign each new, revised, or replacement RFC a unique number. That way, there is not an RFC 1340-A or RFC 1700bis.

The replacement RFCs identify themselves in the RFC index with the Obsoletes RFC# statement. (See RFC 1700's entry in Fig. 1.8.) The obsoleted RFCs entry in the index also identifies (see RFC 1340's entry in Fig. 1.8) that it has been "Obsoleted by RFC#."

1700 S J. Reynolds, J. Postel, “ASSIGNED NUMBERS”, 10/20/1994. (Pages = 230) (Format = .txt) (Obsoletes RFC1340) (STD 2)

1698 I P. Furniss, “Octet Sequences for Upper-Layer OSI to Support Basic Communications Applications”, 10/26/1994. (Pages = 29) (Format = .txt)

1697 PS D. Brower, R. Purvy, A. Daniel, M. Sinykin, J. Smith, “Relational Database Management System (RDBMS) Management Information Base (MIB) using SMIv2”, 08/23/1994. (Pages = 38) (Format = .txt)

1696 PS J. Barnes, L. Brown, R. Royston, S. Waldbusser, “Modem Management Information Base (MIB) using SMIv2”, 08/25/1994. (Pages = 31) (Format = .txt)

1695 PS M. Ahmed, K. Tesink, “Definitions of Managed Objects for ATM Management Version 8.0 using SMIv2”, 08/25/1994. (Pages = 73) (Format = .txt)

1694 DS T. Brown, K. Tesink, “Definitions of Managed Objects for SMDS Interfaces using SMIv2”, 08/23/1994. (Pages = 35) (Format = .txt) (Obsoletes RFC1304)

1693 E T. Connolly, P. Amer, P. Conrad, “An Extension to TCP : Partial Order Service”, 11/01/1994. (Pages = 36) (Format = .txt)

/\

1342 PS K. Moore, “Representation of Non-ASCII Text in Internet Message Headers”, 06/11/1992. (Pages = 7) (Format = .txt) (Obsoleted by RFC1522)

1341 PS N. Borenstein, N. Freed, “MIME (Multipurpose Internet Mail Extensions): Mechanisms for Specifying and Describing the Format of Internet Message Bodies”, 06/11/1992. (Pages = 80) (Format = .txt, .ps) (Obsoleted by RFC1521)

1340 S J. Reynolds, J. Postel, “ASSIGNED NUMBERS”, 07/10/1992. (Pages = 139) (Format = .txt) (Obsoletes RFC1060) (STD 2) (Obsoleted by RFC1700)

1339 E S. Dorner, P. Resnick, “Remote Mail Checking Protocol”, 06/29/1992. (Pages = 5) (Format = .txt)

Figure 1.8 Excerpt from the RFC index.

All newly published RFCs show up in the next catalog of the RFCs. DS.InterNIC.Net makes the complete and up-to-date list of RFCs available in the rfc-index.txt file. Figure 1.8 contains excerpts from that file.

As you can see, the RFC index lists the RFCs beginning with the most recent (at the top of the file). What you cannot see from this excerpt is that the list continues to the oldest RFC (RFC 1, dated April 1969) at the bottom. Since this is a plain ASCII text file, you can open it using almost any word processing application and search for a topic of interest to find the most recent RFC on that subject.

These RFCs fall into six categories that carry a short code after their number in the RFC index, to identify their category. The following list contains those letters and their explanation:

S Standards specify the requirements for protocol(s) or functions.

DS Draft standards are in the process of becoming an official standard.

PS	Proposed standards are currently going through an initial screening process.
E	Experimentals are not being considered for standardization.
H	Historicals cite protocols that have been declared obsolete.
I	Informationals contain information that may be interesting to Internet users.

The RFC index identifies the RFC number, category, author(s), title, creation date, number of pages, format, any relationship to other RFCs, and (when applicable) the standard reference. We have talked about some of these areas, and others are easy to understand on their own. Some further notes will make understanding easier.

Note that all RFCs must be available in text format (.txt), though some authors also provide the RFC in PostScript™ (.ps) format, particularly when they contain graphics.

The IAB maintains a list of the TCP/IP and Internet standards specified by RFCs and other documentation. The Internet Official Protocol Standards RFC (RFC 1920, as of this writing) identifies these RFCs and their STD (STandarD) numbers. Figure 1.9 shows a table of these STD RFCs and their ID numbers. Note that the network interface layer specifications are separate and RFCs do not specify these fully defined industry standards.

RFCs are available from a number of sites. The most up-to-date is DS.InterNIC.Net. You can use the Web and point the browser to http://www.internic.net/rfc to look at an RFC or the RFC index. To retrieve the index or one or more RFCs, point the browser to ftp://www.internic.net/rfc (note the difference between http and ftp). You can also retrieve them by anonymous FTP or by electronic mail.

Note: This book comes with a CD that includes the RFCs.

The first RFCs we need, after the RFC index and the Internet Official Protocol Standards RFC, are the Assigned Numbers RFC (see Fig. 1.10 for RFC 1700's first page) and the For Your Information RFCs (Fig. 1.11).

Assigned Numbers RFC

This RFC is the magic decoder ring that we missed in that last box of Cracker Jacks. In other words, it contains the codes and identifying numbers that the TCP/IP and related protocols use in their daily communications. As you can see from the numbers assigned to this RFC, it has changed many times to add new information and to change the format of the information. With more users on the Internet, we expect changes or updates to occur more often than the 2- to 3-year cycles seen since 1987.

Since this version of the Assigned Numbers RFC does not have a table of contents, we will put the appropriate portions of it at the end of each chapter. We recommend you retrieve this RFC from DS.InterNIC.Net and keep it on a

Protocol	Name	Status	RFC	STD
	Internet Official Protocol Standards	Req	1920	1
	Assigned Numbers	Req	1700	2
	Host Requirements—Communications	Req	1122	3
	Host Requirements—Applications	Req	1123	3
IP	Internet Protocol	Req	791	5
	as amended by:			
	IP Subnet Extension	Req	950	5
	IP Broadcast Datagrams	Req	919	5
	IP Broadcast Datagrams with Subnets	Req	922	5
ICMP	Internet Control Message Protocol	Req	792	5
IGMP	Internet Group Multicast Protocol	Rec	1112	5
UDP	User Datagram Protocol	Rec	768	6
TCP	Transmission Control Protocol	Rec	793	7
TELNET	Telnet Protocol	Rec	854, 855	8
FTP	File Transfer Protocol	Rec	959	9
SMTP	Simple Mail Transfer Protocol	Rec	821	10
SMTP-SIZE	SMTP Service Ext for Message Size	Rec	1870	10
SMTP-EXT	SMTP Service Extensions	Rec	1869	10
MAIL	Format of Electronic Mail Messages	Rec	822	11
CONTENT	Content Type Header Field	Rec	1049	11
NTPV2	Network Time Protocol (Version 2)	Rec	1119	12
DOMAIN	Domain Name System	Rec	1034, 1035	13
DNS-MX	Mail Routing and the Domain System	Rec	974	14
SNMP	Simple Network Management Protocol	Rec	1157	15
SMI	Structure of Management Information	Rec	1155	16
Concise-MIB	Concise MIB Definitions	Rec	1212	16
MIB-II	Management Information Base-II	Rec	1213	17
NETBIOS	NetBIOS Service Protocols	Ele	1001, 1002	19
ECHO	Echo Protocol	Rec	862	20
DISCARD	Discard Protocol	Ele	863	21
CHARGEN	Character Generator Protocol	Ele	864	22
QUOTE	Quote of the Day Protocol	Ele	865	23
USERS	Active Users Protocol	Ele	866	24
DAYTIME	Daytime Protocol	Ele	867	25
TIME	Time Server Protocol	Ele	868	26
TFTP	Trivial File Transfer Protocol	Ele	1350	33
TP-TCP	ISO Transport Service on top of the TCP	Ele	1006	35
ETHER-MIB	Ethernet MIB	Ele	1643	50
PPP	Point-to-Point Protocol (PPP)	Ele	1661	51
PPP-HDLC	PPP in HDLC Framing	Ele	1662	51
IP-SMDS	IP Datagrams over the SMDS Service	Ele	1209	52

Figure 1.9 Standards RFCs.

Network Working Group
Request for Comments: 1700
STD: 2
Obsoletes RFCs: 1340, 1060, 1010, 990, 960,
943, 923, 900, 870, 820, 790, 776, 770,
762, 758,755, 750, 739, 604, 503, 433, 349
Obsoletes IENs: 127, 117, 93
Category: Standards Track

J. Reynolds
J. Postel
ISI
October 1994

ASSIGNED NUMBERS

Status of this Memo

This memo is a status report on the parameters (i.e., numbers and keywords) used in protocols in the Internet community. Distribution of this memo is unlimited.

OVERVIEW

This RFC is a snapshot of the ongoing process of the assignment of protocol parameters for the Internet protocol suite. To make the current information readily available the assignments are kept up-to-date in a set of online text files. This RFC has been assembled by catenating these files together with a minimum of formatting "glue". The authors apologize for the somewhat rougher formatting and style than is typical of most RFCs.

We expect that various readers will notice specific items that should be corrected. Please send any specific corrections via email to <iana@isi.edu>.

/\

Reynolds & Postel [Page 1]

Figure 1.10 Front page of the Assigned Numbers RFC.

system for ready access. Since it is an ASCII text document, it is readable using any standard word or text processor.

For Your Information RFC

The For Your Information (FYI) RFCs are those that the IAB wants you to R-E-A-D first. (Aren't acronyms fun?) They provide Internet users with information about any topics that relate to the Internet. Topics can range from historical memos to answers to frequently asked questions (FAQs).

The FYIs serve a wide audience from beginners to very advanced users. Since all FYIs are informational RFCs, anyone who has some information to share with the Internet community (and the time to write it) may submit it as an RFC and suggest it become an FYI. There are 29 FYIs as of this writing.

Figure 1.11 FYI Index.

0001	FYI on FYI: Introduction to the FYI Notes. G.S. Malkin, J.K. Reynolds. Mar-01-1990. (Format: TXT = 7867 bytes) (Updates FYI0001) (Updated by FYI0001) (Also RFC1150)
0002	FYI on a network management tool catalog: Tools for monitoring and debugging TCP/IP internets and interconnected devices. R. Enger & J. Reynolds. June 1993. (Format: TXT = 308528 bytes) (Updates RFC1147) (Also RFC1470)
0003	FYI on where to start: A bibliography of internetworking information. K.L. Bowers, T.L. LaQuey, J.K. Reynolds, K. Roubicek, M.K. Stahl, A. Yuan. Aug-01-1991. (Format: TXT = 67330 bytes) (Updates FYI0003) (Updated by FYI0003) (Also RFC1175)
0004	FYI on Questions and Answers - Answers to commonly asked "New Internet User" Questions. A. Marine, J. Reynolds, & G. Malkin. March 1994. (Format: TXT = 98753 bytes) (Obsoletes RFC1177, RFC1206, RFC1325) (Also RFC1594)
0005	Choosing a name for your computer. D. Libes. Aug-01-1991. (Format: TXT = 18472 bytes) (Updates FYI0005) (Updated by FYI0005) (Also RFC1178)
0006	FYI on the X window system. R.W. Scheifler. Jan-01-1991. (Format: TXT = 3629 bytes) (Updates FYI0006) (Updated by FYI0006) (Also RFC1198)
0007	FYI on Questions and Answers: Answers to commonly asked "experienced Internet user" questions. G.S. Malkin, A.N. Marine, J.K. Reynolds. Feb-01-1991. (Format: TXT = 33385 bytes) (Updates FYI0007) (Updated by FYI0007) (Also RFC1207)
0008	Site Security Handbook. J.P. Holbrook, J.K. Reynolds. Jul-01-1991. (Format: TXT = 259129 bytes) (Updates FYI0008) (Updated by FYI0008) (Also RFC1244)
0009	Who's Who in the Internet: Biographies of IAB, IESG and IRSG Members. G. Malkin. May 1992. (Format: TXT = 92119 bytes) (Obsoletes RFC1251) (Also RFC1336)
0010	There's Gold in them thar Networks! or Searching for Treasure in all the Wrong Places. J. Martin. January 1993. (Format: TXT = 71176 bytes) (Obsoletes RFC1290) (Also RFC1402)
0011	A Revised Catalog of Available X.500 Implementations. A. Getchell & S. Sataluri, Editors. May 1994. (Format: TXT = 124111 bytes) (Obsoletes RFC1292) (Also RFC1632)
0012	Building a Network Information Services Infrastructure. D. Sitzler, P. Smith, A Marine. February 1992. (Format: TXT = 29135 bytes) (Also RFC1302)
0013	Executive Introduction to Directory Services Using the X.500 Protocol. C. Weider, J. Reynolds. March 1992. (Format: TXT = 9392 bytes) (Also RFC1308)
0014	Technical Overview of Directory Services Using the X.500 Protocol. C. Weider, J. Reynolds, S. Heker. March 1992. (Format: TXT = 35694 bytes) (Also RFC1309)
0015	Privacy and Accuracy Issues in Network Information Center Databases. J. Curran, A. Marine. August 1992. (Format: TXT = 8858 bytes) (Also RFC1355)
0016	Connecting to the Internet - What Connecting Institutions Should Anticipate. ACM SIGUCCS. August 1992. (Format: TXT = 53449 bytes) (Also RFC1359)

0017	The Tao of IETF - A Guide for New Attendees of the Internet Engineering Task Force. The IETF Secretariat & G. Malkin. November 1994. (Format: TXT = 50477 bytes) (Obsoletes RFC1539, RFC1391) (Also RFC1718)
0018	Internet Users' Glossary. G. Malkin. August 1996. (Format: TXT = 123008 bytes) (Obsoletes RFC1983)
0019	FYI on Introducing the Internet— A Short Bibliography of Introductory Internetworking Readings. E. Hoffman & L. Jackson. May 1993. (Format: TXT = 7116 bytes) (Also RFC1463)
0020	FYI on "What is the Internet?". E. Krol & E. Hoffman. May 1993. (Format: TXT = 27811 bytes) (Also RFC1462)
0021	A Survey of Advanced Usages of X.500. C. Weider & R. Wright. July 1993. (Format: TXT = 34883 bytes) (Also RFC1491)
0022	Frequently Asked Questions for Schools. J. Sellers & J. Robichaux. May 1996. (Format: TXT = 150980 bytes) (Obsoletes RFC1578) (Also RFC1941)
0023	Guide to Network Resource Tool. EARN Staff. March 1994. (Format: TXT = 235112 bytes) (Also RFC1580)
0024	How to Use Anonymous FTP. P. Deutsch, A. Emtage & A. Marine. May 1994. (Format: TXT = 27258 bytes) (Also RFC1635)
0025	A Status Report on Networked Information Retrieval: Tools and Groups. J. Foster. August 1994. (Format: TXT = 375469 bytes) (Also RFC1689, RTR0013)
0026	K-12 Internetworking Guidelines. J. Gargano, D. Wasley. November 1994. (Format: ASCII,PS = 66659,662030 bytes) (Also RFC1709)
0027	Tools for DNS debugging. A. Romao. November 1994. (Format: TXT = 33500 bytes) (Also RFC1713)
0028	Netiquette Guidelines. S. Hambridge. October 1995. (Format: TXT = 46185 bytes) (Also RFC1855)
0029	Catalogue of Network Training Materials. J. Foster, M. Isaacs & M. Prior. October 1996. (Format: TXT = 78941 bytes) (Also RFC2007)

Figure 1.11 *(Conclusion)*

Unlike the RFC index, the FYI index (Fig. 1.11) lists the RFCs in order of their FYI number, from 1 through 29. Like the RFC index (though in a different order), each entry contains the title, author(s), date (month and year only), format, size (in bytes instead of pages), and any relationship to other RFCs. The FYI index adds a reference to the RFC that the IAB chose to also be an FYI.

Chapter

2

TCP/IP Stack Overview

Layer Responsibilities

No direct correlation exists between the TCP/IP protocol suite and the OSI protocol suite. However, many people understand protocol stacks by using the OSI protocol suite as a model. Therefore, we have put the two stacks here (Fig. 2.1) for a rough comparison.

The process layer is roughly equal to the OSI Presentation and Application layers. The host-to-host layer matches multiple functions of the OSI Transport and Session layers. The Internet layer functions, including Address Resolution Protocol (ARP) and ICMP, span the OSI Data Link, Network, and Transport layers. The OSI Physical and Data Link Control layers correspond to the network interface layer in TCP/IP.

Users may think the TCP/IP protocols work as a result of a particular request from the user and all the rest of the protocols in the suite just "do what they are supposed to do" to support network communications. Because it is a little more complicated than that, let's look at the functions each layer supports by creating an analogy (using the U.S. Postal Service) to understand how TCP/IP conveys information from one person to another.

TCP/IP	OSI
Process	Application Presentation
Host-to-host	Session Transport
Internet	Network
Network interface	Data link Physical

Figure 2.1 TCP/IP stack aligned with OSI model.

FTP	Telnet	SMTP	HTTP	BGP4	DNS	BootP	TFTP	SNMP	RIP

Figure 2.2 Process layer messages.

Process layer

In our analogy, Amy selects the right application to move her idea from her processor (brain) to a form that she can send to Clint, in this case writing her thoughts on paper (Fig. 2.2).

In TCP/IP, she picks the process layer application that does that job. For example, to send electronic mail the application is SMTP. To quickly transfer a single file, she might select Trivial File Transfer Protocol (TFTP). (See Fig. 2.3.)

Host-to-host layer

Once Amy's ideas are on paper, she must (in our analogy) choose reliability or speed as the key to transporting her ideas to Clint. The Postal Service can provide either or both, but our analogy to TCP/IP restricts us to one or the other.

In TCP/IP, the application the user chooses decides which transport layer protocol it will use. Software authors write their applications to use either

Process	Function
FTP	File Transfer Protocol
Telnet	Terminal-to-server application
SMTP	Simple Mail Transfer Protocol
HTTP	Hypertext Transaction Protocol, WWW
BGP4	Border Gateway Protocol, Version 4
DNS	Domain Name System
BootP	Boot Protocol
DHCP	Dynamic Host Configuration Protocol
TFTP	Trivial File Transfer Protocol
SNMP	Simple Network Management Protocol
RIP	Routing Information Protocol

Figure 2.3 Processes and functions.

FTP	Telnet	SMTP	HTTP	BGP4	DNS	BootP	TFTP	SNMP	RIP
TCP					UDP				

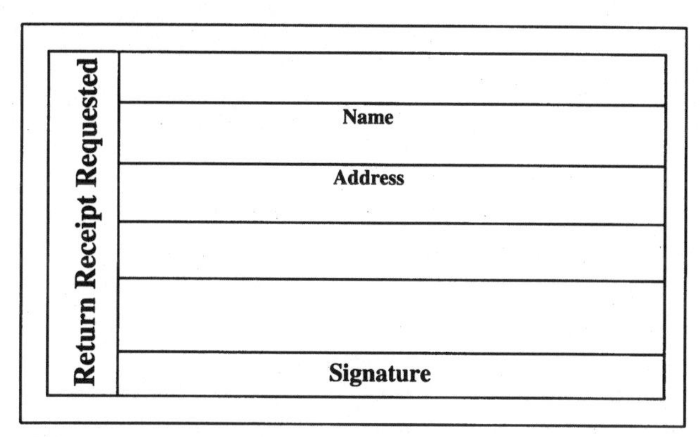

Figure 2.4 Host-to-host layer segments.

Transmission Control Protocol or User Datagram Protocol. As a connection-based protocol, TCP offers far greater reliability along with acknowledging received data. UDP is connectionless (best effort) and so less reliable. It can send data with the first UDP datagram while TCP must complete a three-segment handshake.

Although some applications [like Domain Name System (DNS)] can run over either protocol, they cannot use both simultaneously. As you can see in Fig. 2.4, the applications tie to different protocols in the host-to-host layer.

The programmer typically bases the Transmission Control Protocol (TCP) or User Datagram Protocol (UDP) decision on the source and target locations. In the same network, the user assumes reliability and selects speed. In connecting to another network the user cannot be sure of reliability, and that becomes the decision point. Once the application makes the choice (TCP in the case of Simple Mail Transfer Protocol [SMTP]), TCP adds its header (more about that in the TCP chapter) to the front of the application message. The result is a TCP segment.

In our analogy, Amy chooses reliability and calls Clint to coordinate the pending arrival of ideas. The Postal Service provides reliability through a return receipt service that has Clint sign for the package at his end, and the return receipt card goes back to Amy to verify delivery. Similarly, TCP will account for the bytes received by sending an acknowledgment to the originating system.

By contrast (if UDP could carry SMTP), UDP sends the data and hopes for the best. The only acknowledgment would have to come from the application at the other end.

Internet layer

TCP, after adding its header data to the user's message, passes the TCP segment to the Internet layer (along with instructions) for processing. The Internet layer adds its header (see Chap. 6) to the TCP segment to create an Internet Protocol (IP) datagram. An IP datagram is a routable chunk of data that includes the IP header and whatever information it is transferring (its payload).

The Internet layer (Fig. 2.5) performs three main jobs: fragmentation, addressing, and routing.

Fragmentation keeps from sending more data to the network interface layer than it can handle in one datagram. Routers often require fragmentation as they send datagrams from one interface protocol to another, in other words, token ring to Ethernet. If Amy's package weighs too much, the Postal Service may require her to separate it into several smaller packages.

Addressing has two aspects at this layer. First, the source IP entity must find the physical address that matches the given target IP address. The Address Resolution Protocol performs that function. For example, Amy knows Clint's name and the street name of his address but not his house number. The person delivering Amy's letter would have to call everyone on Clint's street to find Clint and get the number.

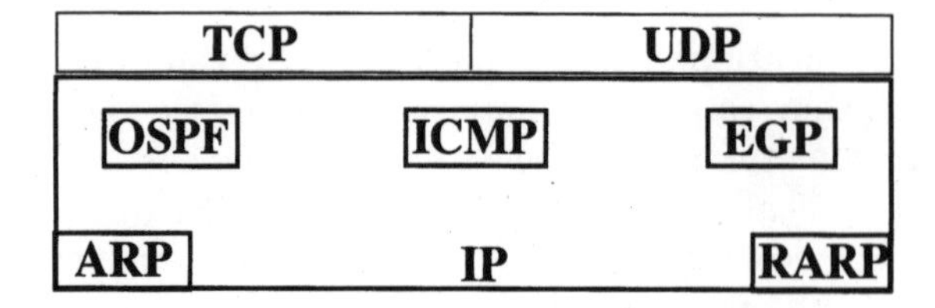

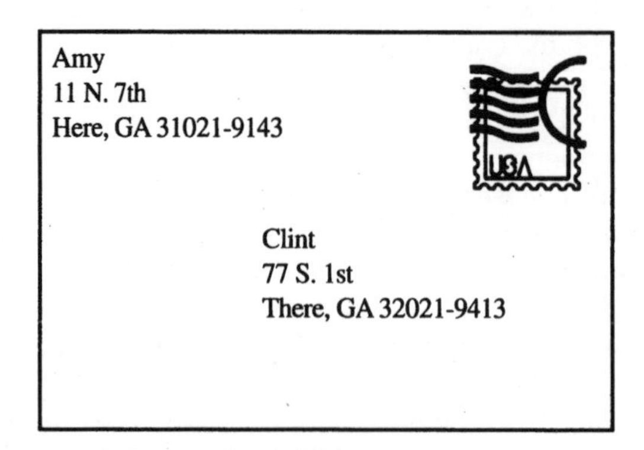

Figure 2.5 Internet layer datagrams.

Second, IP adds the source and target IP addresses to the IP header so the routers in between the source and target hosts will know where to send the datagram (or which address to inform if it cannot deliver the message). In our Postal Service analogy, the zip code offers that ability through the postal network.

Network interface layer

IP passes the datagram (or separate fragments) down the stack to the network interface layer (Fig. 2.6). An IP datagram with a network interface layer header attached to the front and an error check to the rear may be an Ethernet or token-ring frame in some networking, but it is a *packet* in TCP/IP. The term probably came from X.25 packets that carried data from the early days of TCP/IP development. IP does not care what network interface layer protocol is running as long as it can recognize the protocol and communicate with it.

In our analogy, Amy drops her parcel in a mailbox or at a post office to begin its journey through the network to Clint. Diagrams of networks often represent the network as a vague cloud because there are so many unknown points of switching, routing, source systems, and target systems to identify

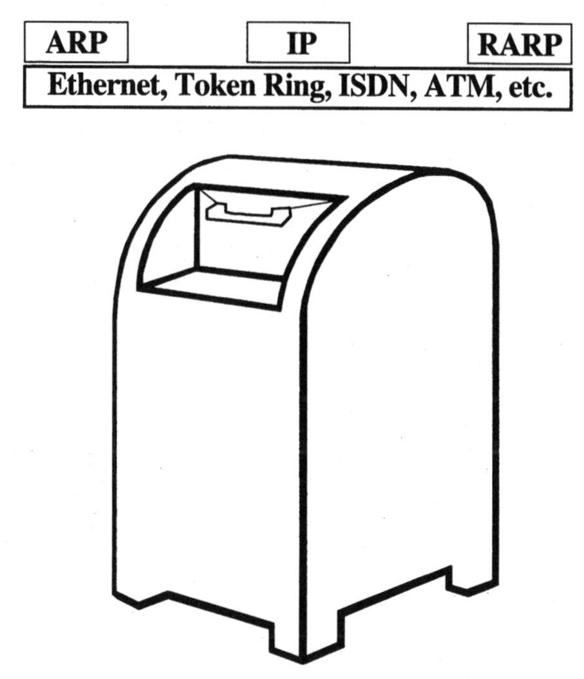

Figure 2.6 Network interface layer packets.

each one of them. In the most complex networks, TCP and IP work together with the various network interface layers to speed information accurately toward its target.

When the packet arrives, the target hardware interface verifies the destination address and the error checking before passing the datagram to the identified protocol in the Internet layer. After completing all of its verification and reassembly tasks, IP passes the segment to the entity specified in the Internet protocol header. These tasks include:

- Verifying the IP checksum
- Reassembling datagrams (putting the original information together from the fragments)
- Checking the target IP address
- Noting the host-to-host layer protocol and other needed information

In our analogy, TCP is the host-to-host entity. TCP verifies the checksum, identifies the source and target ports, and sends an acknowledgment back to the TCP entity in the originating host. Then it passes the message to the named application (SMTP in our example) for processing.

The analogy uses the same process of physical and protocol or personal address verification, assembling any fragments, signing the receipt, and returning it to the sender. Clint then uses the application of reading to put Amy's information into his processor (brain).

Some Assembly Required

As data drops into messages, and those messages into segments, and segments into datagrams, and datagrams into packets out onto the network, the TCP/IP protocols provide ongoing *logical* connectivity. With so much at stake, direct connectivity or changes in the physical network interface must have no bearing on delivery. The process layer application on the source (client) talks with the process layer application on the target (server) and this logic continues down the two stacks.

The matching entities on the two hosts talk to each other using the layers below them to carry the session between the two systems. The layers must use the same protocols (language) to talk with each other during a session. For example, UDP on one host does not talk with TCP on the other. We will see that we can track these conversations and use them to troubleshoot the network.

The network-ordered layout of the headers (the order that they appear on the network), in the Fig. 2.7 packets, show the combined host data and application header is a variable length (if present). (The challenge with graphically displaying this is that each application has its own header.)

Ethernet	IP	UDP	Application	CRC
14 bytes	20 to 60 bytes	8 bytes	variable length	4 bytes

Ethernet	IP	TCP	Application	CRC
14 bytes	20 to 60 bytes	20 or 24 bytes	variable length	4 bytes

Ethernet	IP	ICMP	CRC
14 bytes	20 to 60 bytes	9+ bytes	4 bytes

Figure 2.7 TCP/IP headers.

From there (working to our left in Fig. 2.7), the header sizes are more predictable (see the appropriate chapters for each header layout). For comparison, the UDP header could be in the packet instead of the TCP header. The UDP header is always 8 bytes long. The IP header is a minimum of 20 bytes (default). It only expands up to 60 bytes to allow for options like recording the route or specifying the routers the datagram will pass through on its way to the target system. (See Fig. 2.8.)

The maximum transmission unit (MTU) size of the network interface layer packet sets the limit of the payload. The payload includes all headers and data in between the packet header and the error checking. For example, token ring may carry up to 16,000 bytes per packet, while Ethernet can only carry 1500 bytes per packet.

This varied length (and the necessary flexibility) is the reason the Internet layer offers fragmentation and why the application header and its data allow variable length. We will discuss fragmentation further in Chap. 6.

Source		Target
Process	<Application>	Process
Host-to-host	<Same transport>	Host-to-host
Internet	<IP>	Internet
Network interface	<Same protocol>	Network interface
	Network	

Figure 2.8 Source stack to and from the target stack.

Chapter

3

Network Interface Layer

At the end of the last chapter, we talked about the maximum transmission unit size of the network interface layer. Let's continue that discussion by looking into the details of the protocols that inhabit this layer. While TCP/IP uses these protocols to travel along the network from one point to another, its flexibility allows a different network interface layer protocol for each segment of media that the IP datagram travels. With that in mind, we will lightly review these protocols for familiarity.

We are purposely limiting our discussion to the most popular protocols for carrying TCP/IP. Since this is a book on TCP/IP instead of network interface protocols, we will look at Ethernet II, 802.3, 802.5 (token ring), Serial Line Internet Protocol, and Point-to-Point Protocol. There are many well-done books and RFCs on other protocols in case you need to know more about them.

The most popular of the network interface protocols is Ethernet II or DEC Intel Xerox (D.I.X.) Ethernet. As we saw in Fig. 2.7, the Ethernet II header is 14 bytes. It provides two addresses, a target and a source, along with a code to identify what it is carrying in its payload.

Ethernet II Addresses

Ethernet II addresses contain six 8-bit bytes (also called octets). That makes an Ethernet II address length of 48 bits that protocol analyzers display as 12 hexadecimal characters. They may indicate a specific target interface (unicast), a group of target interfaces with something in common (multicast), or all possible target interfaces within a portion of the network (broadcast).

The Ethernet II unicast addresses are easily recognizable by an even value in their first (most significant) byte since their least significant (multicast) bit contains a binary 0. Every Ethernet source hardware address must be a unicast address. Both the source and target addresses on the unicast row of Fig. 3.1 have that even value of zero (00).

Multicast addresses contain a multicast bit set to 1. That bit is the least

Packet	Target address	Source address
Unicast	00 00 c0 a0 51 24	00 00 0c 7d 4d 2c
Multicast	01 00 5e 00 00 00	Target only
Broadcast	ff ff ff ff ff ff	Target only

Figure 3.1 Ethernet II address table.

significant bit of the most significant (first) byte of the address (see Fig. 3.2). We can recognize an Ethernet multicast address as one whose first byte contains an odd value (01, 03, 05, 07, 09, 0b, 0d, 0f, through ff). The multicast address should not be a source address in an Ethernet packet header.

The ultimate multicast address includes all possible addresses in an environment. We identify these all-inclusive target hardware addresses as broadcast addresses. They contain all binary 1s, which protocol analyzers display as hexadecimal Fs "ff ff ff ff ff ff." The broadcast address should not be a source address in an Ethernet packet header.

The first three bytes (pairs of hexadecimal characters) of a unicast address contain the vendor address component. The last three bytes carry the serial number of that vendor's card. Together the 6 bytes become the network interface card (NIC) address (not Network Information Center) or media access control (MAC) address (not Macintosh).

Some vendors have registered their code with the Institute of Electrical and Electronic Engineers (IEEE). As a result, these addresses have their global address bit set to 1. This is the seventh bit of the most significant (first) byte of the vendor address field.

Many vendors have not registered their code, though they are careful not to use registered codes. Others are not as careful: A code may be the same on two or more vendor's NICs. If we install these cards (with the same MAC address) on the same LAN segment, the results will be unpredictable.

A collected (but not all registered) list of vendor address components appears in App. A at the end of this chapter. Some of those codes also appear in the Assigned Numbers RFC.

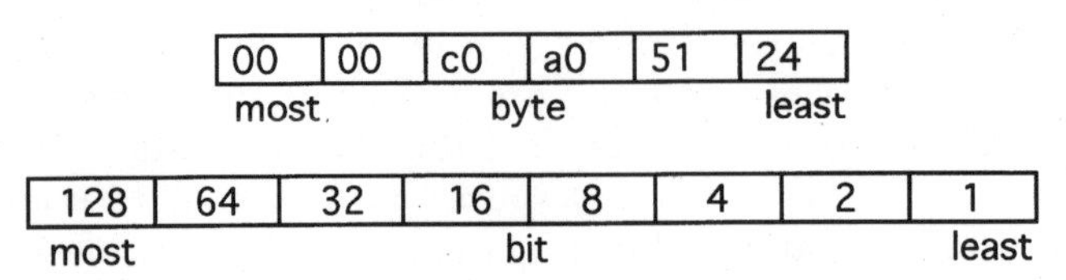

Figure 3.2 Byte and bit significance.

6 bytes	6 bytes	2 bytes	46–1500 bytes	4 bytes
Target hardware address	Source hardware address	Protocol type	Data	Cyclic redundancy check
00 00 c0 a0 51 24	00 00 0c 4d 7d 2c	08 00	IP	01 7f

Figure 3.3 Ethernet header layout.

Ethernet header fields

The target hardware address field is the first field in a packet. This way, the NICs know which packets to process as they watch the network. They will only keep packets whose target address matches their unicast address, one of their multicast addresses or the broadcast address.

The source hardware address field identifies the specific hardware card that originated the Ethernet frame. Remember, the specifications state that software drivers may not set a source address to a multicast or broadcast value in normal network operation. (After all, how could a packet come from many addresses at the same time?)

As you can see in Fig. 3.3, the protocol type field acts as a shipping label to identify what protocol at the next (IP) layer is to receive the contents of this packet (its payload) at the target end of the transmission.

The defined protocol data (payload) is next. We will discuss it further later in this chapter. The last field is a cyclic redundancy check that provides error detection for the packet. It uses a formula to calculate a value based on the actual header and data the packet carries. The receiving end system performs the same calculation to verify that it received the data the source sent.

802.3 Is Not Ethernet

There are times when it could be handy to have two different network interface layer protocols operating on the same physical network. For example, a small company that has only one physical cable for their network might not want to spend extra money for a bridge to separate the network traffic. Their solution is to send two different network protocols (Ethernet II and IEEE 802.3) over the same physical cabling.

They may want to separate their network traffic so that some workers will not have access to certain servers or the Internet by running part of the workstations with Ethernet II and the others with 802.3. Both will work on the same media since they are both Carrier Sense, Multiple Access with Collision Detection (CSMA/CD) protocols. They will not communicate with

6 bytes	6 bytes	2 bytes
Target hardware address	Source hardware address	Protocol type
00 00 c0 a0 51 24	00 00 0c 4d 7d 2c	08 00

6 bytes	6 bytes	2 bytes
Target hardware address	Source hardware address	Data length
00 00 c0 a0 51 24	00 00 0c 4d 7d 2c	05 db

Figure 3.4 Ethernet II header vs. IEEE 802.3 header.

each other because their headers contain different fields. The protocol that the server or Internet connection runs will dictate which stations have access to these resources.

As you can see from the comparison layout (Fig. 3.4) of the headers for IEEE 802.3 and Ethernet II, the fields at the 13th and 14th bytes of the two headers are different. In the Ethernet header, those two bytes contain the identification of the header's payload that we call the protocol type. A list of the valid protocol types is in the Assigned Numbers RFC (RFC 1700 as of this writing). We will find it in a section the RFC labels as "Ether type" (see App. B).

Rather than having you flip to the end of the chapter to look at it right now, Fig. 3.5 shows the first few entries from the Ether type section. Hex values from "00 00" through "05 dc" identify the packet as IEEE 802.3 (they are the length field values). The hex values above "05 dc" identify the packet as Ethernet and contain the protocol type field values. By making the values mutually exclusive, Xerox, IEEE, and the Internet Assigned Numbers Authority (IANA) make sure we can easily distinguish the correct Ethernet type.

To help in decoding and understanding the Ethernet II header, we include a listing of the Ether type code section from RFC 1700 (the current Assigned Numbers RFC), sorted alphabetically by name, in App. C.

Packet information

Ethernet II and 802.3 protocol rules set limitations on the size of the data that the header may carry and the overall size of the packet. The MTU specifies that the largest data that the software driver may enclose in an Ethernet II or 802.3 packet is 1500 bytes (Fig. 3.6).

Hex	Protocol
0000-05dc	IEEE 802.3 Length Field
0600	Xerox NS IDP
0660	DLOG
0661	DLOG
0800	Internet IP (IPv4)
0801	X.75 Internet
0802	NBS Internet
0803	ECMA Internet
0804	Chaosnet
0805	X.25 Level 3
0806	ARP
0807	XNS Compatability
0808	Frame Relay ARP
081C	Symbolics Private
0900	Ungermann-Bass Net Debugr
86DD	IPv6

Figure 3.5 Ethernet type examples.

If an Ethernet or 802.3 NIC discovers a packet larger than 1518 bytes (14-byte header, 1500 bytes of data, and a 4-byte CRC to detect errors) it considers it a "giant" packet and ignores it because it is too large. IP prevents this by fragmenting the data into smaller pieces that fit the MTU for travel along the network and reassembling those pieces into the original datagram at the destination system.

If an Ethernet or 802.3 NIC discovers a packet smaller than 64 bytes (the header, 46 bytes of data, and the CRC) it considers it a "runt" packet and ignores it because it is too small. The Ethernet (or 802.3) driver adds padding to the data to bring it up to at least 46 bytes before sending the packet onto the network. The destination system removes the padding before passing the data up the receiving stack.

Protocol	MTU
Ethernet	1500
IEEE 802.3	1500
IEEE 802.2 with SNAP	1492
SLIP	1006*
PPP	1500
Token Ring	4096*
FDDI	4096*
ATM	48*

*Typical; may vary.

Figure 3.6 Size limits.

6 bytes	6 bytes	2 bytes	3 bytes	5 bytes	38-1492 bytes
Target hardware address	Source hardware address	Data length	802.2 LLC	SNAP	Data
00 c0 93 1e 31 4c	00 00 0c 1b ff 37	00 7b	AA AA 03	00 00 00 08 00	IP

1 byte	1 byte	1 byte
DSAP	SSAP	Control

802.2 Logical Link Control Header

3 bytes	2 bytes
Organization identifier	Protocol type
00 00 00	08 00

Subnetwork Access Protocol Header

Figure 3.7 802.3 header layout.

IEEE 802.3

When 802.3 carries TCP/IP, we will see two more headers: an 802.2 Logical Link Control (LLC) header and a Subnetwork Access Protocol (SNAP) header (Fig. 3.7).

The 802.2 header has three single-byte fields that contain the hex values of "AA AA 03" when carrying TCP/IP. The SNAP header has two fields: a vendor address field and the Ether type field. The protocol driver sets the vendor address field to all zeros for TCP/IP. The Ether type contains the same information as the protocol type field in Ethernet. Those values are typically "08 00" for IP version 4, "08 06" for Address Resolution Protocol, or "80 35" for Reverse Address Resolution Protocol (RARP).

The same runt and giant limits apply to 802.3 as they do to Ethernet, though with the eight more bytes of the 802.2 and SNAP headers. These limits restrict the data field to 1492 bytes.

Token Ring or IEEE 802.5

One difference between 802.3 and 802.5 (carrying TCP/IP) is the additional two bytes on the front of a header that is otherwise very similar to 802.3. These are the access control byte and the frame control byte.

As Fig. 3.8 indicates, the access control byte uses

- A 3-bit priority field
- A single token bit to say if data is present
- A single monitor bit to make sure the receiving station got the data
- Three reservation bits that work with the priority bits to control ring access

1 byte	1 byte	6 bytes	6 bytes	2 bytes	3 bytes	5 bytes	n bytes	1 byte	1 byte
Access control	Frame control	Target address	Source address	Data length	802.2 LLC	SNAP	IP Data	End delimiter	Frame status

802.5 Header

3 bits	1 bit	1 bit	3 bits
Priority	Token	Monitor	Reservation

Access Control Byte

2 bits	6 bits
Frame type	Control

Frame Control Byte

1 bit	1 bit	2 bits	1 bit	1 bit	2 bits
Address recognized	Frame copied	Reserved	Address recognized	Frame copied	Reserved

Frame Status Byte

Figure 3.8 IEEE 802.5 header layout.

The frame control byte uses a 2-bit field to identify whether the frame contains logical link control data or is a media access control frame. If it is a MAC frame, the 6 control bits identify the type of MAC frame. These values are shown in Fig. 3.9.

Another difference between 802.3 and 802.5 is the bit order in the addresses. 802.3 transmits the least significant bit first. Token ring transmits the most significant bit first. This places the multicast address bit and the global address bit as the first two bits of the target address onto the media.

Another difference between 802.5 and 802.3 is the size of the data field. Since it is possible to send an empty token onto the ring, there is no minimum data requirement. The maximum is 16,000 bytes, though most implementations use 4096 or 2048 bytes as their MTU.

The remaining difference is the frame status byte at the end of packet. The last byte, after the frame check sequence and the end frame delimiter, contains values that indicate one of three possible circumstances:

Frame type	Bit pattern
Claim token	00 000011
Duplicate address test	00 000000
Active monitor present	00 000101
Standby monitor present	00 000110
Beacon	00 000010
Purge	00 000100

Figure 3.9 MAC frame control values.

- The destination station is not active on the ring
- The station is active but has not copied the packet
- The destination station copied the packet

Serial Line Interface Protocol

Serial Line Interface Protocol (SLIP) provides a common method for transmitting TCP/IP over serial data lines using RS-232 or V.35 physical connections. The challenge with SLIP is that there are few standards and many flavors that are not compatible with each other, particularly when using public domain SLIP software.

The SLIP header is very simple, as is evident in the Berkeley UNIX version in Fig. 3.10. It uses a 1-byte, c0 (hex), SLIP End (the beginning "end" of the header) to start the header and up to 3 bytes to complete the header. This variation allows for the possibility that the data may end with a byte that matches the SLIP End character's value of c0. That would require the addition of a SLIP Escape byte of db hex and a SLIP Final 1 (dc hex).

If the end of the SLIP payload was the exact hex string c0 db dc, the protocol would change to SLIP Final 2 (dd hex). The SLIP End with Final 2 contains the hex characters of c0 db dd. In either case, the amount of data between the SLIP Ends typically ranges from 0 to 1006 bytes (though larger amounts are allowed).

Notice that there is no error checking provision and no method to define the protocol that SLIP is carrying. Without a protocol identification, the two systems communicating over the SLIP connection must run the same single protocol, in this case TCP/IP.

1 byte	0–1006 bytes	1 byte
SLIP end	Data	SLIP end
c0	IP	c0

SLIP Header

1 byte	0–1006 bytes	3 bytes
SLIP end	Data	SLIP end
c0	IP	c0 db dc

SLIP Header with SLIP Final 1

1 byte	0–1006 bytes	3 bytes
SLIP end	Data	SLIP end
c0	IP	c0 db dd

SLIP Header with Slip Final 2

Figure 3.10 SLIP header layout.

1 byte	1 byte	1 byte	2 bytes	0–1500 bytes	2 bytes	1 byte
Flag	Address	Control	Protocol type	Data	Frame check	Flag
7E	FF	03	00 21	IP	7d 1e	7E

Figure 3.11 PPP header layout.

Point-to-Point Protocol

Point-to-Point Protocol (PPP) offers a more standardized method of transmitting TCP/IP over serial data lines. The developers designed it to remove many of SLIP's limitations such as data size, protocol identification, and error checking. For example, the PPP field gives us the ability to send multiple protocols across the link while SLIP can handle only IP.

The PPP header starts and ends with a flag character: 7E (hex). (See Fig. 3.11.) PPP then sets the (target) address field to FF (hex), or all binary 1s, to indicate the broadcast value. On a direct or point-to-point link, the only choice is the other end of the line. The control field uses the same 03 (hex) as the 802.2 header to identify unnumbered information.

While PPP has many similarities to Ethernet, it is also different. Both the Ethernet protocol type and the PPP field are 2 bytes, and both get their values from the Assigned Numbers RFC (RFC 1700 as of this writing). PPP uses 00 21 (hex) to identify the Internet protocol instead of Ethernet's (and others') use of 08 00 (hex).

The amount of data allowed is 0 to 1500 bytes. This makes PPP more compatible with Ethernet and 802.3 than most versions of SLIP. After the data is a 2-byte frame check sequence that covers the PPP header and its data (payload).

Appendix A Ethernet Vendor Address Components

Code	Organization
00000C	Cisco
00000E	Fujitsu
00000F	NeXT
000010	Hughes LAN Systems (formerly Sytek)
000011	Tektronix
000015	Datapoint Corporation
000018	Webster Computer Corporation Appletalk/Ethernet Gateway
00001A	AMD
00001B	Eagle Technology
00001D	Cabletron

Code	Organization
000020	DIAB (Data Intdustrier AB)
000021	SC&C
000022	Visual Technology
000023	ABB Automation AB, Dept. Q
000029	IMC
00002A	TRW
00003C	Auspex
00003D	AT&T
000044	Castelle
000046	ISC-Bunker Ramo, an Olivetti Company
000049	Apricot Ltd.
00004B	APT Appletalk WAN router
00004F	Logicraft 386-Ware P.C. Emulator
000051	Hob Electronic Gmbh & Co. KG
000052	ODS
000055	AT&T
00005A	SK (Syskonnect)
00005A	Xerox 806
00005D	RCE
00005E	U.S. Department of Defense (IANA)
00005F	Sumitomo
000061	Gateway Communications
000062	Honeywell
000065	Network General
000069	Silicon Graphics
00006B	MIPS
00006E	Artisoft, Inc.
000077	Interphase or MIPS, Motorola
000078	Labtam Australia
000079	Net Ware
00007A	Ardent
00007B	Research Machines
00007D	Cray Research ,Inc
00007F	Linotronic
000080	Dowty Network Services
000081	Synoptics
000084	ADI Systems Inc.

Code	Organization
000086	Gateway Communications Inc.
000089	Cayman Systems Gatorbox
00008A	Datahouse Information Systems
00008E	Jupiter
000093	Proteon
000094	Asante MAC
000095	Sony/Tektronix
000097	Epoch
000098	Cross Com
00009F	Ameristar Technology
0000A0	Sanyo Electronics
0000A2	Wellfleet
0000A3	Network Application Technology (NAT)
0000A4	Acorn
0000A5	Compatible Systems Corporation
0000A6	Network General
0000A7	Network Computing Devices (NCD) X-terminals
0000A8	Stratus Computer, Inc.
0000A9	Network Systems
0000AA	Xerox Xerox machines
0000AC	Apollo
0000AF	Nuclear Data Acquisition Interface Modules (AIM)
0000B0	RND (RAD Network Devices)
0000B1	Alpha Microsystems Inc.
0000B3	CIMLinc
0000B4	Edimax
0000B5	Datability Terminal Servers
0000B7	Dove Fastnet
0000BB	TRI-DATA Systems Inc. Netway products, 3274 emulators
0000BC	Allen-Bradley
0000C0	SMC (Std. Microsystems Corp.)
0000C6	HP Intelligent Networks Operation
0000C8	Altos
0000C9	Emulex Terminal Servers
0000CC	Densan Co., Ltd.
0000D0	Develcon Electronics, Ltd.
0000D1	Adaptec, Inc. "Nodem" product

Code	Organization
0000D3	Wang Labs
0000D4	PureData
0000D7	Dartmouth College (NED Router)
0000D8	Novell
0000DD	Gould
0000DE	Unigraph
0000E2	Acer Counterpoint
0000E3	Integrated Micro Products Ltd
0000E6	Aptor Produits De Comm Indust
0000E8	Accton Technology Corporation
0000E9	ISICAD, Inc.
0000ED	April
0000EE	Network Designers Limited
0000EF	Alantec
0000F0	Samsung
0000F3	Gandalf Data Ltd. - Canada
0000F4	Allied Telesis, Inc.
0000F6	A.M.C. (Applied Microsystems Corp.)
0000F8	DEC
0000FD	High Level Hardware
000102	BBN
000143	IEEE 802
000163	NDC (National Datacomm Corporation)
000168	W&G (Wandel & Goltermann)
0001C8	Thomas Conrad Corp.
000852	Technically Elite Concepts
000855	Fermilab
001700	Kabel
004088	Mobuis NuBus
00400B	Crescendo
00400C	General Micro Systems, Inc.
00400D	LANNET Data Communications
004010	Sonic Mac Ethernet interfaces
004014	Comsoft Gmbh
004015	Ascom
00401F	Colorgraph Ltd
004027	Sigma

Code	Organization
00402A	Canoga-Perkins
00402B	TriGem
00402F	XDI
004030	GK Computer
004033	Addtron Technology Co., Ltd.
00403C	Forks, Inc.
004041	Fujikura Ltd.
00404C	Hypertec Pty Ltd.
004050	Ironics, Incorporated
00405B	Funasset Limited
004066	Hitachi Cable, Ltd.
004068	Extended Systems
00406E	Corollary, Inc.
004074	Cable and Wireless
004076	AMP Incorporated
00407F	Agema Infrared Systems AB
00408C	Axis Communications AB
00408E	CXR/Digilog
004092	ASP Computer Products, Inc.
004095	Eagle Technologies
00409D	DigiBoard Ethernet-ISDN bridges
00409E	Concurrent Technologies Ltd.
0040A6	Cray Research Inc.
0040AE	Delta Controls, Inc.
0040B4	3COM K.K.
0040B6	Computerm Corporation
0040C1	Bizerba-Werke Wilheim Kraut
0040C2	Applied Computing Devices
0040C3	Fischer and Porter Co.
0040C5	Micom Communications Corp.
0040C6	Fibernet Research, Inc.
0040C8	Milan Technology Corp.
0040D4	Gage Talker Corp.
0040DF	Digalog Systems, Inc.
0040E7	Arnos Instruments & Computer
0040E9	Accord Systems, Inc.
0040F1	Chuo Electronics Co., Ltd.

Code	Organization
0040F4	Cameo Communications, Inc.
0040F9	Combinet
0040FB	Cascade Communications Corp.
00608C	3Com
008004	Antlow Computers, Ltd.
008005	Cactus Computer Inc.
008006	Compuadd Corporation
008007	Dlog NC-Systeme
00800F	SMC (Standard Microsystem Corp.)
008010	Commodore
008017	PFU
008019	Dayna Communications “Etherprint” product
00801A	Bell Atlantic
00801B	Kodiak Technology
008021	Newbridge Networks Corporation
008023	Integrated Business Networks
008024	Kalpana
008029	Microdyne Corporation
00802D	Xylogics, Inc. Annex terminal servers
00802E	Plexcom, Inc.
008033	Formation
008034	SMT-Goupil
008035	Technology Works
008037	Ericsson Business Comm.
008038	Data Research & Applications
00803B	APT Communications, Inc.
00803E	Synernetics
00803F	Hyundai Electronics
008042	Force Computers
00804C	Contec Co., Ltd.
00804D	Cyclone Microsystems, Inc.
008051	ADC Fibermux
008052	Network Professor
00805B	Condor Systems, Inc.
00805C	Agilis
008060	Network Interface Corporation
008062	Interface Co.

Code	Organization
008069	Computone Systems
00806A	ERI (Empac Research Inc.)
00806C	Cegelec Projects Ltd
00806D	Century Systems Corp.
008074	Fisher Controls
00807B	Artel Communications Corp.
00807C	FiberCom
008086	Computer Generation Inc.
008087	Okidata
00808A	Summit
00808B	Dacoll Limited
00808C	Frontier Software Development
008092	Japan Computer Industry, Inc.
008096	Human Designed Systems X terminals
00809D	Datacraft Manufactur'g Pty Ltd
00809F	Alcatel Business Systems
0080A1	Microtest
0080A3	Lantronix
0080AD	Telebit
0080AE	Hughes Network Systems
0080AF	Allumer Co., Ltd.
0080B2	NET (Network Equipment Technologies)
0080C0	Penril
0080C2	IEEE 802.1 Committee
0080C7	Xircom, Inc.
0080C8	D-Link
0080C9	Alberta Microelectronic Centre
0080CE	Broadcast Television Systems
0080D0	Computer Products International
0080D3	Shiva Appletalk-Ethernet interface
0080D4	Chase Limited
0080D6	Apple
0080D7	Fantum Electronics
0080D8	Network Peripherals
0080DA	Bruel & Kjaer
0080E3	Coral
0080F1	Opus

Code	Organization
0080F7	Zenith Communications Products
0080FB	BVM Limited
00AA00	Intel
00B0D0	Computer Products International
00C001	Diatek Patient Managment
00C004	Japan Business Computer Co. Ltd
00C016	Electronic Theatre Controls
00C01A	Corometrics Medical Systems
00C01C	Interlink Communications Ltd.
00C01D	Grand Junction Networks, Inc.
00C020	Arco Electronic, Control Ltd.
00C024	Eden Sistemas De Computacao SA
00C025	Dataproducts Corporation
00C027	Cipher Systems, Inc.
00C028	Jasco Corporation
00C02B	Gerloff Gesellschaft Fur
00C02C	Centrum Communications, Inc.
00C02D	Fuji Photo Film Co., Ltd.
00C030	Integrated Engineering B. V.
00C031	Design Research Systems, Inc.
00C032	I-Cubed Limited
00C034	Dale Computer Corporation
00C040	ECCI
00C042	Datalux Corp.
00C044	Emcom Corporation
00C048	Bay Technical Associates
00C04E	Comtrol Corporation
00C051	Advanced Integration Research
00C05C	Elonex PLC
00C066	Docupoint, Inc.
00C06D	Boca Research, Inc.
00C071	Areanex Communications, Inc.
00C078	Computer Systems Engineering
00C091	Jabil Circuit, Inc.
00C093	Alta Research Corp.
00C097	Archipel SA
00C098	Chuntex Electronic Co., Ltd.

Code	Organization
00C09D	Distributed Systems Int'l, Inc
00C0A0	Advance Micro Research, Inc.
00C0A2	Intermedium A/S
00C0A8	GVC Corporation
00C0AC	Gambit Computer Communications
00C0AD	Computer Communication Systems
00C0B0	GCC Technologies, Inc.
00C0B8	Fraser's Hill Ltd.
00C0BD	Inex Technologies, Inc.
00C0BE	Alcatel - Sel
00C0C2	Infinite Networks Ltd.
00C0C4	Computer Operational
00C0CA	Alfa, Inc.
00C0CB	Control Technology Corporation
00C0D1	Comtree Technology Corporation
00C0D6	J1 Systems, Inc.
00C0DC	EOS Technologies, Inc.
00C0E2	Calcomp, Inc.
00C0E7	Fiberdata AB
00C0EA	Array Technology Ltd.
00C0EC	Dauphin Technology
00C0EF	Abit Corporation
00C0F4	Interlink System Co., Ltd.
00C0F6	Celan Technology Inc.
00C0F7	Engage Communication, Inc.
00C0F8	About Computing Inc.
00C0FB	Advanced Technology Labs
00DD00	Ungermann-Bass IBM RT
00DD01	Ungermann-Bass
00DD08	Ungermann-Bass
020406	Bolt Beranek and Newman, Inc. internal use
020701	MICOM/Interlan
026060	3Com
026086	Satelcom MegaPac (UK)
02608C	3Com IBM PC; Imagen; Valid; Cisco; Macintosh
02CF1F	CMC Masscomp; Silicon Graphics; Prime EXL
02E6D3	BTI (Bus-Tech, Inc.)

Code	Organization
080001	Computer Vision
080002	3Com
080003	ACC (Advanced Computer Communications)
080005	Symbolics Symbolics LISP machines
080006	Siemens Nixdorf PC clone
080007	Apple
080008	BBN (Bolt Beranek and Newman, Inc.)
080009	Hewlett-Packard
08000A	Nestar Systems
08000B	Unisys
08000D	ICL (International Computers, Ltd.)
08000E	NCR/AT&T
08000F	SMC (Standard Microsystems Corp.)
080010	AT&T
080011	Tektronix, Inc.
080014	Excelan
080017	NSC (Network System Corp.)
08001A	Data General
08001B	Data General
08001E	Apollo
08001F	Sharp
080020	Sun
080022	NBI (Nothing But Initials)
080023	Matsushita Denso
080025	CDC
080026	Norsk Data (Nord)
080027	PCS Computer Systems GmbH
080028	TI Explorer
08002B	DEC
08002E	Metaphor
08002F	Prime Computer Prime 50-Series LHC300
080030	CERN
080036	Intergraph CAE stations
080037	Fujitsu-Xerox
080038	Bull
080039	Spider Systems
08003B	Torus Systems

Code	Organization
08003E	Motorola VME bus processor modules
080041	DCA (Digital Comm. Assoc.)
080044	DSI (DAVID Systems, Inc.)
080046	Sony
080047	Sequent
080048	Eurotherm Gauging Systems
080049	Univation
08004C	Encore
08004E	BICC
080051	Experdata
080056	Stanford University
080057	Evans & Sutherland
080058	DEC
08005A	IBM
080067	Comdesign
080068	Ridge
080069	Silicon Graphics
08006A	AT&T
08006E	Excelan
080070	Mitsubishi
080074	Casio
080075	DDE (Danish Data Elektronik A/S)
080077	Retix
080079	Silicon Graphics
08007C	Vitalink TransLAN III
080080	XIOS
080081	Crosfield Electronics
080083	Seiko Denshi
080086	Imagen/QMS
080087	Xyplex terminal servers
080089	Kinetics AppleTalk-Ethernet interface
08008B	Pyramid
08008D	XyVision XyVision machines
08008E	Tandem Computer
08008F	Chipcom Corp.
080090	Retix, Inc. Bridges
10005A	IBM

Code	Organization
1000D4	DEC
1000E0	Apple
400003	Net Ware
444649	DFI (Diamond Flower Industries)
475443	GTC
484453	HDS
800010	AT&T
80AD00	CNET Technology Inc.
AA0000	DEC obsolete
AA0001	DEC obsolete
AA0002	DEC obsolete
AA0003	DEC Global physical address for some DEC machines
AA0004	DEC Local logical address for systems running DECNET

Appendix B Ether Types (Protocol Types) by Code

Hex	Protocol
0000-05DC	IEEE 802.3 Length Field
0101-01FF	Experimental
0200	XEROX PUP (*see* 0A00)
0201	PUP Addr Trans (*see* 0A01)]
0400	Nixdorf
0600	XEROX NS IDP
0660	DLOG
0661	DLOG
0800	Internet IP (IPv4)
0801	X.75 Internet
0802	NBS Internet
0803	ECMA Internet
0804	Chaosnet
0805	X.25 Level 3
0806	ARP
0807	XNS Compatability
0808	Frame Relay ARP
081C	Symbolics Private
0888-088A	Xyplex
0900	Ungermann-Bass net debugr
0A00	Xerox IEEE802.3 PUP

Hex	Protocol
0A01	PUP Addr Trans
0BAD	Banyan VINES
0BAE	VINES Loopback
0BAF	VINES Echo
1000	Berkeley Trailer nego
1001-100F	Berkeley Trailer encap/IP
1600	Valid Systems
4242	PCS Basic Block Protocol
5208	BBN Simnet
6000	DEC Unassigned (Exp.)
6001	DEC MOP Dump/Load
6002	DEC MOP Remote Console
6003	DEC DECNET Phase IV Route
6004	DEC LAT
6005	DEC Diagnostic Protocol
6006	DEC Customer Protocol
6007	DEC LAVC, SCA
6008-6009	DEC Unassigned
6010-6014	3Com Corporation
6558	Trans Ether Bridging
6559	Raw Frame Relay
7000	Ungermann-Bass download
7002	Ungermann-Bass dia/loop
7020-7029	LRT
7030	Proteon
7034	Cabletron
8003	Cronus VLN
8004	Cronus Direct
8005	HP Probe
8006	Nestar
8008	AT&T
8010	Excelan
8013	SGI diagnostics
8014	SGI network games
8015	SGI reserved
8016	SGI bounce server
8019	Apollo Domain

Hex	Protocol
802E	Tymshare
802F	Tigan, Inc.
8035	Reverse ARP
8036	Aeonic Systems
8038	DEC LANBridge
8039-803C	DEC Unassigned
803D	DEC Ethernet Encryption
803E	DEC Unassigned
803F	DEC LAN Traffic Monitor
8040-8042	DEC Unassigned
8044	Planning Research Corp.
8046	AT&T
8047	AT&T
8049	ExperData
805B	Stanford V Kernel exp.
805C	Stanford V Kernel prod.
805D	Evans & Sutherland
8060	Little Machines
8062	Counterpoint Computers
8065	Univ. of Mass. @ Amherst
8066	Univ. of Mass. @ Amherst
8067	Veeco Integrated Auto.
8068	General Dynamics
8069	AT&T
806A	Autophon
806C	ComDesign
806D	Computgraphic Corp.
806E-8077	Landmark Graphics Corp.
807A	Matra
807B	Dansk Data Elektronik
807C	Merit Internodal
807D-807F	Vitalink Communications
8080	Vitalink TransLAN III
8081-8083	Counterpoint Computers
809B	Appletalk
809C-809E	Datability
809F	Spider Systems Ltd.

Hex	Protocol
80A3	Nixdorf Computers
80A4-80B3	Siemens Gammasonics Inc.
80C0-80C3	DCA Data Exchange Cluster
80C4	Banyan Systems
80C5	Banyan Systems
80C6	Pacer Software
80C7	Applitek Corporation
80C8-80CC	Intergraph Corporation
80CD-80CE	Harris Corporation
80CF-80D2	Taylor Instrument
80D3-80D4	Rosemount Corporation
80D5	IBM SNA Service on Ether
80DD	Varian Associates
80DE-80DF	Integrated Solutions TRFS
80E0-80E3	Allen-Bradley
80E4-80F0	Datability
80F2	Retix
80F3	AppleTalk AARP (Kinetics)
80F4-80F5	Kinetics
80F7	Apollo Computer
80FF-8103	Wellfleet Communications
8107-8109	Symbolics Private
8130	Hayes Microcomputers
8131	VG Laboratory Systems
8132-8136	Bridge Communications
8137-8138	Novell, Inc.
8139-813D	KTI
8148	Logicraft
8149	Network Computing Devices
814A	Alpha Micro
814C	SNMP
814D	BIIN
814E	BIIN
814F	Technically Elite Concept
8150	Rational Corp
8151-8153	Qualcomm
815C-815E	Computer Protocol Pty Ltd

Hex	Protocol
8164-8166	Charles River Data System
817D-818C	Protocol Engines
818D	Motorola Computer
819A-81A3	Qualcomm
81A4	ARAI Bunkichi
81A5-81AE	RAD Network Devices
81B7-81B9	Xyplex
81CC-81D5	Apricot Computers
81D6-81DD	Artisoft
81E6-81EF	Polygon
81F0-81F2	Comsat Labs
81F3-81F5	SAIC
81F6-81F8	VG Analytical
8203-8205	Quantum Software
8221-8222	Ascom Banking Systems
823E-8240	Advanced Encryption System
827F-8282	Athena Programming
8263-826A	Charles River Data System
829A-829B	Inst Ind Info Tech
829C-82AB	Taurus Controls
82AC-8693	Walker Richer & Quinn
8694-869D	Idea Courier
869E-86A1	Computer Network Tech
86A3-86AC	Gateway Communications
86DB	SECTRA
86DE	Delta Controls
86DF	ATOMIC
86E0-86EF	Landis & Gyr Powers
8700-8710	Motorola
876B	TCP/IP Compression
876C	IP Autonomous Systems
876D	Secure Data
8A96-8A97	Invisible Software
9000	Loopback
9001	3Com(Bridge) XNS Sys Mgmt
9002	3Com(Bridge) TCP-IP Sys
9003	3Com(Bridge) loop detect

FF00	BBN VITAL-LanBridge cache
FF00-FF0F	ISC Bunker Ramo
FFFF	Reserved

Appendix C Ether Types by Name

Organization	Code(s)
3Com Corporation	6010-6014
3Com(Bridge) loop detect	9003
3Com(Bridge) TCP-IP Sys	9002
3Com(Bridge) XNS Sys Mgmt	9001
Advanced Encryption System	823E-8240
Aeonic Systems	8036
Allen-Bradley	80E0-80E3
Alpha Micro	814A
Apollo Computer	80F7
Apollo Domain	8019
AppleTalk AARP (Kinetics)	80F3
Appletalk	809B
Applitek Corporation	80C7
Apricot Computers	81CC-81D5
ARAI Bunkichi	81A4
ARP	0806
Artisoft	81D6-81DD
Ascom Banking Systems	8221-8222
AT&T	8008, 8046, 8047, 8069
Athena Programming	827F-8282
ATOMIC	86DF
Autophon	806A
Banyan Systems	80C4, 80C5
Banyan VINES	0BAD
BBN Simnet	5208
BBN VITAL-LanBridge cache	FF00
Berkeley Trailer encap/IP	1001-100F
Berkeley Trailer nego	1000
BIIN	814D, 814E
Bridge Communications	8132-8136
Cabletron	7034
Chaosnet	0804

Organization	Code(s)
Charles River Data System	8164-8166, 8263-826A
ComDesign	806C
Computer Network Tech	869E-86A1
Computer Protocol Pty Ltd	815C-815E
Computgraphic Corp.	806D
Comsat Labs	81F0-81F2
Counterpoint Computers	8062, 8081-8083
Cronus Direct	8004
Cronus VLN	8003
Dansk Data Elektronik	807B
Datability	80E4-80F0, 809C-809E
DCA Data Exchange Cluster	80C0-80C3
DEC Customer Protocol	6006
DEC DECNET Phase IV Route	6003
DEC Diagnostic Protocol	6005
DEC Ethernet Encryption	803D
DEC LAN Traffic Monitor	803F
DEC LANBridge	8038
DEC LAT	6004
DEC LAVC, SCA	6007
DEC MOP Dump/Load	6001
DEC MOP Remote Console	6002
DEC Unassigned	6008-6009, 8039-803C, 803E, 8040-8042
DEC Unassigned (Exp.)	6000
Delta Controls	86DE
DLOG	0660, 0661
ECMA Internet	0803
Evans & Sutherland	805D
Excelan	8010
ExperData	8049
Experimental	0101-01FF
Frame Relay ARP	0808
Gateway Communications	86A3-86AC
General Dynamics	8068
Harris Corporation	80CD-80CE
Hayes Microcomputers	8130
HP Probe	8005

Organization	Code(s)
IBM SNA Service on Ether	80D5
Idea Courier	8694-869D
IEEE 802.3 Length Field	0000-05DC
Inst Ind Info Tech	829A-829B
Integrated Solutions TRFS	80DE-80DF
Intergraph Corporation	80C8-80CC
Internet IP (IPv4)	0800
Internet IP (IPv6)	86DD
Invisible Software	8A96-8A97
IP Autonomous Systems	876C
ISC Bunker Ramo	FF00-FF0F
Kinetics	80F4-80F5
KTI	8139-813D
Landis & Gyr Powers	86E0-86EF
Landmark Graphics Corp.	806E-8077
Little Machines	8060
Logicraft	8148
Loopback	9000
LRT	7020-7029
Matra	807A
Merit Internodal	807C
Motorola	8700-8710
Motorola Computer	818D
NBS Internet	0802
Nestar	8006
Network Computing Devices	8149
Nixdorf	0400
Nixdorf Computers	80A3
Novell, Inc.	8137-8138
Pacer Software	80C6
PCS Basic Block Protocol	4242
Planning Research Corp.	8044
Polygon	81E6-81EF
Proteon	7030
Protocol Engines	817D-818C
PUP Addr Trans	0A01
PUP Addr Trans (*see* 0A01)	0201

Organization	Code(s)
Qualcomm	819A-81A3, 8151-8153
Quantum Software	8203-8205
RAD Network Devices	81A5-81AE
Rational Corp	8150
Raw Frame Relay	6559
Reserved	FFFF
Retix	80F2
Reverse ARP	8035
Rosemount Corporation	80D3-80D4
SAIC	81F3-81F5
SECTRA	86DB
Secure Data	876D
SGI bounce server	8016
SGI diagnostics	8013
SGI network games	8014
SGI reserved	8015
Siemens Gammasonics Inc.	80A4-80B3
SNMP	814C
Spider Systems Ltd.	809F
Stanford V Kernel exp.	805B
Stanford V Kernel prod.	805C
Symbolics Private	081C, 8107-8109
Taurus Controls	829C-82AB
Taylor Instrument	80CF-80D2
TCP/IP Compression	876B
Technically Elite Concept	814F
Tigan, Inc.	802F
Trans Ether Bridging	6558
Tymshare	802E
Ungermann-Bass dia/loop	7002
Ungermann-Bass download	7000
Ungermann-Bass net debugr	0900
Univ. of Mass. @ Amherst	8065, 8066
Valid Systems	1600
Varian Associates	80DD
Veeco Integrated Auto.	8067
VG Analytical	81F6-81F8

Organization	Code(s)
VG Laboratory Systems	8131
VINES Echo	0BAF
VINES Loopback	0BAE
Vitalink Communications	807D-807F
Vitalink TransLAN III	8080
Walker Richer & Quinn	82AC-8693
Wellfleet Communications	80FF-8103
X.25 Level 3	0805
X.75 Internet	0801
XEROX NS IDP	0600
XEROX PUP (*see* 0A00)	0200
Xerox IEEE802.3 PUP	0A00
XNS Compatability	0807
Xyplex	81B7-81B9, 0888-088A

Chapter

4

IP Networks, Subnets, and Hosts

Overview

We usually write IP addresses in a byte-based decimal format called dotted decimal notation (see Fig. 4.1). We say them without the decimals in the same way we leave out the parentheses or hyphens in telephone numbers. For example, in Fig. 4.1 we say the IP address as "one ninety-one two fifty-five one eighteen one twenty-three."

IP addresses are 4 bytes in length. By looking at the value in the first byte of the address, the IP software can know the number of available, unique IP addresses in that network. It can also use that information to determine which bits are network bits and which bits the network administrator controls locally. The first byte value indicates which class (group of networks) contains the address.

With four classes in active use (Classes A, B, C, and D), these determinations are necessary for address resolution and accurate delivery. As we noted earlier, RS.InterNIC.Net assigns the network portion of the IP address. It does this based on the number of systems the given organization will require.

The locally controlled bits can identify the interfaces, often individual systems, on a network. Some network administrators use part of the local bits to make more manageable pieces called subnets. This way, network managers can create logical groups of users they can monitor and manage. The advantage comes in narrowing the area to troubleshoot in case of a network problem.

Decimal	191	.	255	.	118	.	123
Binary	10111111		11111111		01110110		01111011
Logical	Network		Network		Subnet		Interface

Figure 4.1 Decimal and binary IP address examples.

This local control of certain bits means that there can be network, subnet, and host fields in an IP address. Some rules apply to the IP address and those fields as represented in a binary format:

- No field of an interface's IP address may contain all 1s or all 0s.
- All 1s in the host field of a target IP address indicates an IP broadcast to all interfaces on that network or subnet.
- All 0s in the host portion of a target IP address identifies a subnet or a network rather than a specific interface.

If the last two rules seem to violate the first rule, remember that the key is that no *field* (network, subnet, or interface) of a (physical) *interface's* IP address may contain all 1s or all 0s (in the binary format). For example, the second byte of the IP address in Fig. 4.1 is only part of the 2-byte network field.

Class A

If the first bit of an IP address is 0, the software sees it as a Class A address. Their first byte values are from 1 to 127. This seems to offer 127 Class A networks, but 127.0.0.1 is the universal address for IP loopback testing. So, we actually have only 126 Class A addresses available.

With RS.InterNIC.Net assigning the first byte (8 bits), we recognize the remaining three bytes (24 bits) as locally administered. In Fig. 4.2, we designate the network bits with an N and the locally administered bits with an L. If these 24 L bits were only host addresses, we could have 16,777,214 host addresses (with all 0 and 1 addresses eliminated). Class A addresses allow for the largest number of unique addresses on a single network. Since there are so few, the InterNIC rarely assigns them.

With so many possible interface addresses, RS.InterNIC.Net only assigns Class A networks to large organizations such as Ford Motor Company or Massachusetts Institute of Technology. To manage a network of that size, the network administrator frequently separates the network into smaller subnetworks (subnets), which we will examine later in this chapter.

As part of the concern over running out of IP addresses, Class A network addresses have been very difficult to get lately. So difficult that a large

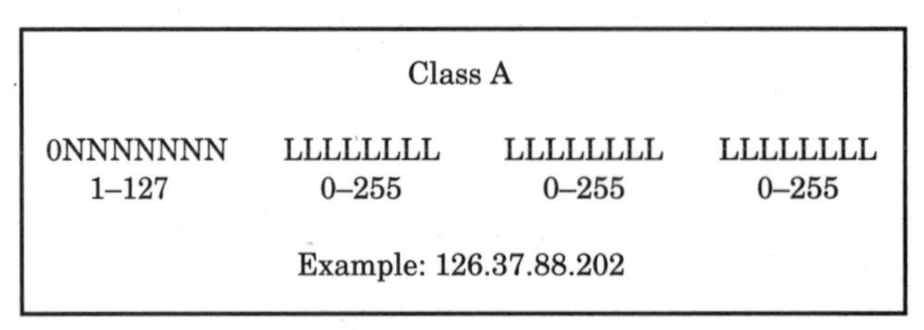

Figure 4.2 Class A IP address format.

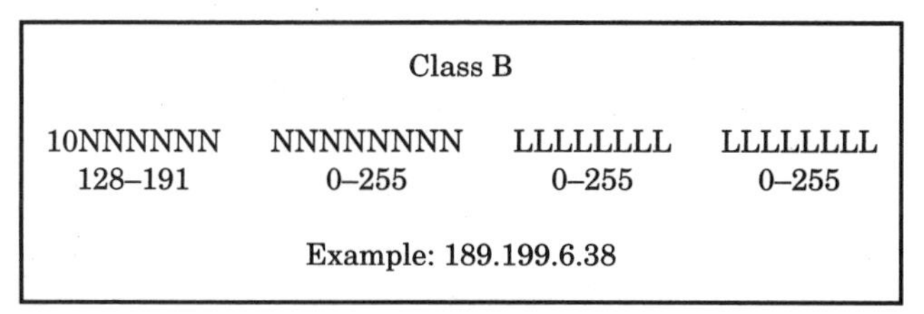

Figure 4.3 Class B IP address format.

organization like AMOCO received 23 Class B networks instead of one Class A.

Class B

When the first bit of an IP address is 1 and the second bit is 0, the computer recognizes it as a Class B network address (Fig. 4.3). Class B IP addresses carry a first-byte value from 128 through 191. Class B network assignments occupy the first two bytes. That offers 16,384 possible Class B addresses with 2 bytes of local space each. Without subnetting, a local space of that size offers 65,534 addresses in each Class B network.

Managing a network of that size, the administrator frequently separates the network into subnets. This can help with network traffic management and help control workgroup access to certain resources. The InterNIC assigns Class B networks to midsize organizations such as colleges and universities. Examples of Class B networked organizations include University of Southern California, Hughes Aircraft Company, Clarkson University, and Proteon, Inc.

If an organization cannot show enough network address need to justify a Class B address, RS.InterNIC.Net may assign multiple Class C networks.

Class C

Setting the first two bits to 1 and the third bit to 0 identifies the IP address as a Class C address (Fig. 4.4). The first byte holds a decimal value from 192 through 223. Class C addresses use the first three bytes. This creates 2,097,152 possible Class C addresses. Each Class C network, with 1 byte of local space, offers 254 interface addresses.

Class C

110NNNNN	NNNNNNNN	NNNNNNNN	LLLLLLLL
192–223	0–255	0–255	0–255

Example: 192.153.186.26

Figure 4.4 Class C IP address format.

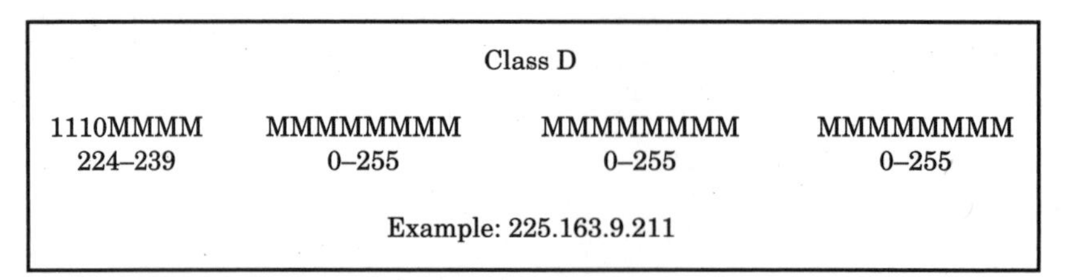

Class D			
1110MMMM	MMMMMMMM	MMMMMMMM	MMMMMMMM
224–239	0–255	0–255	0–255

Example: 225.163.9.211

Figure 4.5 Class D IP address format.

Class C networks do not need subnetting for management unless there are smaller workgroups in diverse locations. More often, organizations will subnet Class C networks to restrict access to specific resources. For example, an access provider may let designated subnets have access to the Internet and restrict access to the NSFnet for security or other reasons.

To streamline the routing process (passing datagrams from the source system through one or more routers to the target system), the InterNIC allocated blocks of Class C addresses to Internet Service Providers to assign to their customers. This changes the source of IP addresses from RS.InterNIC.Net to the ISPs.

Class D

When the first three bits have a value of 1 and the fourth bit is 0, the IP address is a Class D address (Fig. 4.5). The first byte decimal value is from 224 through 239.

Class D addresses are the IP layer multicast addresses. We can reach groups of IP-addressed nodes by assigning the same multicast address to all members of the group. These group member interfaces also have their own individual, Class A, B, or C interface IP address. In other words, an interface will have a unique IP address and may also have one or more Class D multicast addresses by being part of selected multicast group(s).

There are more than 265 million possible multicast addresses. These group addresses can provide access to geographically diverse groups or peer functions in large organizations, for example, a certain level of commanders in the military.

The RFCs designate only Class D addresses for groups of users. That means Class D networks do not have network, subnet, or interface fields.

IP Communications Logic Process

When any IP-addressed interface wants to communicate with any other IP-addressed interface, the software follows a logic process of up to three steps. After finding the target IP address involved, the originating software checks the class of the target IP address to determine which bits are network bits in that address.

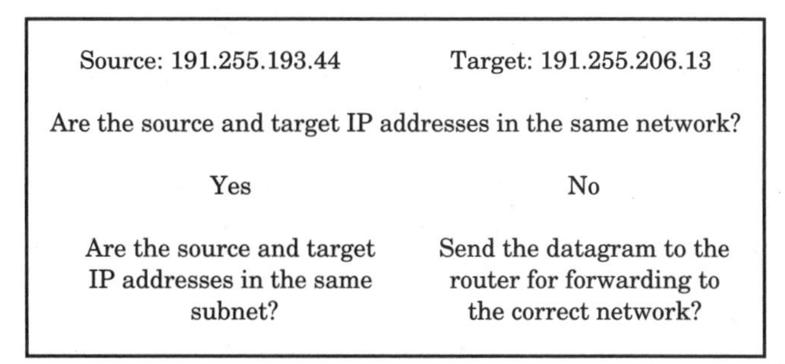

Figure 4.6 The IP communications logic process begins.

Using that information, it completes the first step of the process, answering the question, "Are the source and target IP addresses in the same network?" Matching the network portion of both addresses quickly reveals the answer.

- If *no,* IP sends the datagram to a router for delivery to the right network.
- If *yes,* the system must take the next step to decide if the IP addresses are in the same subnet.

The first byte's value of 191 identifies that the source and target addresses in Fig. 4.6 are Class B addresses. The InterNIC assigns the values of the first two bytes in Class B addresses. The values in the first two bytes of the source (191.255) and the target (191.255) match, so they must be in the same Class B network. With the *yes* answer, we must decide if they are in the same subnet. To move to that second step, we must examine subnetting.

Subnet

A *subnet* is a subset of a network that supports a workgroup, department, section, or some other portion of an organization. In network terms, it is a locally defined grouping of addresses that the network manager or administrator wants to treat as a logical group.

The primary reason to create a subnet is to make the overall network easier to manage by separating it into smaller pieces. By doing this, the rest of the network will be free of interruptions from the problems in a subnet. Network managers and technicians can isolate the problem area and make any needed repairs or adjustments more quickly.

This is not rocket science. We have all been doing this with fixed length fields most of our lives. Telephone numbers operate in the same manner. Consider the U.S. telephone number (222) 333-4444. The 222 area code is the equivalent of the network portion of an address. The 333 exchange is equivalent to the subnet portion of an IP address, and the 4444 extension is equivalent to the host or interface portion of the address.

When to subnet

Network administrators will find that there are several events that may trigger the decision to subnet. These include the following:

- Workgroups are geographically remote from each other and there are enough WAN links that routers (instead of bridges) provide the connectivity.
- Functional areas or management wants to separate that portion of the network from the rest of the organization.
- Certain workgroups' network traffic volumes interfere with other workgroups. For example, research and development uses computer-aided design (CAD), which is very network intensive (high network traffic volume) and conflicts with marketing's heavy printer use. Though the two departments share the same building, putting them on the same network could cause more traffic than the network can bear. If configured on separate subnets and separate sides of a router, however, their traffic will not interfere with each other and they can communicate with each other and the rest of the organization smoothly.
- Management has realigned the functional areas (right-sizing), and the resulting areas will have separate networking requirements. Perhaps a realignment combines two groups into one building, one area, or one manager, though the functions remain separate.
- When two different media protocols connect with each other they should connect through a router. Some bridges can translate token-ring traffic to send it out onto an Ethernet network. This is usually not the best way to go, however. The best method for connecting those two protocols is to install a router with a token-ring NIC and an Ethernet NIC.

  The guideline of having each media protocol on a separate subnet has two exceptions. First, the different media protocols could be on separate Class C networks rather subnetting a Class C network. Second, media protocols with a common base can run on the same subnet without talking with each other. For example, IEEE 802.3, IEEE 802.2, and Ethernet II are all CSMA/CD and run over the same physical media types (10Base2 or ThinNet, 10Base5 or ThickNet, 10BaseT or twisted-pair cabling, 10BaseF or fiber-optic cabling, etc.).
- The network traffic on a bridged Ethernet segment is constantly at, or above, 30 percent. The token-ring traffic volume threshold is 50 percent.
- Anytime a router connects two or more network segments for any of the reasons we gave, it is time to use subnetting. Subnetting without a router is a fancy numbering system that may help locate IP addresses, but it does nothing to separate traffic. A router without subnetting is expecting separate IP-addressed networks (probably Class C) on each of its interfaces. If the interfaces are on separate cables with no subnetting, the router will not know which interface is to receive the data traffic.

Subnetting IP networks

As we have seen, the main reason for subnetting is to better manage the overall network. To help in managing the network, subnetting creates groups of systems to

- Separate local traffic from the rest of the network
- Physically separate systems from the rest of the network or other groups
- Restrict access to certain network resources

To recognize the subnet, each system in the subnet must have the same subnet mask. This mask is not like a Halloween mask, in that instead of hiding, this mask reveals how we should look at the IP addresses in the subnet. The subnet mask identifies which bits are network bits, subnet bits, and interface bits. In this way, it acts as a filter through which we see the parts of the IP address.

To do this, the mask must have the same number of bits as the IP address (32), and we often name it in the same format as the IP address. Though the mask has the same size and layout as an IP address, its first byte always has a value of 255 (decimal) or ff (hex), which is not possible for an IP-addressed interface.

In the mask (more clearly seen in the binary format in Fig. 4.7), binary 1s indicate the position of the network and subnet portion of the IP address, and binary 0s identify those bits that can represent unique (end station) interfaces.

In our previous U.S. telephone number analogy, the invisible, fixed field length, subnet mask, which we have all been taking for granted, would be 111 111 0000 to identify three digits of network, three digits of subnet, and four digits of interface. IP subnetting has the advantage of letting us adjust the size of the subnet and host fields that telephone numbers limit in size.

Subnet calculations

When a network administrator decides to subnet, the first two decisions are

- How many subnets might the organization need from this network?
- What is the greatest number of interfaces that must be in the largest subnet?

11111111	11111111	11111100	00000000	Binary
255	255	252	0	Decimal
ff	ff	fc	00	Hex
nnnnnnnn	nnnnnnnn	SSSSSShh	hhhhhhhh	Meaning

Figure 4.7 Subnet mask.

$(2^n) - 2$ = the number of subnets or hosts in a subnet
n = the number of bits used in the mask

For example,
With 4 bits in the host field of the mask,
$(2^4) - 2 = (16) - 2 = 14$ hosts

With 3 bits in the subnet field of the mask,
$(2^3) - 2 = (8) - 2 = 6$ subnets

Figure 4.8 Subnet formula.

We combine the answers with the limits of the class of the network that we are subnetting and the rules set for IP addressing. For example, with the rule that no field (network, subnet, or host) may contain all 1s or all 0s (binary), there can be no mask with a subnet field (or a host field) of one bit.

To calculate the number of hosts or subnetworks in any IP-addressed network, apply the formula shown in Fig. 4.8: Take 2 to the power of the number of bits used in the subnet mask for the field desired and subtract 2 from the result (for all 1s and all 0s).

The subnet example shows 3 subnet bits in the mask. The calculation is 2 to the third power $(2 \cdot 2 \cdot 2)$, which equals 8. Subtracting 2, for all 0s and all 1s (binary), leaves a total of six usable subnets.

Let's look at this from another angle. Figure 4.9 shows all eight of the possible combinations of 3 bits in a subnet field. It also provides the ranges of

Bits	Subnet(s)	Sample IP address ranges
111	All	223.254.253.255 (broadcast only)
100	128	223.254.253.129 through 223.254.253.159
010	64	223.254.253.65 through 223.254.253.95
110	192	223.254.253.193 through 223.254.253.223
001	32	223.254.253.33 through 223.254.253.63
101	160	223.254.253.161 through 223.254.253.191
011	96	223.254.253.97 through 223.254.253.127
000	None	223.254.253.0

Figure 4.9 Three-bit possibilities.

usable addresses for the six available subnets in a sample Class C network. We know that the functional specification disallows the subnet values of 000 (to identify the network) and 111 (to identify a broadcast to all subnets). Note that some router vendors include the software capability to make subnet 000 available for use in violation of the RFC standards.

The subnet bit pattern in Fig. 4.9 does not seem to follow any pattern. It follows addressing recommendations that suggest assigning the subnet-field bits as if the field were a mirror (reverse) of itself. The order of the subnets would be the order in Fig. 4.9. The same recommendations call for assigning the interface numbers (within each subnet) from the right. The advantage is to work toward the imaginary line dividing the subnet field and the interface field. This lets the administrator lay out the addresses and move the subnet mask 1 bit either way to account for too few subnets or too few hosts.

Building the mask

After setting the size requirements for both the subnet and host fields, the network administrator needs to choose the mask that is appropriate for the class (A, B, or C) of IP network (Fig. 4.10).

In our example, the network administrator needs

- Eleven separate locations for 11 subnets.
- Fifty-four interfaces in the largest subnet, including the printer and the router.

We could spend the next few pages trying different calculations to get the right combination of subnet and host bits to match. Instead, you can look at Class A, B, and C subnet tables (Figs. 4.11, 4.12, and 4.13). Using the Class A (or B) table, we can see that a 4-bit mask will provide 14 subnets. To make room for 55 (54 hosts plus one interface for the router that must be part of the subnet) interfaces, we need at least six host bits.

In the Class A and B examples above, there is plenty of room, with enough left for future growth, in the number of supportable interfaces. The Class C example uses the 6 host bits leaving 2 subnet bits that only allow two subnets with up to 62 addresses in each. Obviously we need multiple (about six) Class C addressed networks to complete the requirement with minimal room for growth.

Class A	11111111.11110000.00000000.00000000	255.240.0.0
Class B	11111111.11111111.11100000.00000000	255.255.224.0
Class C	11111111.11111111.11111111.11000000	255.255.255.192

Figure 4.10 Class A, B, and C masks.

Subnet bits	Subnet mask	Subnets	Hosts	Subnet broadcast
0	255.0.0.0	0	16,777,214	net.255.255.255
2	255.192.0.0	2	4,194,302	net.(subnet+63).255.255
3	255.224.0.0	6	2,097,150	net.(subnet+31).255.255
4	255.240.0.0	14	1,048,574	net.(subnet+15).255.255
5	255.248.0.0	30	524,286	net.(subnet+7).255.255
6	255.252.0.0	62	262,142	net.(subnet+3).255.255
7	255.254.0.0	126	131,070	net.(subnet+1).255.255
8	255.255.0.0	254	65,534	net.subnet.255.255
9	255.255.128.0	510	32,766	net.(subnet+127).255
10	255.255.192.0	1,022	16,382	net.(subnet+63).255
11	255.255.224.0	2,046	8,190	net.(subnet+31).255
12	255.255.240.0	4,094	4,094	net.(subnet+15).255
13	255.255.248.0	8,190	2,046	net.(subnet+7).255
14	255.255.252.0	16,382	1,022	net.(subnet+3).255
15	255.255.254.0	32,766	510	net.(subnet+1).255
16	255.255.255.0	65,534	254	net.subnet.255
17	255.255.255.128	131,070	126	net.(subnet+127)
18	255.255.255.192	262,142	62	net.(subnet+63)
19	255.255.255.224	524,286	30	net.(subnet+31)
20	255.255.255.240	1,048,574	14	net.(subnet+15)
21	255.255.255.248	2,097,150	6	net.(subnet+7)
22	255.255.255.252	4,194,302	2	net.(subnet+3)

Figure 4.11 Class A subnet table.

Subnet bits	Subnet mask	Subnets	Hosts	Subnet broadcast
0	255.255.0.0	0	65,534	net.net.255.255
2	255.255.192.0	2	16,382	net.net.(subnet+63).255
3	255.255.224.0	6	8,190	net.net.(subnet+31).255
4	255.255.240.0	14	4,094	net.net.(subnet+15).255
5	255.255.248.0	30	2,046	net.net.(subnet+7).255
6	255.255.252.0	62	1,022	net.net.(subnet+3).255
7	255.255.254.0	126	510	net.net.(subnet+1).255
8	255.255.255.0	254	254	net.net.subnet.255
9	255.255.255.128	510	126	net.net.(subnet+127)
10	255.255.255.192	1,022	62	net.net.(subnet+63)
11	255.255.255.224	2,046	30	net.net.(subnet+31)
12	255.255.255.240	4,094	14	net.net.(subnet+15)
13	255.255.255.248	8,190	6	net.net.(subnet+7)
14	255.255.255.252	16,382	2	net.net.(subnet+3)

Figure 4.12 Class B subnet table.

Subnet bits	Subnet mask	Subnets	Hosts	Subnet broadcast
0	255.255.255.0	0	254	net.net.net.255
2	255.255.255.192	2	62	net.net.net.(subnet+63)
3	255.255.255.224	6	30	net.net.net.(subnet+31)
4	255.255.255.240	14	14	net.net.net.(subnet+15)
5	255.255.255.248	30	6	net.net.net.(subnet+7)
6	255.255.255.252	62	2	net.net.net.(subnet+3)

Figure 4.13 Class C subnet table.

Applying the mask

The network administrator uses the mask as a guide to design the subnetting plan. Once the plan is in place, she assigns IP addresses to the individual interfaces or delegates a subnet to a subnet manager. The plan (as we suggest below) may not identify the subnets in order, but the network administrator should make the actual assignments in numeric order to let the routers keep their routing tables smaller.

A good guideline to begin the subnet plan is to identify the subnets from the highest order bits (left) and the hosts from the lowest order bits (right). This lets you adjust the mask bits left or right as needed, right up to the last interface assignments for the greatest possible flexibility.

In the example in Fig. 4.14, the Class B address mask designates 62 subnets with up to 1022 interfaces in each. The first subnet we would assign under that rule is subnet 128 (with a 6-bit subnet field of 100000 binary) and the first interface is 1 (with a binary, 10-bit host field of 00.00000001). The overall address would be 191.255.128.1 hex or 10111111.11111111.10000000.00000001 binary. The next interface in that subnet is 2.

The next assigned subnet, following the bit mirror guideline above, sets the second highest subnet bit (64 value) or subnet 64 (with a 6-bit, binary subnet field of 010000). The third subnet, under this guideline of staying to the left with subnet bit usage, would be subnet 192.

	Binary	Decimal
Network	10111111.11111111.00000000.00000000	191.255. 0.0
Mask	11111111.11111111.11111100.00000000	255.255.252.0
First subnet	10111111.11111111.10000000.00000000	191.255.128.0
Second subnet	10111111.11111111.01000000.00000000	191.255. 64.0
Third subnet	10111111.11111111.11000000.00000000	191.255.192.0
Mask values	nnnnnnnn.nnnnnnnn.sssssshh.hhhhhhhh	

Figure 4.14 Applying the mask.

To look at this another way, the last interface in that first subnet would be 191.255.131.254. The network value stays at 191.255. The subnet value stays at 128. The interface uses all the host bits, except the last one in the fourth byte to prevent an all 1s (binary) address. Since two of the host bits are in the third byte, we add those values (the last two, 2 and 1, on the right) to the 128 subnet value to complete the third byte IP address value of 131.

Once we have mapped all the subnet possibilities, we can put them in numeric order before assigning them to user subnets. That will make it easier for us to keep track of which subnets we have assigned and what we have left.

Interfaces lost

When the manager or administrator decides to subnet, the organization must pay the price in available interfaces. Using the Class C network examples shown in Fig. 4.15 and some simple math, we find the total number of interfaces under subnets is always less than the number of interfaces without subnetting. That is the trade-off for the ability to manage a network more easily.

The lost addresses are the all 0s and all 1s (binary) host addresses for each subnet and the all 1s and all 0s (binary) values for the subnet field itself. In the no-subnetting example in Fig. 4.12, the only unusable (shaded) addresses are the all 0s in the top-left corner and the all 1s in the bottom-right corner.

In the 2-bit subnet field example in the figure, we lose the 32 addresses in the all 0s subnet (shaded at the top of the example). In the 01 subnet, we lose the all 0s interface (the first square on the fifth row) and the all 1s interface (the last square on the eighth row). In the 10 subnet, we lose the all 0s interface (the first or shaded square on the ninth row) and the all 1s interface (the last square on the 12th row). We also lose the 32 addresses (shaded in the last four rows of the example) in the all 1s subnet.

The graphics in Fig. 4.15 make it easy to see that the "price" of subnetting varies with the numbers of subnet bits and the class of network used. The least costly masks for each class of networking balance exactly the number of bits in both the subnet field and the host field (four of each in this Class C example).

IP Communications Logic Process (Step Two)

Now that we have a basic understanding of the subnetting process, we can move to the second step of the logic process (Fig. 4.16). This step requires that both the source and target IP addresses live in the same network.

By applying the subnet mask to the source and target IP addresses, we can quickly determine if the two are in the same subnet. In the example above, the Class B mask indicates that the first two bytes may contain only network bits. The last byte, according to the decimal version of the mask, contains only host bits. This means we must focus our attention on the third byte of the mask.

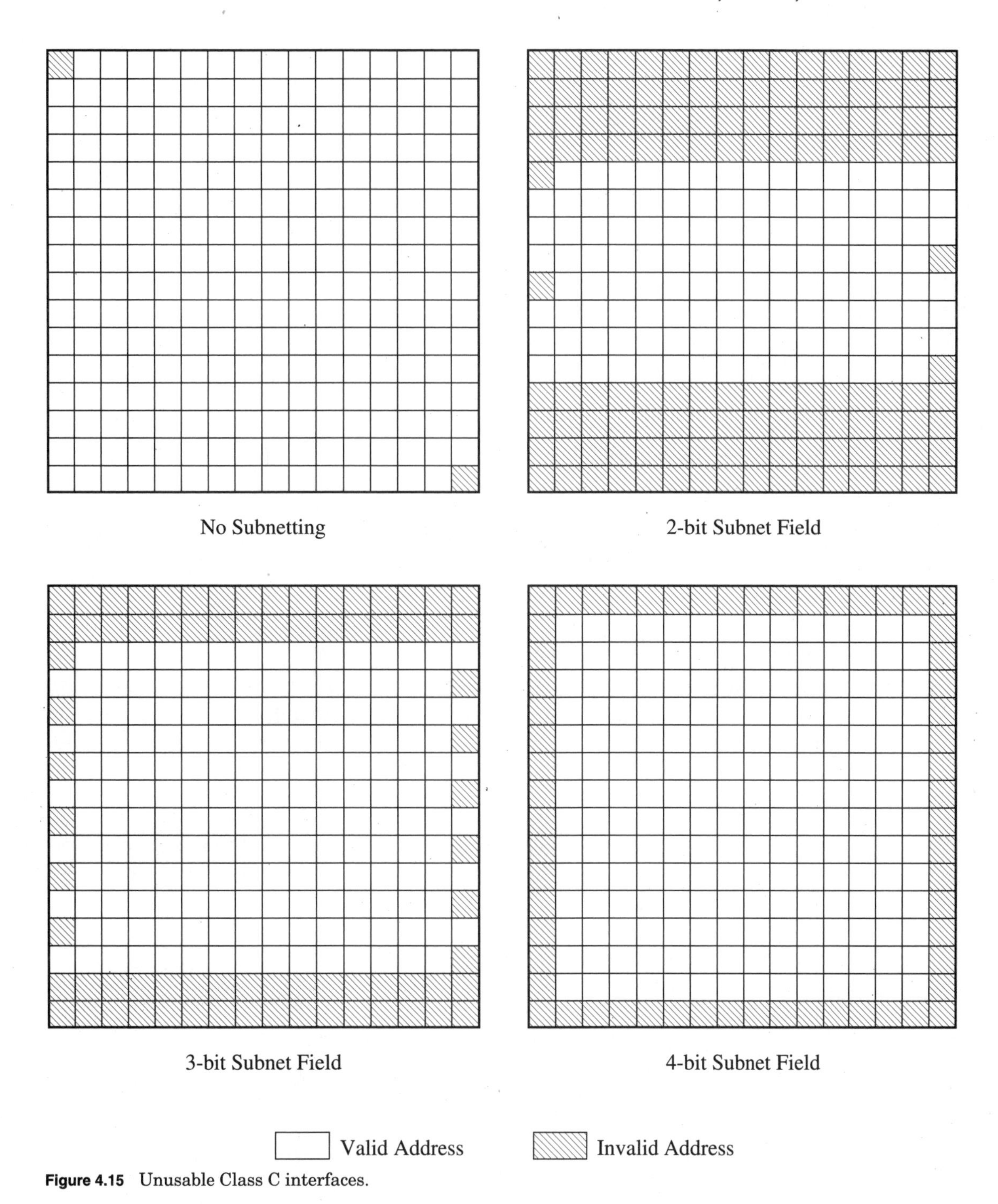

Figure 4.15 Unusable Class C interfaces.

Source: 191.255.193.44 Target: 191.255.206.13

Are the source and target IP addresses in the same network?

Yes

Are the source and target IP addresses in the same subnet?

Subnet mask: 255.255.240.0

Source binary: 11111111.11111111.11000001.00101100
Binary mask: 11111111.11111111.11110000.00000000
Target binary: 11111111.11111111.11001110.00001101

Figure 4.16 Step two of the logic process.

We could perform a logical AND operation, but there is a much simpler way to answer the question. The binary mask lets the address translator isolate the subnet bits from the host bits. The binary version of the third byte of the mask shows that there are 4 bits of subnet field and 4 bits of host field. Examining the third byte of the source and target IP addresses reveals that the 4 subnet bits (per the mask) contain the same value. That indicates that these two IP addresses are in the same subnet.

IP Communications Logic Process (Step Three)

By now we know that both the source and target IP addresses exist in the same network and in the same subnet. Had the IP addresses been in different subnets, then IP would send the datagram to the router serving the source system's subnet for forwarding to the correct remote subnet's router (and on to the target IP address).

The third step of the logic process asks another question that we must answer before IP can release the datagram to the local subnet (Fig. 4.17).

Source: 191.255.193.44 Target: 191.255.206.13

Are the source and target IP addresses in the same network?

Yes

Are the source and target IP addresses in the same subnet?

Yes	No
Do I have the physical address for the target IP address?	Send the datagram to the router to get it to the correct subnet?

Figure 4.17 Step three of the logic process.

Does the source system have the physical address of the target IP-addressed system's interface?

Just as the Postal Service needs a (physical) street address, the TCP/IP host must have the exact physical or MAC address. Without it, the NIC would have to send each piece of data as a broadcast, interrupting all the other hosts on that network in an attempt to reach the desired system. It is then a matter of matching the addresses, which is a good topic for the next chapter.

Chapter

5

Address Matching

Overview

When an IP-addressed interface on a system wants to communicate with another IP-addressed interface, it follows an IP logic process. Before the IP logic process can begin, the source system must have an IP address for the target interface. The user can enter this target as an IP address or as a name that the system will use to find the matching IP address.

When the source software gets a name, the sending application or process layer must translate to an IP address. Each TCP/IP vendor has its own method for translating names, though many will check the local system's hosts file and the Domain Name System (DNS). These vendors usually make the decision of which system (hosts, DNS, or other) to ask first, a configuration parameter.

The hosts file is like a personal telephone directory or little black book. Each entry contains an IP address and the name (or names) the local user has for that IP-addressed system. Figure 5.1 shows an example of a hosts file.

Most vendors also allow multiple alias names for each IP address. This lets the user specify any one of the names to reach the desired IP-addressed interface.

```
191.255.129.66      maggie, margaret
223.255.255.220     bill, jane
191.255.101.197     joe, cola
126.49.12.234       jim, preach
126.49.12.231       tom, dell
223.255.255.2       pat, patrick
191.255.3.78        amy, clint
223.255.255.111     david, nose
126.99.1.246        gin, pete
```

Figure 5.1 Hosts file.

Since this file is on the user's system, that user can customize it for her own needs. In other words, the names are those that the user will recognize and contact. Some organizations create a basic hosts file and let users add their own names.

Domain Name System

DNS is a distributed database that TCP/IP applications use to convert names that users enter into IP addresses. By distributed, we mean that no single Internet site or domain name server knows all the information. Each local server has the information on its area or zone and shares that with other servers as needed.

The application receives a command, such as: telnet unix.class. neurolink.com and turns to its resolver to find the matching IP address. As Fig. 5.2 shows, the resolver is not part of the operating system: It is part of the application.

If that check of the local hosts file finds no match (or the requesting system's configuration directs it), the resolver contacts the local domain name server to find the matching IP address. The resolver contacts as many domain name servers as it needs to locate the correct IP address. It stops only when it has a fully qualified domain name (FQDN) that ends in a period and the matching IP address. In our example the FQDN is *unix.class. neurolink.com.*

Domain Name System tree

The root or core of the DNS tree, shown in Fig. 5.3, is an artificial point where the various branches of the DNS come together. We never name this position but assume it to be the label after the final period in an FQDN. A label is the series of characters between the delimiting periods or left of the left-most period.

RFCs 952, 1035, and 1123 set certain specifications for these labels:

- Upper- and lower-case domain names and labels are equal (i.e., CCScom is the same as ccscom).

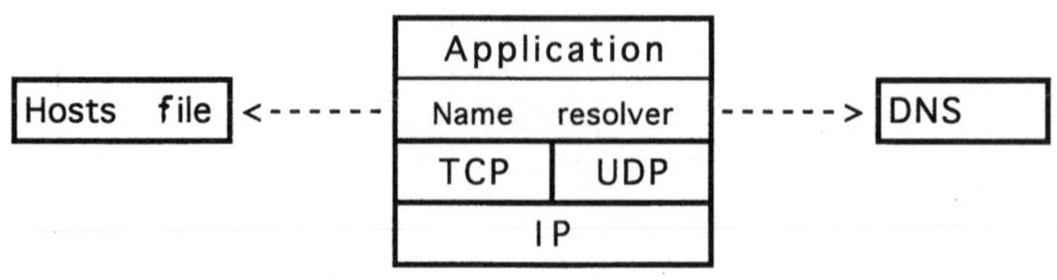

Figure 5.2 The name resolver.

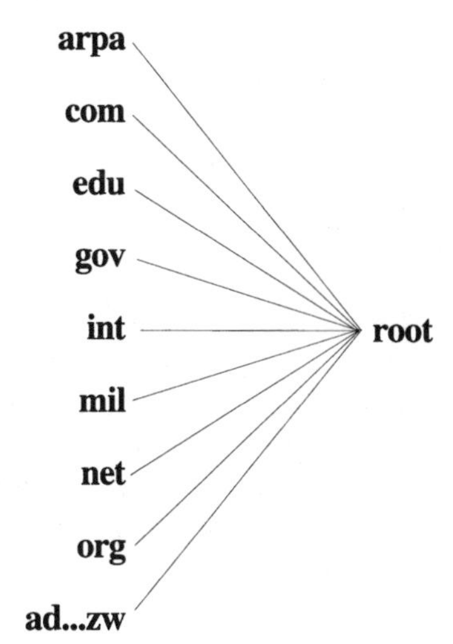

Figure 5.3 Partial domain name tree.

- Each label may have up to 63 characters, though local policies and "netiquette" dictate much shorter labels.
- Each label may contain only the 26 ASCII letters, the 10 digits, and the hyphen or dash (-).
- Each label must start with a letter or a digit per RFC 1123's update.
- No label should contain all digits, although some do.

By usage, we know that every node must have a unique domain name, though we can use labels more than once. This brings up another interesting challenge. Without detailed knowledge of the subject domain, we cannot determine if the far-left label identifies a zone of the DNS or a host computer.

As a solution to that situation, the arpa domain offers address-to-name mapping. For example, the resolver searches for the name that matches 192.136.118.123 by looking for the 123.118.136.192.in-addr.arpa name. The software writes the IP address in reverse since IP addresses are more specific from the right whereas, as Fig. 5.4 shows, a domain name is more specific to

123.118.136.192.in-addr.arpa
unix.class.neurolink.com
mit.edu
whitehouse.gov
nato.int
bull.pearl.pac.navy.mil
ds.internic.net
cert.org
cnri.reston.va.us

Figure 5.4 Sample domain names.

the left. This lets remote servers verify the origin domain name of users who are requesting access.

We call the remaining three-character domains the top-level domains (TLD). The number of domains within each organization may vary, though each organization falls under one of the top-level domains or the two-character, ISO-coded country domains.

Organizational domain grouping

The current top-level domains separate organizations into seven categories. This cuts down on the time a resolver needs to find the right network address. Grouping similar organizations means the resolver does not have to look through all the possible Internet domains in tracking down a university file server for an FTP session. Figure 5.5 and the following list detail the top-level domains.

- The .com domain identifies commercial organizations.
- The .edu domain provides support for educational institutions such as colleges and universities.
- Local school districts use the .gov domain like other U.S. civilian government agencies at all levels (county, state, and federal), including whitehouse.gov (which we would use to send e-mail to the president).
- .int is for use by international organizations such as nato.int.
- Like the .gov domain, specifications restrict the .mil domain to the United States. These systems support the military side of the U.S. federal government.
- Network providers use the .net domain to set themselves apart from other organizational groups.
- The .org domain specifies nonprofit organizations.

DS.InterNIC.Net administers all of these top-level domains, including the ISO-coded domains for the many countries that participate in the Internet. (See App. A at the end of this chapter.)

Commercial organizations	.com
Educational institutions	.edu
Nonmilitary government agencies	.gov
International organizations	.int
U.S. military	.mil
Networks	.net
Nonprofit organizations	.org

Figure 5.5 Top-level domains.

AT	Austria	FR	France	JP	Japan
AU	Australia	GR	Greece	KR	Korea (South)
CA	Canada	HK	Hong Kong	NZ	New Zealand
CR	Costa Rica	HR	Croatia	SE	Sweden
DE	Germany	IE	Ireland	SG	Singapore
DK	Denmark	IL	Israel	TW	Taiwan
ES	Spain	IN	India	UK	United Kingdom
FI	Finland	IT	Italy	US	United States

Figure 5.6 Sample country domains.

Country domains

The two-character codes (with samples in Fig. 5.6) are the ISO's abbreviated names for nations. Many countries also form second-layer domains inside their country code similar to the three-digit generic codes. For example, the United Kingdom and others use .co for commercial organizations and .ac for academic institutions, which gives universities a domain name that ends with .ac.uk and companies a domain name that ends in .co.uk.

Though many organizations in the United States use the three-character TLDs, more and more choose to use the .us country domain. State, country, parish, and local governmental agencies are among those who have made this choice (at times under pressure). The only restricted generic domains in the United States are .gov and .mil. (See RFC 1480 on the enclosed CD for more detail on the .us domain.) A list of the ISO country codes is in App. A.

Name server

While DS.InterNIC.Net supports the top-level domains, they delegate organizational domains (and any internal zones) to each organization for maintenance. A zone is a separately administered portion of the domain name tree. Each zone has at least one domain name server, though multiple servers prevent a single point of failure. When the NIC assigns an IP address to an organization, it asks for the name of their primary name server and the date it will be up and running.

Figure 5.7 shows a sample organization's DNS server deployment. The network administrator divided this organization into six zones. The DNS server in each zone is the primary server for that zone. Each server also acts as the secondary (backup) server to one or more neighboring zones. The secondary servers use a zone transfer, which typically happens every two to three hours, to update resource records.

One of these primary servers is also the organization's primary server (let's say the top-left server in the figure). That organizational primary server has its backups at one or more of the adjacent zone's server. It also has a secondary server outside the organization at another organization. The sec-

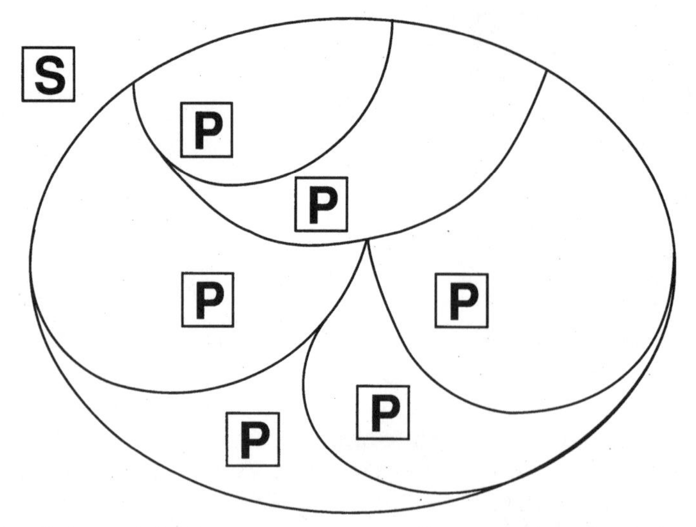

Figure 5.7 Primary and secondary DNS zones.

ondary server is a separate, redundant server that gets its data from the primary name server.

Note that the various zones' DNS servers may be close to each other geographically while supporting separate zones. When we add a network node, the DNS manager adds resource record(s) to the primary server. As we will discover, resource records provide different pieces of information that the DNS uses.

Resource records

Resource records store DNS information in RFC-specified formats. Figure 5.8 lists some of the DNS resource records that the name server uses. Internet users recommend new records periodically. The IAB has implemented some that users previously recommended, and has also declared obsolete some that it had previously implemented.

A	Alias name(s) and 32-bit IP address
CNAME	Canonical name
HINFO	Information about the host
MX	Mail exchange
NS	Name server
NULL	Storage space
PTR	Pointer
SOA	Start of zone authority
WKS	Well-known service

Figure 5.8 Common DNS resource records.

Additional detail on the records listed in Fig. 5.8 follows.

The alias (A) record can store multiple names for a single IP address.

The canonical name (CNAME) records contain the real name of the requested host so the DNS can use aliases.

Information about the host (HINFO) resource records contain two pieces of information: the central processing unit (CPU) type and the operating system being used in the specified system.

The DNS uses the mail exchange (MX) record in multiple scenarios:

- Alternate host mail delivery
- Virtual host IDs for a common mail server name
- Internet mail exchange for non-Internet systems
- Mail filtering for firewalled organizations

Name server (NS) records specify the name server for a requested zone.

The NULL record is a storage space of 65,536 bytes.

Pointer (PTR) records contain the in-addr.arpa information for an IP address to name resolution.

The start of zone authority (SOA) record provides maintenance data about this zone to other servers.

The Well-Known Service (WKS) record is a table of services provided by this zone. Its codes are in the Assigned Numbers RFC and are listed in App. B.

Down the Stack

Once the resolver has matched the user-provided host name and the target IP address, it passes the address down the stack to the IP layer. The IP layer can then begin the logic process that we looked into earlier for sending the data.

After determining that the source and target IP addresses are in the same network and the same subnet, the logic process continues in Fig. 5.9.

IP Communications Logic Process (Conclusion)

When one host wants to communicate with another host on the same piece of network cable (or the same side of a router), it must know the exact physical address. Just as it is not enough for the Postal Service to know a person is located on a given street to be able to deliver a letter, IP also needs to match the IP address to a physical (MAC or network interface) address.

In a TCP/IP-networked system (any device with an IP address), the software stores the MAC address in a volatile RAM space known as the Address

Source: 191.255.193.44	Target: 191.255.206.13
Are the source and target IP addresses in the same subnet?	
Yes	
Do I have the MAC address for the target IP address?	
Yes	No
Send the datagram to the target physical address.	Send an ARP request for the target physical address and retry.

Figure 5.9 IP communications logic process (step 3).

Resolution Protocol (ARP) cache. This gives an IP virtually instant access to the physical address that it must pass to the network interface layer for delivery of the datagram.

Address Resolution Protocol Cache

The ARP cache stores the IP address, the hardware address, and a timer for each entry as shown in the example in Fig. 5.10 (without the identifying header labels). Each entry identifies an IP-addressed device that has communicated with this system. Even a ping is enough to require an ARP request and reply exchange, regardless of who initiated the conversation.

There is no standard (RFC or otherwise) or requirement for the maximum size of the ARP timer. This is another example of the vendor's implementation control. Years of observation show most timers in a range from 30 to 900 seconds while some, for their own reasons, go beyond that. For example, Sun uses 20 minutes (1200 seconds) and Cisco uses 4 hours (14,400 seconds).

IP address	MAC address	Timer
191.255.101.66	00c093a7b1ea	61
191.255.101.220	0000c01f448d	109
191.255.101.197	00000c8d37c2	247
191.255.101.234	0080ad783f0b	444
191.255.101.231	0080ad783b09	473
191.255.101.22	0000c017468e	556
191.255.101.11	0000c09e455d	602
191.255.101.46	0080ad7dbf22	865

Figure 5.10 ARP cache example.

Some TCP/IP vendors have settled on a 900-second (15-minute) timer to begin the count toward zero and the entry's time out. The counter begins its descent after it receives the last IP datagram from that addressed node. If another conversation starts, the timer reverts to its original top value (in this case the 900-second mark).

When a timer reaches 0, the ARP cache entry's validity has expired. The vendor can choose to have the cache continue to carry the information and identify that entry's timer as expired with a value of 0. When this happens, any attempt to communicate with that device (with the expired entry) will require another ARP request and reply to update the cache and return the timer to its top value.

Layout Format and Decoding Tips

You will recognize the same style of protocol analyzer hexadecimal (hex) layout in Fig. 5.11 throughout this book. Let's look at some guidelines that will help aid understanding of the protocol decoding process.

The characters used in hexadecimal (base 16) are 0, 1, 2, 3, 4, 5, 6, 7, 8, 9, a, b, c, d, e, and f. Each hex character has a decimal equivalent. The numeric values (0 to 9) translate to themselves. We must convert the alpha values to the matching decimal amount, as shown in Fig. 5.12.

A byte or octet (8 bits) in hex is two characters. The left character is equal to multiples of 16. This works in the same way that a second digit in our decimal (base 10) counting system is multiples of 10. For example, hex 4e is equal to decimal 78 and decimal 255 equals hex ff. As an aid to Fig. 5.12, we will help decode the hex values with you.

To convert from a set of hexidecimal characters to a decimal number, find the hex characters in the appropriate column and look to the next cell to the right for the matching decimal value. For example, converting the hex value of 0 5 d c to a single decimal number takes four steps:

1. Decide what format the result will be (IP address, single number, etc.).
2. Break up the hex into the correct parts for decoding, i.e.,

0000 500 d0 c

ff	ff	ff	ff	ff	ff	00	00	-	c0	93	19	00	08	06	00	01
08	00	06	04	00	01	00	00	-	c0	93	19	00	c0	88	76	01
00	00	00	00	00	00	c0	88	-	76	32	00	00	55	00	00	dc
00	6c	00	d6	00	00	00	a3	-	00	00	00	41				

Figure 5.11 ARP hexadecimal layout.

Hex	Decimal	Hex	Decimal	Hex	Decimal	Hex	Decimal
0000	0	000	0	00	0	0	0
1000	4096	100	256	10	16	1	1
2000	8192	200	512	20	32	2	2
3000	12288	300	768	30	48	3	3
4000	16384	400	1024	40	64	4	4
5000	20480	500	1280	50	80	5	5
6000	24576	600	1536	60	96	6	6
7000	28672	700	1792	70	112	7	7
8000	32768	800	2048	80	128	8	8
9000	36864	900	2304	90	144	9	9
a000	40960	a00	2560	a0	160	a	10
b000	45056	b00	2816	b0	176	b	11
c000	49152	c00	3072	c0	192	c	12
d000	53248	d00	3328	d0	208	d	13
e000	57344	e00	2584	e0	224	e	14
f000	61440	f00	3840	f0	240	f	15

Figure 5.12 Hex to decimal values.

3. Collect the matching decimal values, i.e.,

   0 1280 208 12

4. Add the decimal values, i.e.,

   0 + 1280 + 208 + 12 = 1500

This protocol analyzer displays two groups of 8 bytes in hex, separating the groups with a hyphen. It checks the 4-byte checksum or frame check sequence and shows the balance of only those packets that pass this calculation. It discards any packets in which it detects an error.

Address Resolution Protocol

Address resolution protocol (ARP) helps the local source system find the hardware address of the target system when it has the corresponding IP address. This is only a concern when the two stations share the same network and subnet.

The packet in Fig. 5.11 begins with a 14-byte Ethernet header. Using the broadcast target address (six pairs of ff) makes all the NICs on the subnet take in the packet to pass to the next layer for an IP address check.

This would be very inefficient if we used it all the time. It would interrupt every system on the segment of cable, for every piece of every conversation beginning or ending on this network segment. Since the software only uses a broadcast for the ARP request, it is just another example of the flexibility of TCP/IP.

2 bytes	2 bytes	1 byte	1 byte	2 bytes	6 bytes*	4 bytes*	6 bytes*	4 bytes*
Hardware type	Protocol type	Hardware length	Protocol length	Operation	Source hardware address	Source protocol address	Target hardware address	Target protocol address
00 01	08 00	06	04	00 01	00 00 c0 93 19 00	c0 88 76 01	00 00 00 00 00 00	c0 88 76 32

*As designated by the hardware length or protocol length fields.

Figure 5.13 ARP header layout.

The source address 00 00 c0 93 19 00 matches the network interface sending out this packet. As we know, it also tells us (by its first 3 bytes) the card was sold by Western Digital or SMC.

The Ethernet protocol type of 08 06 identifies the Ethernet payload as ARP. Since it is the last field of the Ethernet header, the ARP header starts in the next byte (the next to the last byte, from the left, on the first line of Fig. 5.11). Let's examine a sample ARP message that we captured from the network, so we can understand what ARP is doing and why.

We will use the RFC-specified ARP layout (Fig. 5.13) to follow the fields in our hex display. That way we can discover how each field does its part in the overall scheme of supporting IP's need for a MAC address that matches the IP address we have for the target system.

Hardware type

This field (the last 2 bytes on the first line of the hex code in Fig. 5.11) indicates the hardware type that the physical layer of the network is using. As you can see from the part of RFC 1700's Address Resolution Protocol Parameters section we have provided in Fig. 5.14, the hex code of 00 01 identifies this hardware type as Ethernet. (See App. C for the entire Address Resolution Protocol Parameters section.)

Type	Description	Type	Description
1	Ethernet (10 Mb)	9	Lanstar
2	Experimental Ethernet (3 Mb)	10	Autonet short address
3	Amateur radio AX.25	11	LocalTalk
4	Proteon ProNET token ring	12	LocalNet
5	Chaos	13	Ultra Link
6	IEEE 802 networks	14	SMDS
7	ARCNET	15	Frame relay
8	Hyperchannel	16	ATM

Figure 5.14 ARP hardware types.

Hex	Description
0000-05DC	IEEE 802.3 Length Field
0600	Xerox NS IDP
0800	IPv4
0806	ARP
0BAD	Banyan Systems
8035	RARP
8137-8138	Novell, Inc.

Figure 5.15 ARP protocol types.

Protocol type

This field (the first two pairs of hex characters on the second line of the Fig. 5.11 hex display) shows the type of network protocol address the local network (or subnet) is using. The protocol type portion of the Address Resolution Protocol Parameters section of the Assigned Numbers RFC refers us to the Ethernet Numbers of Interest section (though RFC 1700 renamed that section Ether Types) that has the code and the translation for that field. In this case, rather than telling us what the Ethernet header is carrying, the code only identifies the type of protocol we are trying to match with a hardware address. A short sample of values is shown in Fig. 5.15, while the whole list is given in Apps. B and C at the end of Chap. 3.

Since we explored the Ethernet protocol type field in Chap. 3, you can refer to the end of that chapter to find the Ether types section of RFC 1700. In that table, you will find the 08 00 entry from Fig. 5.11 and its identification of IP.

Hardware length

The third byte on the second line of the Fig. 5.11 hex display shows the number of bytes used in the hardware address. As explained in the excerpt from RFC 1700 in Fig. 5.16, this value is 48 bits, expressed as 12 hexadecimal digits or 6 bytes. The ARP header sets both the source and target hardware address lengths here to be the same 6-byte length.

Ethernet hardware addresses are 48 bits, expressed as 12 hex digits.

These 12 hex digits consist of the left 6 digits, which should match the vendor of the Ethernet interface, and the right 6 digits that specify the interface serial number for that interface vendor.

Ethernet addresses should be written hyphenated by octets (12-34-56-78-9A-BC).

These addresses are physical station addresses, not multicast nor broadcast, so the second hex digit (reading from the left) will be even, not odd.

Figure 5.16 Ethernet vendor address components excerpt from RFC 1700.

Operation Code (op)
1 Request
2 Reply

Figure 5.17 ARP operation codes.

Protocol length

This field (the fourth byte on the second line of the hex example in Fig. 5.11) sets the number of bytes that each of the protocol addresses, source and target, will contain in the ARP. Since the protocol type field contains the hex value of 08 00 for IP, and the current (version 4) length of IP addresses is 4 bytes according to RFCs 760 and 791, this field must contain a value of 04 (hex), which is also 4 in decimal value.

Operation

The fifth and sixth bytes on the second line of the hex display in Fig. 5.11 identify the operation field, which identifies the operation ARP is attempting. As you can see from the RFC 1700 text in Fig. 5.17, the 00 01 hex value tells us that we are looking at an ARP request. The alternative ARP entry for this field is 00 02 hex for an ARP reply. The ARP is only successful if the station sending the ARP request receives an ARP reply.

Source hardware address

The physical, also called the hardware or MAC address of the interface sending the ARP message is in the next field (the 2 bytes to the left of the hyphen, and the first four bytes to the right of the hyphen, on the second line of the hex display in Fig. 5.11). According to the hardware length field, the source hardware address field must contain 6 bytes.

It is interesting to see that these 6 bytes occur in the same position on the first and second lines of the hex display. The top line is from the Ethernet header. This apparent duplication of the source address is the result of the Ethernet header's size and the protocol analyzer's arrangement of the hex characters. While this happens with ARP over Ethernet, not all packets will appear this way.

When the Ethernet entity passes the ARP data up the stack to the ARP entity, it does not include the Ethernet header. Without the Ethernet header available, the ARP must include the duplicate address information so the ARP entity can still match the hardware and protocol addresses.

With access to the proper information, the Ethernet vendor address component section of the Assigned Numbers RFC shown in App. A of Chap. 3, we can decode this address to determine that the network interface device is a product of Western Digital (now SMC). Since many of the Ethernet vendor address component codes are not in the Assigned Numbers RFC, it is a very good idea to have an inventory of the hardware and matching protocol addresses to deal with the challenge of duplicate addresses in the same subnet.

We can create that inventory manually by keeping track of the Ethernet cards as we install them in systems, or by using an SNMP-based network management station (NMS) such as HP OpenView or NetView. Most of the SNMP NMSs have an autodiscovery capability. It retrieves the ARP cache from every system it can find on the network(s) it is told to search. The drawback is that the systems must be operating and must have recently communicated with another IP system like a router, a server, or a workstation.

Source protocol address

The next field contains the IP address of the station that is sending this example message. Since the protocol type field indicated the protocol to be IP, the ARP will provide the source IP address here. The protocol length field specifies that the protocol addresses contain 4 bytes. To find them, we look for the last four bytes on the second line of the hex display in Fig. 5.11.

In the example, the entry (c0 88 76 01) translates to 192.136.118.1, the IP address of the device that sent the ARP request. When we decode the 4-hex bytes to decimal, we do so one at a time so that we can write the value in the standard dotted-decimal-notation we recognize as an IP address.

We chose this packet as the example to show what would happen when a router needed to find the node in the subnet so it could pass the packet on from another subnet. Remember, we have already checked the target address to see if it is in the same network and the same subnet.

The example in Fig. 5.18 shows the source sends an ARP request to router 1 and receives an ARP reply before it sends the IP datagram to router 1. The routers and gateways forward the IP datagram by using their routing tables and configuration instructions. (There will be more about routing in Chap. 9.)

Router 2 sends an ARP request to find the target hardware address so it can forward the IP datagram that the source originated. It is important to note that routers and gateways do *not* forward ARP requests or replies. ARP messages stay in their subnet. Even in this example, Router 2's interface is part of the subnet that is the target for this ARP request.

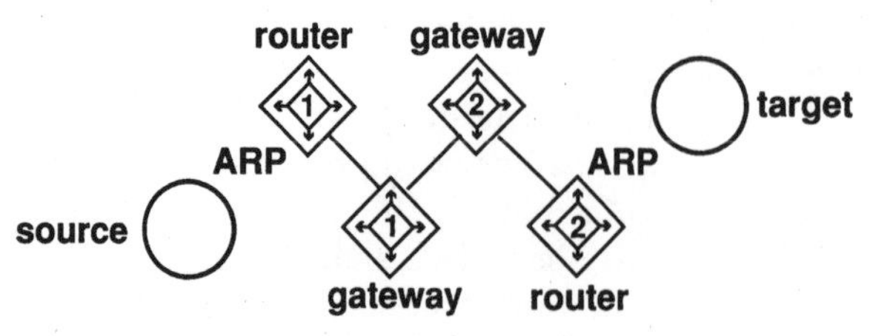

Figure 5.18 ARP and routers.

As we will see in Chap. 9, an ARP request is only necessary when the router's connection is to a network (or subnet) that may have multiple MAC addresses. This is also why router-to-router direct connections through serial ports do not need MAC or IP addresses on those serial ports. Since routers only use the network access layer as a trucking company to carry the data from the router to another router or a host (server or workstation), IP only needs the ARP to find the local hardware address.

Target hardware address

The target hardware address (the first six bytes on the third line of the hex display in Fig. 5.11) is under the same restrictions as the source hardware address: The address is Ethernet (to comply with the ARP hardware type field) and it contains 6 bytes (to comply with the ARP hardware length field).

As the reason for an ARP request is to find the physical address of the device to receive the message, vendors fill this field in different ways. The four possible methods of completing this field in an ARP request are:

- Six bytes of zeros (00 00 00 00 00 00) as we see in Figs. 5.11 and 5.13
- Six bytes of ff (ff ff ff ff ff ff)
- Six bytes from the sending host's buffer (00 02 0f 0e 17 00) that look like a real address but are simply a buffer dump
- The current or recently expired hardware address from the ARP cache that matches the message's target protocol address

It is important to note that in an ARP request these 6 bytes simply take up the right amount of space.

Target protocol address

The last field of the ARP contains the protocol address (specified as IP in the protocol type field) of the station that is the target for the message. Like the source protocol address field, it contains 4 bytes. In the example in Fig. 5.11, the field consists of 2 bytes, on each side of the hyphen, in the middle of the third line of the hex display.

The c0 99 b9 32 hex value indicates the IP address of 192.153.185.50 as the target of this ARP request (and the host whose need for an ARP cache update made this ARP request necessary).

LAN fill

Since Ethernet-based ARP requests and replies do not have enough characters for the minimum 46 bytes of user data that the Ethernet specification requires, the Ethernet driver must add bytes to make up the difference. The

$$46 - [8 + (2 \cdot \text{hl}) + (2 \cdot \text{pl})]$$

Figure 5.19 ARP Ethernet padding formula.

formula in Fig. 5.19 tells us how much padding the Ethernet driver will need to add to meet that minimum.

In this formula, we can subtract (from the minimum 46 bytes) the combined total amount of the ARP. The ARP part of the formula includes

- Eight bytes of the fixed length fields of hardware type, protocol type, hardware length (hl), protocol length (pl), and operation
- Two times the value in the hl field (one for the source and one for the target)
- Two times the value in the protocol length field (one for the source and one for the target)

In the example, we have

- Eight bytes of fixed length fields
- Two times an hl of 6 for 12 more bytes
- Two times the pl of 4 for 8 bytes

Adding those up gives an ARP length of 28 bytes. We will require 18 bytes of padding to meet the 46-byte Ethernet minimum.

ARP Implementation

While all IP nodes must support ARP from the other nodes in the same subnet, TCP/IP vendors handle ARP requests and replies in their own way. When the ARP request arrives at a system other than the one with the matching IP address, the receiving system usually throws it away. Figure 5.20 shows the correct ARP exchange.

1. To initiate a conversation with a server in the same subnet, the user workstation's TCP/IP suite sends an ARP request to retrieve the server's hardware address.
2. When the server receives the ARP request, it updates its own ARP cache by adding the source hardware and source protocol addresses. It also sets the new entry's timer before responding to the ARP requester.

	User station		Server	
1	ARP request	→		Broadcast
2			Record source hardware and protocol addresses, set timer	
3		←	ARP reply	Unicast
4	Record source hardware and protocol addresses, set timer			
5	IP datagram	→		Unicast
6		←	IP datagram	Unicast

Figure 5.20 Correct ARP exchange.

3. The ARP reply carries the server's hardware and protocol addresses in the source fields since it is initiating the ARP reply.
4. When the user's system receives the ARP reply, it updates its own ARP cache by adding the source hardware and source protocol addresses. It also sets the new entry's timer before it sends its first IP datagram.

Steps 5 and 6 are the beginning (and maybe the end) of the communication between the two systems as the IP datagrams can now travel in unicast Ethernet headers.

Some hosts respond before saving the sender's addresses in their ARP cache. In that case, the host sends the ARP reply as a broadcast instead of the unicast that it should use since the ARP does not yet officially know the address. That leads to more broadcasts and delay problems. To simplify the explanation of that faulty ARP exchange, let's note the following steps in Fig. 5.21.

1. The user's system, having discovered that the server it wants to call is in the same subnet, sends a broadcast ARP request.

2. The server answers the ARP request. It does not save the user's hardware and protocol addresses in its ARP cache. Its ARP reply must be a broadcast as it has no ARP cache entry for the target protocol address of the user station.

3. The user's system records the source hardware and protocol addresses from the ARP reply in its ARP cache and turns on the timer.

4. The user's system sends a unicast IP datagram to the target hardware address in its ARP cache.

	User station		Server	
1	ARP request	→		Broadcast
2		←	ARP reply	Broadcast
3	Record source hardware and protocol addresses, set timer			
4	IP datagram	→		Unicast
5		←	ARP request	Broadcast
6	ARP reply	→		Unicast
7			Record source hardware and protocol addresses, set timer	
8	IP datagram retransmit	→		Unicast
9		←	IP datagram	Unicast

Figure 5.21 Faulty ARP exchange.

5. Since it has to have an ARP cache entry for the user's system, the server sends a broadcast ARP request to find the hardware address of the system that is trying to communicate with it.
6. The user's system sends a unicast ARP reply to tell (or remind) the server who is calling and where to send the data.
7. The server records the source hardware and protocol addresses from the ARP reply in its ARP cache and starts the timer.
8. Since it hits a timeout (a failure to hear a response from its IP datagram), the client retransmits the IP datagram that the client sent in step 4.
9. The server responds to the IP datagram with one of its own to continue the communication.

When the hosts at both ends properly process the ARPs by entering the source addresses in the receiving ARP caches, the ARP sequence would

- Not need steps 4, 5, or 6 of our faulty sample
- Move step 7 between steps 1 and 2
- Include only one interrupting broadcast instead of three

Now, at the end of this chapter covering address matching, let's look at the changes that happened in the ARP cache while we were looking at the process. Figure 5.22 shows renewed communication with 191.255.101.66 and 191.255.101.46 while the timer for 191.255.101.220 has expired. The rest of the entries seem to have lost only 120 seconds. It seems Einstein is right: time is relative.

IP address	MAC address	Timer
191.255.101.66	00co93a7b1ea	890
191.255.101.220	0000c01f448d	0
191.255.101.197	00000c8d37c2	127
191.255.101.234	0080ad783f0b	324
191.255.101.231	0080ad783b09	353
191.255.101.22	0000c017468e	436
191.255.101.11	0000c09e455d	482
191.255.101.46	0080ad7dbf22	785

Figure 5.22 Sample ARP cache.

Appendix A ISO Country Codes

Country	Code
Afghanistan	AF
Albania	AL
Algeria	DZ
American Samoa	AS
Andorra	AD
Angola	AO
Anguilla	AI
Antarctica	AQ
Antigua and Barbuda	AG
Argentina	AR
Armenia	AM
Aruba	AW
Australia	AU
Austria	AT
Azerbaijan	AZ
Bahamas	BS
Bahrain	BH
Bangladesh	BD
Barbados	BB
Belarus	BY
Belgium	BE
Belize	BZ
Benin	BJ
Bermuda	BM

Country	Code
Bhutan	BT
Bolivia	BO
Bosnia and Herzegowina	BA
Botswana	BW
Bouvet Island	BV
Brazil	BR
British Indian Ocean Territory	IO
Brunei Darussalam	BN
Bulgaria	BG
Burkina Faso	BF
Burundi	BI
Cambodia	KH
Cameroon	CM
Canada	CA
Cape Verde	CV
Cayman Islands	KY
Central African Republic	CF
Chad	TD
Chile	CL
China	CN
Christmas Island	CX
Cocos (Keeling) Islands	CC
Colombia	CO
Comoros	KM
Congo	CG
Cook Islands	CK
Costa Rica	CR
Cote d'Ivoire	CI
Croatia (local name: Hrvatska)	HR
Cuba	CU
Cyprus	CY
Czech Republic	CZ
Denmark	DK
Djibouti	DJ
Dominica	DM
Dominican Republic	DO
East Timor	TP

Country	Code
Ecuador	EC
Egypt	EG
El Salvador	SV
Equatorial Guinea	GQ
Eritrea	ER
Estonia	EE
Ethiopia	ET
Falkland Islands (Malvinas)	FK
Faroe Islands	FO
Fiji	FJ
Finland	FI
France	FR
France, Metropolitan	FX
French Guiana	GF
French Polynesia	PF
French Southern Territories	TF
Gabon	GA
Gambia	GM
Georgia	GE
Germany	DE
Ghana	GH
Gibraltar	GI
Greece	GR
Greenland	GL
Grenada	GD
Guadeloupe	GP
Guam	GU
Guatemala	GT
Guinea	GN
Guinea-Bissau	GW
Guyana	GY
Haiti	HT
Heard and McDonald Islands	HM
Honduras	HN
Hong Kong	HK
Hungary	HU
Iceland	IS

Country	Code
India	IN
Indonesia	ID
Iran (Islamic Republic of)	IR
Iraq	IQ
Ireland	IE
Israel	IL
Italy	IT
Jamaica	JM
Japan	JP
Jordan	JO
Kazakhstan	KZ
Kenya	KE
Kiribati	KI
Korea, Democratic People's Republic of	KP
Korea, Republic of	KR
Kuwait	KW
Kyrgyzstan	KG
Lao People's Democratic Republic	LA
Latvia	LV
Lebanon	LB
Lesotho	LS
Liberia	LR
Libyan Arab Jamahiriya	LY
Liechtenstein	LI
Lithuania	LT
Luxembourg	LU
Macau	MO
Macedonia, the former Yugoslav Republic of	MK
Madagascar	MG
Malawi	MW
Malaysia	MY
Maldives	MV
Mali	ML
Malta	MT
Marshall Islands	MH
Martinique	MQ
Mauritania	MR

Country	Code
Mauritius	MU
Mayotte	YT
Mexico	MX
Micronesia, Federated States of	FM
Moldova, Republic of	MD
Monaco	MC
Mongolia	MN
Montserrat	MS
Morocco	MA
Mozambique	MZ
Myanmar	MM
Namibia	NA
Nauru	NR
Nepal	NP
Netherlands	NL
Netherlands Antilles	AN
New Caledonia	NC
New Zealand	NZ
Nicaragua	NI
Niger	NE
Nigeria	NG
Niue	NU
Norfolk Island	NF
Northern Mariana Islands	MP
Norway	NO
Oman	OM
Pakistan	PK
Palau	PW
Panama	PA
Papua New Guinea	PG
Paraguay	PY
Peru	PE
Philippines	PH
Pitcairn	PN
Poland	PL
Portugal	PT
Puerto Rico	PR

Country	Code
Qatar	QA
Reunion	RE
Romania	RO
Russian Federation	RU
Rwanda	RW
Saint Kitts and Nevis	KN
Saint Lucia	LC
Saint Vincent and the Grenadines	VC
Samoa	WS
San Marino	SM
Sao Tome and Principe	ST
Saudi Arabia	SA
Senegal	SN
Seychelles	SC
Sierra Leone	SL
Singapore	SG
Slovakia (Slovak Republic)	SK
Slovenia	SI
Solomon Islands	SB
Somalia	SO
South Africa	ZA
South Georgia and the South Sandwich Islands	GS
Spain	ES
Sri Lanka	LK
St. Helena	SH
St. Pierre and Miquelon	PM
Sudan	SD
Suriname	SR
Svalbard and Jan Mayen Islands	SJ
Swaziland	SZ
Sweden	SE
Switzerland	CH
Syrian Arab Republic	SY
Taiwan (Province of China)	TW
Tajikistan	TJ
Tanzania, United Republic of	TZ
Thailand	TH

Country	Code
Togo	TG
Tokelau	TK
Tonga	TO
Trinidad and Tobago	TT
Tunisia	TN
Turkey	TR
Turkmenistan	TM
Turks and Caicos Islands	TC
Tuvalu	TV
Uganda	UG
Ukraine	UA
United Arab Emirates	AE
United Kingdom	GB
United States	US
United States Minor Outlying Islands	UM
Uruguay	UY
Uzbekistan	UZ
Vanuatu	VU
Vatican City State (Holy See)	VA
Venezuela	VE
Viet Nam	VN
Virgin Islands (British)	VG
Virgin Islands (U.S.)	VI
Wallis and Futuna Islands	WF
Western Sahara	EH
Yemen	YE
Yugoslavia	YU
Zaire	ZR
Zambia	ZM
Zimbabwe	ZW

Appendix B Protocol and Service Names for DNS WKS Records (from RFC 1700, Assigned Numbers, October 1994)

Protocol and Service Names

These are the Official Protocol Names as they appear in the Domain Name System WKS records and the NIC Host Table. Their use is described in [RFC952].

A protocol or service may be up to 40 characters taken from the set of uppercase letters, digits, and the punctuation character hyphen. It must start with a letter, and end with a letter or digit.

ARGUS	ARGUS Protocol
ARP	Address Resolution Protocol
AUTH	Authentication Service
BBN-RCC-MON	BBN RCC Monitoring
BL-IDM	Britton Lee Intelligent Database Machine
BOOTP	Bootstrap Protocol
BOOTPC	Bootstrap Protocol Client
BOOTPS	Bootstrap Protocol Server
BR-SAT-MON	Backroom SATNET Monitoring
CFTP	CFTP
CHAOS	CHAOS Protocol
CHARGEN	Character Generator Protocol
CISCO-FNA	CISCO FNATIVE
CISCO-TNA	CISCO TNATIVE
CISCO-SYS	CISCO SYSMAINT
CLOCK	DCNET Time Server Protocol
CMOT	Common Mgmnt Info Ser and Prot over TCP/IP
COOKIE-JAR	Authentication Scheme
CSNET-NS	CSNET Mailbox Nameserver Protocol
DAYTIME	Daytime Protocol
DCN-MEAS	DCN Measurement Subsystems Protocol
DCP	Device Control Protocol
DGP	Dissimilar Gateway Protocol
DISCARD	Discard Protocol
DMF-MAIL	Digest Message Format for Mail
DOMAIN	Domain Name System
ECHO	Echo Protocol
EGP	Exterior Gateway Protocol
EHF-MAIL	Encoding Header Field for Mail
EMCON	Emission Control Protocol
EMFIS-CNTL	EMFIS Control Service
EMFIS-DATA	EMFIS Data Service
FCONFIG	Fujitsu Config Protocol
FINGER	Finger Protocol
FTP	File Transfer Protocol

FTP-DATA	File Transfer Protocol Data
GGP	Gateway Gateway Protocol
GRAPHICS	Graphics Protocol
HMP	Host Monitoring Protocol
HOST2-NS	Host2 Name Server
HOSTNAME	Hostname Protocol
ICMP	Internet Control Message Protocol
IGMP	Internet Group Management Protocol
IGP	Interior Gateway Protocol
IMAP2	Interim Mail Access Protocol version 2
INGRES-NET	INGRES-NET Service
IP	Internet Protocol
IPCU	Internet Packet Core Utility
IPPC	Internet Pluribus Packet Core
IP-ARC	Internet Protocol on ARCNET
IP-ARPA	Internet Protocol on ARPAnet
IP-CMPRS	Compressing TCP/IP Headers
IP-DC	Internet Protocol on DC Networks
IP-DVMRP	Distance Vector Multicast Routing Protocol
IP-E	Internet Protocol on Ethernet Networks
IP-EE	Internet Protocol on Exp. Ethernet Nets
IP-FDDI	Transmission of IP over FDDI
IP-HC	Internet Protocol on Hyperchannnel
IP-IEEE	Internet Protocol on IEEE 802
IP-IPX	Transmission of 802.2 over IPX Networks
IP-MTU	IP MTU Discovery Options
IP-NETBIOS	Internet Protocol over NetBIOS Networks
IP-SLIP	Transmission of IP over Serial Lines
IP-WB	Internet Protocol on Wideband Network
IP-X25	Internet Protocol on X.25 Networks
IRTP	Internet Reliable Transaction Protocol
ISI-GL	ISI Graphics Language Protocol
ISO-TP4	ISO Transport Protocol Class 4
ISO-TSAP	ISO TSAP
LA-MAINT	IMP Logical Address Maintenance
LARP	Locus Address Resolution Protocol
LDP	Loader Debugger Protocol
LEAF-1	Leaf-1 Protocol

LEAF-2	Leaf-2 Protocol
LINK	Link Protocol
LOC-SRV	Location Service
LOGIN	Login Host Protocol
MAIL	Format of Electronic Mail Messages
MERIT-INP	MERIT Internodal Protocol
METAGRAM	Metagram Relay
MIB	Management Information Base
MIT-ML-DEV	MIT ML Device
MFE-NSP	MFE Network Services Protocol
MIT-SUBNET	MIT Subnet Support
MIT-DOV	MIT Dover Spooler
MPM	Internet Message Protocol (Multimedia Mail)
MPM-FLAGS	MPM Flags Protocol
MPM-SND	MPM Send Protocol
MSG-AUTH	MSG Authentication Protocol
MSG-ICP	MSG ICP Protocol
MUX	Multiplexing Protocol
NAMESERVER	Host Name Server
NETBIOS-DGM	NETBIOS Datagram Service
NETBIOS-NS	NETBIOS Name Service
NETBIOS-SSN	NETBIOS Session Service
NETBLT	Bulk Data Transfer Protocol
NETED	Network Standard Text Editor
NETRJS	Remote Job Service
NI-FTP	NI File Transfer Protocol
NI-MAIL	NI Mail Protocol
NICNAME	Who Is Protocol
NFILE	A File Access Protocol
NNTP	Network News Transfer Protocol
NSW-FE	NSW User System Front End
NTP	Network Time Protocol
NVP-II	Network Voice Protocol
OSPF	Open Shortest Path First Interior GW Protocol
PCMAIL	PCmail Transport Protocol
POP2	Post Office Protocol—Version 2
POP3	Post Office Protocol—Version 3

PPP	Point-to-Point Protocol
PRM	Packet Radio Measurement
PUP	PUP Protocol
PWDGEN	Password Generator Protocol
QUOTE	Quote of the Day Protocol
RARP	A Reverse Address Resolution Protocol
RATP	Reliable Asynchronous Transfer Protocol
RE-MAIL-CK	Remote Mail Checking Protocol
RDP	Reliable Data Protocol
RIP	Routing Information Protocol
RJE	Remote Job Entry
RLP	Resource Location Protocol
RTELNET	Remote Telnet Service
RVD	Remote Virtual Disk Protocol
SAT-EXPAK	Satnet and Backroom EXPAK
SAT-MON	SATNET Monitoring
SEP	Sequential Exchange Protocol
SFTP	Simple File Transfer Protocol
SGMP	Simple Gateway Monitoring Protocol
SNMP	Simple Network Management Protocol
SMI	Structure of Management Information
SMTP	Simple Mail Transfer Protocol
SQLSRV	SQL Service
ST	Stream Protocol
STATSRV	Statistics Service
SU-MIT-TG	SU/MIT Telnet Gateway Protocol
SUN-RPC	SUN Remote Procedure Call
SUPDUP	SUPDUP Protocol
SUR-MEAS	Survey Measurement
SWIFT-RVF	Remote Virtual File Protocol
TACACS-DS	TACACS-Database Service
TACNEWS	TAC News
TCP	Transmission Control Protocol
TCP-ACO	TCP Alternate Checksum Option
TELNET	Telnet Protocol
TFTP	Trivial File Transfer Protocol
THINWIRE	Thinwire Protocol

TIME	Time Server Protocol
TP-TCP	ISO Transport Service on top of the TCP
TRUNK-1	Trunk-1 Protocol
TRUNK-2	Trunk-2 Protocol
UCL	University College London Protocol
UDP	User Datagram Protocol
NNTP	Network News Transfer Protocol
USERS	Active Users Protocol
UUCP-PATH	UUCP Path Service
VIA-FTP	VIA Systems-File Transfer Protocol
VISA	VISA Protocol
VMTP	Versatile Message Transaction Protocol
WB-EXPAK	Wideband EXPAK
WB-MON	Wideband Monitoring
XNET	Cross Net Debugger
XNS-IDP	Xerox NS IDP

References

[RFC952] Harrenstien, K., M. Stahl, and E. Feinler, "DoD Internet Host Table Specification," RFC 952, SRI, October 1985.

URL = ftp://ftp.isi.edu/in-notes/iana/assignments/service-names

Appendix C Address Resolution Protocol Parameters (from RFC 1700, Assigned Numbers, October 1994)

Address Resolution Protocol Parameters

The Address Resolution Protocol (ARP) specified in [RFC826] has several parameters. The assigned values for these parameters are listed here.

Reverse Address Resolution Protocol Operation Codes The Reverse Address Resolution Protocol (RARP) specified in [RFC903] uses the "Reverse" codes below.

Dynamic Reverse ARP The Dynamic Reverse Address Resolution Protocol (DRARP) uses the "DRARP" codes below. For further information, contact: David Brownell (suneast!helium!db@Sun.COM).

Inverse Address Resolution Protocol The Inverse Address Resolution Protocol (IARP) specified in [RFC1293] uses the "InARP" codes below.

Assignments:

Number	Operation Code (op)	Reference
1	REQUEST	[RFC826]
2	REPLY	[RFC826]
3	request Reverse	[RFC903]
4	reply Reverse	[RFC903]
5	DRARP-Request	[David Brownell]
6	DRARP-Reply	[David Brownell]
7	DRARP-Error	[David Brownell]
8	InARP-Request	[RFC1293]
9	InARP-Reply	[RFC1293]
10	ARP-NAK	[Mark Laubach]

Number	Hardware Type (hrd)	Reference
1	Ethernet (10Mb)	[JBP]
2	Experimental Ethernet (3Mb)	[JBP]
3	Amateur Radio AX.25	[PXK]
4	Proteon ProNET Token Ring	[JBP]
5	Chaos	[GXP]
6	IEEE 802 Networks	[JBP]
7	ARCNET	[JBP]
8	Hyperchannel	[JBP]
9	Lanstar	[TU]
10	Autonet Short Address	[MXB1]
11	LocalTalk	[JKR1]
12	LocalNet (IBM PCNet or SYTEK LocalNET)	[JXM]
13	Ultra link	[RXD2]
14	SMDS	[GXC1]
15	Frame Relay	[AGM]
16	Asynchronous Transmission Mode (ATM)	[JXB2]
17	HDLC	[JBP]
18	Fibre Channel	[Yakov Rekhter]
19	Asynchronous Transmission Mode (ATM)	[Mark Laubach]
20	Serial Line	[JBP]
21	Asynchronous Transmission Mode (ATM)	[MXB1]

Protocol Type (pro) Use the same codes as listed in the section called "Ethernet Numbers of Interest" (all hardware types use this code set for the protocol type).

References

[RFC826] Plummer, D., "An Ethernet Address Resolution Protocol or Converting Network Protocol Addresses to 48-bit Ethernet Addresses for Transmission on Ethernet Hardware," STD 37, RFC 826, MIT-LCS, November 1982.

[RFC903] Finlayson, R., T. Mann, J. Mogul, and M. Theimer, "A Reverse Address Resolution Protocol," STD 38, RFC 903, Stanford University, June 1984.

[RFC1293] Bradley, T., and C. Brown, "Inverse Address Resolution Protocol," RFC 1293, Wellfleet Communications, Inc., January 1992.

URL = ftp://ftp.isi.edu/in-notes/iana/assignments/arp-parameters

Chapter

6

Internet Protocol

After the address matching, which we covered in Chap. 5, is complete, IP takes over to get the data to the target system. To do that, it will package the data, provided by one of the protocols it supports, into a datagram. This datagram includes the data payload and the IP header. While IP is a connectionless protocol, what may appear to be a liability works in its favor to create the effect of a self-healing network.

Self-Healing Networks

A self-healing network has no hard-coded path between the source and the target hosts. This means that each IP datagram can take whatever route is currently available between the two hosts. By letting each datagram choose its own connectionless path between locations, IP is doing its part in supporting TCP/IP's flexibility. We call that connectionless IP network *self-healing* because no single failure or group of failures in the Internet will break the connection between hosts unless all the paths are broken.

The term also means that the configuration of the routes can change without a central control to direct those changes. Removing the need for a central control gives the network the ability to handle moves easily, including adds and changes to the population of hosts and their connections to each other. As a result, the network has no single point of failure, with greater system flexibility and fault tolerance.

IP Header

As we found in Chaps. 3 and 5, each of the protocols in the TCP/IP suite uses a series of bytes (known as a header) to perform its required functions. The IP header, shown in Fig. 6.1, is no different in that respect. Its required functions are to determine logical addressing, segmentation, data length, quality of service, higher layer protocol identification, routing, diagnostics, and options.

4 bits	4 bits	3 bits	5 bits	2 bytes	2 bytes	2 bytes	1 byte	1 byte	2 bytes	4 bytes	4 bytes
Vers.	Header length	Precedence	TOS	Total IP length	Datagram ID no.	Fragment area	Time to live	Protocol carried	Header checksum	Source IP address	Target IP address
4	5	0	0	05 b4	a2 b7	00 00	40	11	88 a8	co 99 b8 01	co 99 b9 03

Figure 6.1 IP header layout.

By exploring the IP header, we see the way the fields of the header fulfill each of these requirements. These fields can contain a single bit or multiple octets (8-bit bytes). As we walk through the header, we will look into each of these fields to see what part each plays in completing IP's assigned tasks.

The example in Fig. 6.2 is typical of the hex display of a protocol analyzer. The center hyphen (between the eighth and ninth bytes or octets) makes it easier to identify verbally which pair of hex characters we are discussing.

Just as we did in the previous chapter's look at ARP, we will step through a hex example of an IP datagram, paying special attention to its header fields. The Ethernet header, which is the first 14 bytes (pairs of hex characters), identifies the protocol it is carrying as IP (08 00 hex) according the Ether type section of the Assigned Numbers RFC (discussed in Chap. 3's appendixes). That means that the IP header will start in the next byte.

IP Header Decode

The first byte of the IP header (the next-to-the-last byte, on the first line of the hex display in Fig. 6.2) combines two 4-bit fields: the version number and the header length.

Version

This first hex character sets the version of IP that created the header. This becomes more important when the next generation of IP begins connecting for the first time. If the versions do not match and both source and target software do not support a common version, the two sides cannot communicate. The current version of IP has been 4 since 1981, so this will be 4 until the next generation changes the header.

```
00  00  c0  a0  51  24  00  c0  -  93  21  88  a7  08  00  45  00
00  5a  dc  28  00  00  ff  01  -  88  08  c0  99  b8  01  c0  99
b8  03  08  00  d0  34  3b  19  -  01  00  18  2b  7a  83  9e  00
70  2d  08  09  0a  0b  0c  0d  -  0e  0f  10  11  12  13  14  15
```

Figure 6.2 Protocol analyzer display of an IP header.

Header length

The second hex character sets the IP header length as a number of 32-bit data words or four 8-bit bytes (i.e., 32 bits divided by 8). An IP header, without options, defaults to a value of 5 or ($5 \times 4 = 20$ bytes).

Some analysts look for the hex byte of 45 to find the start of the IP header. Longer IP headers and other TCP/IP instances of hex 45 are possible. It is better to look for the IP header to follow the network access layer's header ID of IP (08 00). This (looking for the Ether type of 08 00 hex to find IP) is also true when options are present and the first byte reads 4f hex rather than 45 hex.

Type-of-Service Byte

The second byte of the IP header carries the 3-bit precedence of data field and the four type-of-service bits: delay, throughput, reliability, and cost. Many technicians call it the TOS (type-of-service) byte. The last bit of the byte carries a value of 0 to indicate that the RFCs reserve it for future use.

These fields describe the routing service the source requests of each router through which the datagram passes. Each datagram can have a different request than other datagrams originating from the same source host.

Some vendors do not implement processing of the values in the fields, and those who do implement the values do not do so for all routing protocols. If a datagram that contains a requested precedence or type of service arrives at a router that does not support that requested function, the router must forward the datagram as if the software had set no value. The value remains in the header.

These fields provide the quality-of-service function. Let's look at what that can mean.

Precedence of the data

This field informs the receiving IP gateways and routers, along the network path that it follows, of the importance of the data it carries. The specifications offer many of the values for DARPA or military use, but several are of potential use for clearing network congestion problems. Figure 6.3 provides the

Binary	Field meaning
111	National network control
110	Internetwork control
101	CRTIC/ECP
100	Flash override
011	Flash
010	Immediate
001	Priority
000	Routine

Figure 6.3 Precedence field decode table.

precedence field codes. Some values allow network management tools to explore problems in a network that is experiencing errors. Most commercial software packages are just beginning to use this field.

When the software uses the precedence of data, it labels normal traffic as routine. When a router's buffer gets close to being full and the software decides to clear the buffer to start over, the router can dump the routine data to free up buffer space. It then forwards the remaining higher precedence traffic.

The other use of this field is for users who want access even though the circuits are full with traffic. They must have a higher precedence code than at least one of the existing users. The theory is that the higher precedence user has more important data to transmit. The lower level user is then bumped off the circuit in favor of the higher precedence user. This bumping process works up to the flash precedence level. When all circuits are busy with flash traffic, a user must have a flash override code for access. That user immediately reverts to flash level once she has access.

Military regulations specify the use of these codes although some government agencies also use these precedence codes. For example, the network (or staff) must deliver a flash message with 15 minutes (coordinated universal time) of the original transmission. The average user cannot use that code because it identifies a message that is of national security importance.

Figure 6.4 depicts the types of service bits used for each precedence field. The following bits are used in this figure:

Delay bit. Setting this bit to 1 requests a route with the least amount of propagation delay. If IP routers along the path support the delay bit, the datagram takes the shortest route and the fastest communications path(s) to the target-addressed system.

Throughput bit. If this bit is 1, supporting IP routers have the datagram travel over the path(s) with the highest throughput (greatest number of bytes per second). If FTP is trying to move files with the greatest efficiency it may set this bit to 1.

Reliability bit. This lets the application request that the datagram travel over the route with the least chance of data loss. When the software sets

Precedence field	Type-of-service bits				
	Delay	Throughput	Reliability	Cost	Reserved
0 0 0	1	0	0	0	0
0 0 0	0	1	0	0	0
0 0 0	0	0	1	0	0
0 0 0	0	0	0	1	0

Figure 6.4 Type-of-service layout.

the bit to a value of 1, the reliability bit still requires that IP routers support its use over the network.

Cost bit. RFC 1349 added the use of a cost bit that would allow the data to travel over the route that costs the sending organization the least amount of money to use.

Special note on TOS bits. The delay, throughput, reliability, and cost bits are mutually exclusive. Therefore, you can only set one of the bits on for each datagram. If the software sets more than 1 bit, the target software may return an ICMP Parameter Problem (logical) error message. For more information see the section on ICMP messages in Chap. 10.

Total IP Length Field

This 2-byte field indicates the total length of the datagram including the IP header and all the data behind it. With 16 bits available, the maximum datagram size is 65,535 bytes. Originally that number was higher than is practical, given the buffer space and error rates of the communications circuits. With new technology, higher speed lines, and improved software capabilities, this is no longer true. In RFC 1323, Van Jacobson explains the challenges and proposes an excellent solution.

Note that the total IP length field does not include any padding that the network access layer may add to meet minimum requirements. This is a very helpful field for determining the length of data being carried by IP. Simple subtraction of the IP header's byte length from the total IP length provides the number of bytes of data (made up of other headers and their data).

For example, subtracting the IP header length (20 bytes with no options) and the first 8 bytes of the ICMP header from the total IP length can provide the number of bytes the user is sending as a result of issuing the ping command. In our Fig. 6.2 example, we start with 00 5a hex that is 90 decimal. Subtracting 28 bytes from that 90 bytes leaves 62 bytes that the target of the ping should echo back to us.

Note that the maximum size for this field when the datagram is traveling over Ethernet or IEEE 802.3 is 1500 bytes (05 dc hex) as both protocols have a maximum transmission unit (MTU) size of 1500 bytes.

Datagram ID Number

This 2-byte, host-specific field carries the unique ID number of each datagram this host (or device) sends over (what the RFC calls) "a long time." We can find it as the third and fourth byte on the second line of the hex example in Fig. 6.2. This number most often starts at 00 01 (hex) when TCP/IP software starts operating on a host. The software adjusts this ID by one for every separate datagram that the host's IP layer sends.

The unique number plan worked fine for the ARPAnet running at 56 kbps with the long time of more than three days. A DS1 brings the time down to under three hours, and 10-Mbps Ethernet drops the time needed to wrap around the datagram ID number to just over 28 minutes. The 100-Mbps services, such as FDDI and Ethernet, bring that number to under three minutes. As we look to the future of multigigabit speeds, the wrap-around time may drop to only a few seconds, making "a long time" an odd term. In RFC 1323, Jacobson, Braden, and Borman explain the challenges and propose an excellent solution.

This datagram ID number primarily correlates logical errors and helps in reassembling datagram fragments. When fragmentation occurs, each datagram that is part of a larger message will have the same datagram ID number. This way the target IP-addressed system can reassemble the fragments into a single datagram before forwarding its payload to the next protocol up the stack. This can also occur when the originating host must fragment the datagram to introduce it to the network access layer.

Fragment Area

If IP has an IP payload that is larger than is allowed by the network access layer (NAL) it uses, then the IP software must break that payload into smaller pieces. Whenever IP "plugs in" to an NAL, it learns the value of this field, and the NAL driver tells the IP software the maximum size it will accept. This process of reducing data to a size that is acceptable to the NAL is called *fragmenting*.

Fragmenting occurs when the IP software breaks a large IP payload into smaller pieces. Each piece is small enough to travel on the next network segment. The IP software in another router may need to break those fragments into smaller fragments. Each fragment travels the network(s) to arrive at the target system. Only the targeted, IP-addressed system may reassemble the fragments into the original datagram.

Let's look at how IP makes sure the fragments of data can be properly reconstructed at the target host by examining the breakdown of possible values from these two bytes. See Fig. 6.5 for a fragment field bit layout and Fig. 6.6 for a sample of fragment field hex values.

Reserved	Don't	Status	Fragment offset
0	1	0	0000000000000
0	0	1	0000000000000
0	0	1	0000010111001
0	0	0	0000101110010
0	0	0	0000000000000

Figure 6.5 Fragment area bit layout.

00	5a	dc	28	40	00	ff	01	-	88	08	c0	99	b8	01	c0	99	Don't
00	5a	dc	28	20	00	ff	01	-	88	08	c0	99	b8	01	c0	99	First
00	5a	dc	28	20	b9	ff	01	-	88	08	c0	99	b8	01	c0	99	Middle
00	5a	dc	28	01	72	ff	01	-	88	08	c0	99	b8	01	c0	99	Last
00	5a	dc	28	00	00	ff	01	-	88	08	c0	99	b8	01	c0	99	None

Figure 6.6 Fragment area sample hex values.

Reserved bit

The RFCs reserve the first bit in the first byte of the fragment area, which means it carries a value of 0.

Don't fragment bit

If we tell the software to set this second bit to 1 (binary), the application is demanding that the IP software, in any receiving router, *not* fragment the IP datagram. If a router in the path cannot send the data without fragmenting, it will discard the datagram and transmit an ICMP Type 3 error message back to the sender. (See Chap. 10 for more information on a Type 3 error message.)

Fragment status

If the IP entity sets this third bit to 1, it says there are more fragments to follow. If this bit's value is 0, it identifies the datagram as either the final fragment or the only fragment. With this bit and the fragment offset field, the target IP can tell if it has received all of the data for a particular message.

Fragment offset

This 13-bit field carries the number of 64-bit (8-octet) words by which this data fragment is separate (offset) from the original message. By *offset* we mean the distance (in 64-bit words) from the start of a fragment to the first byte in the original datagram.

As we can see in Fig. 6.7, IP offsets each datagram by the total length of the fragments before it. We can now combine what we have learned about the other pieces of the fragment area into the following rules of thumb:

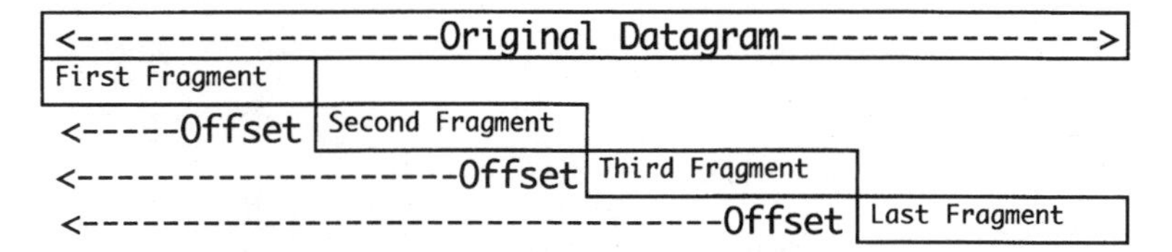

Figure 6.7 Sample offset diagram.

If the fragment status is 1 and the fragment offset is 0, we are looking at the first fragment.

If the fragment status is 1 and the fragment offset is greater than 0, we are looking at a middle fragment. We can arrange middle fragments in order of their offset values.

If the fragment status is 0 and the fragment offset is greater than 0, we are looking at the last fragment of the original datagram.

If the fragment status is 0 and the fragment offset is 0, we are looking at the only fragment or at a datagram that is unfragmented.

Fragment reassembly

As we said earlier, only the target IP-addressed system's IP software can reassemble the fragments into the original datagram. When the first datagram carrying a fragment arrives at the target system, a reassembly timer begins counting down toward 0. If all the datagrams carrying the fragments do not arrive at the target IP before the reassembly timer expires, IP discards the fragments and sends an ICMP Type 3 error message to the originating IP address.

Fragmenting fragments

As we can see in Fig. 6.8, the original packet, traveling on the token-ring network, contains 4096 bytes of data. After the router checks its routing table and determines that the path from itself to the target will use its Ethernet

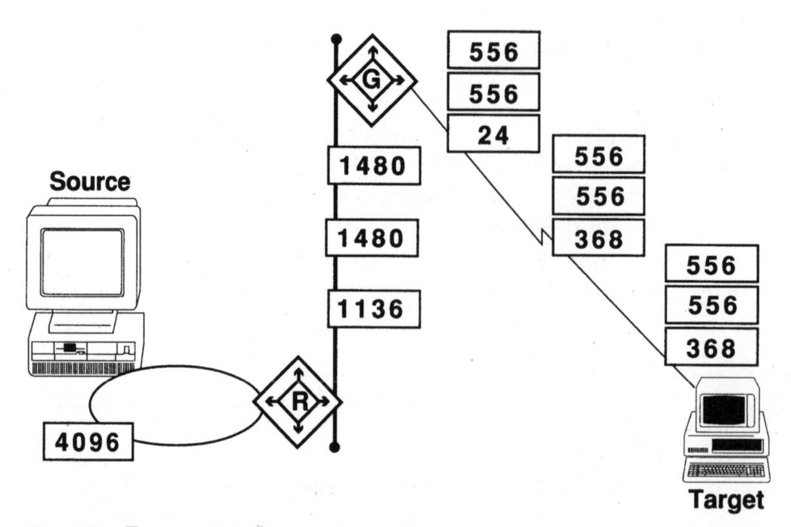

Figure 6.8 Fragmenting fragments.

interface, the router's IP entity has to fragment the datagram down to a size that Ethernet can handle, 1500 bytes or less.

The 1480-byte fragment gets a 20-byte IP header to complete the 1500-byte MTU that Ethernet has as its upper limit. The 1136-byte fragment fits under the 1500-byte MTU and carries the fragment status of 1 and a fragment offset of 370 64-bit words or 2960 bytes (370 · 8 [8-bit bytes] = 2960).

As the fragments arrive at the gateway, their destination IP address directs the gateway to pass them across the WAN link that is without an MTU. The IP specifications set a minimum acceptable data size of 576 bytes including the IP header. That means the gateway has to fragment the arriving fragments into smaller pieces to be sure that they can travel across the link to the target IP-addressed system.

In this case we can see that it is likely that all fragments arrived at the target system and its IP entity reassembled them in time. In the real world, fragmenting fragments present a high potential for problems. Since each fragment goes whatever way it must to reach the target system, losing one means the source has to retransmit the entire original datagram. On the other hand, with so many routes, a minor delay in getting all the fragments to the target system can cause a reassembly timeout and require the source to retransmit the datagram. In short, there is a direct correlation between an increase in fragmentation and an increase in lost datagrams due to lost fragments.

On the positive side, TCP/IP's flexibility stays alive with each fragment being routed independently. Also, reassembling the fragments at the target system lets the routers do what they do best—route datagrams—instead of reassembling datagrams only to refragment them to another size.

An alternative to this fragmentation in routers is to let the source do the fragmenting. That uses a Path MTU Discovery, specified in RFC 1191, to find the largest MTU size that does not require fragmentation anywhere along the path from the source to the target.

Time to Live

The time-to-live (TTL) field, the ninth byte of the IP header, tells how many seconds the datagram can live before a router (or gateway) must deliver it or discard it. That does not mean that the source user can look at her watch to see how long before the datagram dies. Instead, we should consider these seconds as coins in a pouch that we can use to pay for parking or tolls.

Gateways do not know how long it takes data to move across a link. To compensate for that, each IP gateway (or router) takes a one-second (coin) toll from each datagram as it arrives. Experience tells us that while all router vendors support this aspect of the TTL function, some do not support the remaining (following) part.

In this second part, which is not fully supported by all vendors, each router stamps the time the datagram arrives at the router. The system checks the clock again when the datagram leaves the buffer for the interface and the network. If one or more seconds have passed, the router subtracts that number of seconds from the datagram's TTL field as a parking charge (for taking up buffer space). If the counter reaches 0, the router discards that datagram and transmits an ICMP Type 3 error message back to the source IP address.

Without the TTL counter, some data may never arrive; that data could travel the network forever, taking up constant bandwidth. The TTL value in Fig. 6.2 is not indefinite but is the closest we can get, 255 seconds (ff hex) in the seventh byte on the second line of the hex display. That is a very long time to be in a network.

A time-to-live example

The flexibility of TCP/IP, to run over a network whose systems have different CPU clock times, dictates that the time the datagram leaves the workstation makes no difference. When the datagram arrives at the first router, that router clocks in the datagram and subtracts 1 from the time-to-live field. Since RFC 1700 recommends a default TTL for 64 seconds (40 hex), we start with that value in the TTL field.

The IP software in the router checks the datagram's departure time. It finds that the datagram sat in the buffer for less than one second. The datagram departs with (3f hex) 63 decimal seconds in its TTL field.

Arriving at the gateway, the datagram's TTL field loses one more second. Since the gateway is very busy, it takes the IP entity more than 4 seconds to get to the datagram and send it on toward the target system. When the gateway sends the datagram out, the gateway takes 4 more seconds from the TTL field, bringing the TTL to 58 seconds (3a hex). Figure 6.9 portrays an example of a TTL field in a source to target transmission.

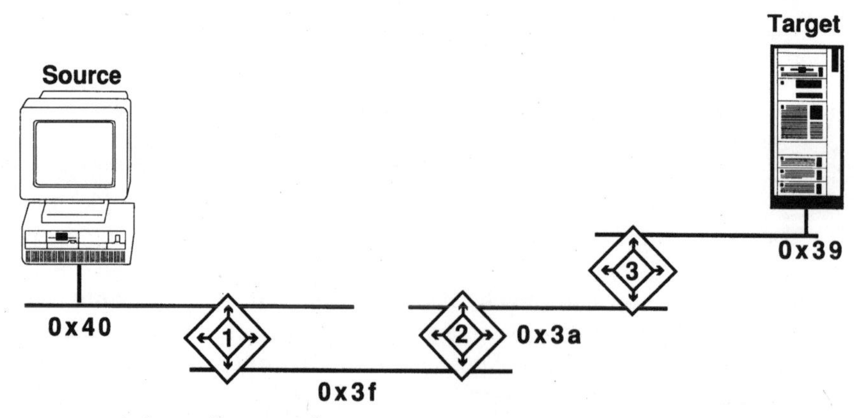

Figure 6.9 A time-to-live example.

Hex	Decimal	Protocol
01	1	ICMP
06	6	TCP
08	8	EGP
11	17	UDP
59	89	OSPF
58	88	IGRP

Figure 6.10 Common protocol numbers.

Protocol Field

This 8-bit field carries the ID number of the higher-level protocol, like a shipping label, that identifies which protocol is to receive the datagram's payload. The most common value found here is 06, which the RFC specification assigns to TCP.

We have identified other common values in Fig. 6.10. Remember, the value found in this field in a protocol analyzer's hex field (or a hex trace of a session) is in hex. We must convert it to a decimal number before looking it up in the Protocol Numbers section of RFC 1700, or reproductions of that section in Figs. 6.11 and 6.12.

IP Header Checksum

Since the protocols above IP have their own error detection, this field provides error checking on the IP header only and does not cover the data that IP is carrying at the end of the header. When options cause the IP to extend the IP header, the checksum covers the extended length of the header. In short, the IP header checksum covers the number of octets or bytes identified in the IP header length field. If the target IP-addressed interface receives a datagram with a failed checksum, the IP software discards the entire datagram.

IP Addresses

The last two fields of the basic IP header are the source and target IP addresses. An interesting thing happens from the IP header up in the TCP/IP stack. When the software identifies the source and target fields, the source appears before the target. Contrast this with the network access layer where the target field appears before the source field, as we saw in Fig. 6.2.

Source IP address

The sender interface's 32-bit Internet address is in 4 hex bytes or octets (pairs of hex characters). In Fig. 6.2, that field began with the third byte, to

Figure 6.11 RFC 1700 protocol numbers section.

PROTOCOL NUMBERS

In the Internet Protocol (IP) [DDN], [RFC791] there is a field, called Protocol, to identify the next level protocol. This is an 8 bit field.

Assigned Internet Protocol Numbers

Decimal	Keyword	Protocol
0		Reserved
1	ICMP	Internet Control Message
2	IGMP	Internet Group Management
3	GGP	Gateway-to-Gateway
4	IP	IP in IP (encasulation)
5	ST	Stream
6	TCP	Transmission Control
7	UCL	UCL
8	EGP	Exterior Gateway Protocol
9	IGP	any private interior gateway
10	BBN-RCC-MON	BBN RCC Monitoring
11	NVP-II	Network Voice Protocol
12	PUP	PUP
13	ARGUS	ARGUS
14	EMCON	EMCON
15	XNET	Cross Net Debugger
16	CHAOS	Chaos
17	UDP	User Datagram
18	MUX	Multiplexing
19	DCN-MEAS	DCN Measurement Subsystems
20	HMP	Host Monitoring
21	PRM	Packet Radio Measurement
22	XNS-IDP	XEROX NS IDP
23	TRUNK-1	Trunk-1
24	TRUNK-2	Trunk-2
25	LEAF-1	Leaf-1
26	LEAF-2	Leaf-2
27	RDP	Reliable Data Protocol
28	IRTP	Internet Reliable Transaction
29	ISO-TP4	ISO Transport Protocol Class 4
30	NETBLT	Bulk Data Transfer Protocol
31	MFE-NSP	MFE Network Services Protocol
32	MERIT-INP	MERIT Internodal Protocol
33	SEP	Sequential Exchange Protocol
34	3PC	Third Party Connect Protocol
35	IDPR	Inter-Domain Policy Routing Protocol
36	XTP	XTP
37	DDP	Datagram Delivery Protocol
38	IDPR-CMTP	IDPR Control Message Transport Proto
39	TP++	TP++ Transport Protocol
40	IL	IL Transport Protocol
41	SIP	Simple Internet Protocol
42	SDRP	Source Demand Routing Protocol
43	SIP-SR	SIP Source Route
44	SIP-FRAG	SIP Fragment
45	IDRP	Inter-Domain Routing Protocol

Decimal	Keyword	Protocol
46	RSVP	Reservation Protocol
47	GRE	General Routing Encapsulation
48	MHRP	Mobile Host Routing Protocol
49	BNA	BNA
50	SIPP-ESP	SIPP Encap Security Payload
51	SIPP-AH	SIPP Authentication Header
52	I-NLSP	Integrated Net Layer Security TUBA
53	SWIPE	IP with Encryption
54	NHRP	NBMA Next Hop Resolution Protocol
55-60	Unassigned	Unassigned
61	Host internal protocol	any host internal protocol
62	CFTP	CFTP
63	Any local network	any local network
64	SAT-EXPAK	SATNET and Backroom EXPAK
65	KRYPTOLAN	Kryptolan
66	RVD	MIT Remote Virtual Disk Protocol
67	IPPC	Internet Pluribus Packet Core
68	Any distributed file sys.	any distributed file system
69	SAT-MON	SATNET Monitoring
70	VISA	VISA Protocol
71	IPCV	Internet Packet Core Utility
72	CPNX	Computer Protocol Network Executive
73	CPHB	Computer Protocol Heart Beat
74	WSN	Wang Span Network
75	PVP	Packet Video Protocol
76	BR-SAT-MON	Backroom SATNET Monitoring
77	SUN-ND	SUN ND PROTOCOL-Temporary
78	WB-MON	WIDEBAND Monitoring
79	WB-EXPAK	WIDEBAND EXPAK
80	ISO-IP	ISO Internet Protocol
81	VMTP	VMTP
82	SECURE-VMTP	SECURE-VMTP
83	VINES	VINES
84	TTP	TTP
85	NSFNET-IGP	NSFNET-IGP
86	DGP	Dissimilar Gateway Protocol
87	TCF	TCF
88	IGRP	IGRP
89	OSPFIGP	OSPFIGP
90	Sprite-RPC	Sprite RPC Protocol
91	LARP	Locus Address Resolution Protocol
92	MTP	Multicast Transport Protocol
93	AX.25	AX.25 Frames
94	IPIP	IP-within-IP Encapsulation Protocol
95	MICP	Mobile Internetworking Control Pro.
96	SCC-SP	Semaphore Communications Sec. Pro.
97	ETHERIP	Ethernet-within-IP Encapsulation
98	ENCAP	Encapsulation Header
99	any private encryption	any private encryption scheme
100	GMTP	GMTP
101-254	unassigned	Unassigned
255	reserved	Reserved

Figure 6.11 *(Conclusion)*

Figure 6.12 IP protocol numbers sorted by keyword.

Keyword	Decimal	Protocol
3PC	34	Third Party Connect Protocol
Any distributed file sys.	68	any distributed file system
Any local network	63	any local network
Any private encryption	99	any private encryption scheme
ARGUS	13	ARGUS
AX.25	93	AX.25 Frames
BBN-RCC-MON	10	BBN RCC Monitoring
BNA	49	BNA
BR-SAT-MON	76	Backroom SATNET Monitoring
CFTP	62	CFTP
CHAOS	16	Chaos
CPHB	73	Computer Protocol Heart Beat
CPNX	72	Computer Protocol Network Executive
DCN-MEAS	19	DCN Measurement Subsystems
DDP	37	Datagram Delivery Protocol
DGP	86	Dissimilar Gateway Protocol
EGP	8	Exterior Gateway Protocol
EMCON	14	EMCON
ENCAP	98	Encapsulation Header
ETHERIP	97	Ethernet-within-IP Encapsulation
GGP	3	Gateway-to-Gateway
GMTP	100	GMTP
GRE	47	General Routing Encapsulation
HMP	20	Host Monitoring
host internal protocol	61	any host internal protocol
I-NLSP	52	Integrated Net Layer Security TUBA
ICMP	1	Internet Control Message
IDPR	35	Inter-Domain Policy Routing Protocol
IDPR-CMTP	38	IDPR Control Message Transport Proto
IDRP	45	Inter-Domain Routing Protocol
IGMP	2	Internet Group Management
IGP	9	any private interior gateway
IGRP	88	IGRP
IL	40	IL Transport Protocol
IP	4	IP in IP (encasulation)
IPCV	71	Internet Packet Core Utility
IPIP	94	IP-within-IP Encapsulation Protocol
IPPC	67	Internet Pluribus Packet Core
IRTP	28	Internet Reliable Transaction
ISO-IP	80	ISO Internet Protocol
ISO-TP4	29	ISO Transport Protocol Class 4
KRYPTOLAN	65	Kryptolan
LARP	91	Locus Address Resolution Protocol
LEAF-1	25	Leaf-1
LEAF-2	26	Leaf-2
MERIT-INP	32	MERIT Internodal Protocol
MFE-NSP	31	MFE Network Services Protocol
MHRP	48	Mobile Host Routing Protocol
MICP	95	Mobile Internetworking Control Pro.

Keyword	Decimal	Protocol
MTP	92	Multicast Transport Protocol
MUX	18	Multiplexing
NETBLT	30	Bulk Data Transfer Protocol
NHRP	54	NBMA Next Hop Resolution Protocol
NSFNET-IGP	85	NSFNET-IGP
NVP-II	11	Network Voice Protocol
OSPFIGP	89	OSPFIGP
PRM	21	Packet Radio Measurement
PUP	12	PUP
PVP	75	Packet Video Protocol
RDP	27	Reliable Data Protocol
Reserved	255	Reserved
RSVP	46	Reservation Protocol
RVD	66	MIT Remote Virtual Disk Protocol
SAT-EXPAK	64	SATNET and Backroom EXPAK
SAT-MON	69	SATNET Monitoring
SCC-SP	96	Semaphore Communications Sec. Pro.
SDRP	42	Source Demand Routing Protocol
SECURE-VMTP	82	SECURE-VMTP
SEP	33	Sequential Exchange Protocol
SIP	41	Simple Internet Protocol
SIP-FRAG	44	SIP Fragment
SIP-SR	43	SIP Source Route
SIPP-AH	51	SIPP Authentication Header
SIPP-ESP	50	SIPP Encap Security Payload
Sprite-RPC	90	Sprite RPC Protocol
ST	5	Stream
SUN-ND	77	SUN ND PROTOCOL-Temporary
SWIPE	53	IP with Encryption
TCF	87	TCF
TCP	6	Transmission Control
TP++	39	TP++ Transport Protocol
TRUNK-1	23	Trunk-1
TRUNK-2	24	Trunk-2
TTP	84	TTP
UCL	7	UCL
UDP	17	User Datagram
Unassigned	55-60	Unassigned
Unassigned	101-254	Unassigned
VINES	83	VINES
VISA	70	VISA Protocol
VMTP	81	VMTP
WB-EXPAK	79	WIDEBAND EXPAK
WB-MON	78	WIDEBAND Monitoring
WSN	74	Wang Span Network
XNET	15	Cross Net Debugger
XNS-IDP	22	XEROX NS IDP
XTP	36	XTP

Figure 6.12 *(Conclusion)*

the right of the hyphen, on the second line of hex. There we find the hex value of c0 99 b8 01, which translates to an IP address of 192.153.184.1 in decimal. We do this by translating each hex byte of the address separately since they appear separately in the dotted decimal version of the IP address.

c0 equals 12 (c hex) times 16 or 192

99 equals 9 times 16 or 144, plus 9, for a total of 153

b8 equals 11 (b hex) times 16 or 176, plus 8, for a total of 184

01 equals 1

Inserting the periods between the bytes thus yields an IP address of 192.153.184.1.

Target IP address

The target host's 32-bit Internet address is the last four hex bytes (pairs of hex characters) in the basic IP header. In Fig. 6.2 they are the last two bytes on the second line of the hex display and the first two bytes on the third line. They contain the hex value of c0 99 b8 03, which translates to a decimal IP address of 192.153.184.3 in decimal by following the same procedure that we saw in the previous paragraph.

The IP header ends here when the IP header length field contains a value of 5. When the IP datagram's header is carrying options, the IP header length field contains a number larger than 5, as we see in Fig. 6.13.

IP Option Fields

Users may have access to several useful options in IP to measure the time delay or routing of network traffic or to specify routing instructions to the routing devices in the network. We can recognize the presence of IP options by the first byte of the IP header carrying a value greater than 45 hex. To use these options, IP must set the correct option class(es) and code(s) in the

00	00	c0	a0	51	24	00	c0	-	93	21	88	a7	08	00	4f	00
00	7c	oc	22	00	00	ff	01	-	a6	9d	c0	99	b9	02	c0	99
b9	64	07	27	04	00	00	00	-	00	00	00	00	00	00	00	00
00	00	00	00	00	00	00	00	-	00	00	00	00	00	00	00	00
00	00	00	00	00	00	00	00	-	00	00	08	00	62	cd	20	00
0f	08	8b	74	67	2f	80	8b	-	c7	00	09	0a	0b	0c	0d	0e

Figure 6.13 Hex packet display with IP options.

option fields. Let's take a look at how the options work by looking into the bits and bytes of those IP option fields.

Copy-through-gate bit

Although the copy-through-gate bit is part of the specification, IP would *only* use the bit in designating the IP addresses of the router that the datagram (or fragments) follows to reach the target system. IP sets this bit to 1 to tell the router, which fragments the datagram, that all the resulting fragments carry the same option fields. That means the IP entity in the router attaches the option fields to every fragment instead of only the first fragment.

Since the fragments can otherwise follow any path to the target system, a user asking the IP to record the route the datagram follows can get multiple routes back from fragments that follow different paths. There is no way in the IP option fields to provide the requester with a list of those multiple paths on the 37-byte limit on the space available for an answer. You can see the reason that the record route option does not use the copy-through-gate bit.

With the bit set to 0, only the first fragment carries the option's data.

IP option values

The 7 bits (after the copy-through-gate bit) in this first byte of the IP option fields identify the option(s) used by their option class and option number as we can see in Fig. 6.14. Terms used in this figure include the following:

Option class. Two bits of the option classes: 00 for network or datagram control and 10 for system measurements and debugging. The RFC specifications reserve 01 and 11 for future use.

Option number. A 5-bit field used to select the option that IP is to invoke. The specifications only define eight of the 32 possible options. With the next version of IP on the horizon, we do not expect any more assignments.

Class	Number	Length	Name
00	00000	n/a	End of option list
00	00001	n/a	No operation
00	00010	11	Security handling
00	00011	Variable	Loose source routing
10	00100	Variable	Collect internet timestamps
00	00111	Variable	Record route
00	01000	4	SATNET send stream ID
00	01001	Variable	Strict source routing

Figure 6.14 IP option values table.

End-of-option list. This marks the end of the data field for variable length options.

No-operation bytes. Provide padding characters so the option fits the 32-bit word header length increments.

Security handling. An option that governs how routers should treat the data.

SATNET Send Stream ID. An obsolete military protocol.

Route-based options

Length. The length field lets us determine the total length of the IP options field. This also designates the maximum size that the options' data can become. In the examples in Fig. 6.15 (and in the current version of IP), that value is 39 decimal or 27 hex. It uses these values since the IP does not count the first byte of the option field, the class and number byte, in the length. After an IP header of 20 bytes, the remaining space in a maximum-length header is 40 bytes. We can determine that length limit by looking at the IP header length field that IP set to f hex in Fig. 6.14. Multiplying its decimal value of 15 by 4, to count the number of bytes instead of 32-bit words, we arrive at a total IP header length of 60 bytes.

Pointer. IP uses the pointer byte to identify the location where the router should insert, or read, the next bytes of data from the options' data field. If an option length field equals the pointer, the options' data space is full, or the IP has followed all the option instructions.

In the examples in Fig. 6.15, the IP set the pointer for the first byte where IP loads the first 4-byte IP address the router is to find in the loose source and strict source options. The pointer in the record route option tells the router where to begin loading the IP addresses of the routers it finds along the route to the target system.

Let's look at some examples of these three options to understand their differences and common areas as well as how the IP uses them in the real world of TCP/IP internetworking.

Class and number byte							
CtG	Class	Number	Length	Pointer	IP addresses	No op. pad	Option
0	00	00011	27	04	One or more	01	Loose source
0	00	01001	27	04	One or more	01	Strict source
0	00	00111	27	04	None or more	01	Record route

Figure 6.15 Route-based options layout.

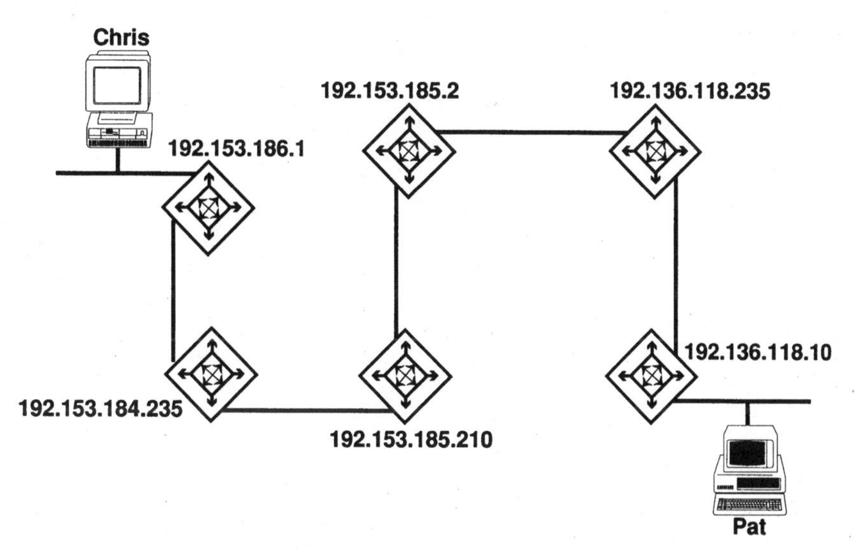

Figure 6.16 Routing options diagram.

Loose source routing. This option (03 hex or 0000 0011 binary) specifies a list of IP addresses (identifying a series of routers) that the datagram follows to the target IP-addressed system. We might use it as a diagnostic path that verifies connectivity. The datagram can pass through other IP routers between the listed IP-addressed routers.

In the example in Fig. 6.16, Chris uses the loose source routing option in sending a datagram to Pat to verify connectivity to some routers in the network. Chris designates the receiving, IP-addressed router interfaces that the routers use as a path toward the target system. This option overrides the routers' routing tables for handling this datagram.

Chris, however, only needs to identify certain interfaces along the route. The main requirement is that the source-specified, IP-addressed interfaces be accessed in the order they appear in the list of IP addresses in the option field. With loose source routing, the datagram passes through any unlisted routers on its way to the next router that the pointer identifies.

Strict source routing. With this diagnostic option, the source specifies the exact route that the datagram must follow. If an IP router cannot send the data to the next option-listed, IP-addressed router, the sending router must discard the packet. That router then sends an ICMP Type 3 error message to the original source address. Network technicians and managers often use this option to test the routing tables of networks for errors.

In the example in Fig. 6.16, Pat worries about the proprietary nature of the information going across the network to Chris. As a precaution against unauthorized eyes seeing the data, Pat invokes the strict source routing option in sending the datagram to Chris. This option verifies each interface of the path that the datagram follows through the network. With strict source routing, the datagram can only use the exact source-designated route through the network. If the datagram cannot follow the route, the first router to recognize that fact sends an ICMP "Destination Unreachable" message back to Pat.

In Fig. 6.16, Pat has forgotten that 192.136.118.200 is in the network path that the datagram must follow back to Chris. As a result, the router's IP entity sends an ICMP message back to Pat, and the router discards the datagram.

Record route. This option allows the IP datagram to collect the addresses of all IP gateways visited between the source and the target IP addresses. Managers and technicians often use this option to troubleshoot routing problems in the network.

In the Fig. 6.16 example, Pat has found that the datagram did not reach Chris and must determine why it failed. To accomplish this, Pat sends a record route option to find the exact route to specify in a subsequent strict source route option. With the record route option, the IP entity in each router adds its IP address to the list as it passes the option-extended datagram through the network.

Not all vendors' TCP/IP packages support the record route option. Therefore, it is important that the target IP-addressed device's software supports this option to be sure that it returns the desired results. In our example, Pat finds the correct path of IP addresses to specify in the next strict source route option that will carry the proprietary data to Chris.

Internet timestamp collection option

The last option collects the timestamps from each IP address that it visits along a route. It lets scientists and engineers get a record of the delay in a time-locked network. A time-locked network has every device's CPU clock ticking to the same beat of a master (usually atomic) clock.

Overflow. The overflow field identifies the number of IP gateways that the datagram found after the previous routers filled the option field.

Operation flags. Operation (op) flags provide additional instructions to the IP. As we can see in the table in Fig. 6.17, the op flags match the values we find in the operation column. For example:

- The timestamps only (time-only) flag will return up to nine timestamps. Each timestamp takes up 4 bytes and the remaining space, after the option overhead, is 36 bytes.
- The timestamps and add IP addresses (add IP) flag collects 4-byte timestamps and adds the 4-byte IP address of the interface that provided the timestamp. We can only collect four timestamps with this option flag.

Byte			Byte	Byte	Byte		4 bytes	4 bytes	
CtG	Class	Number	Length	Pointer	Overflow	Op. flag	IP address	Timestamp	Operation
0	10	00100	27	05	0000	0000	Unused	As needed	Time only
0	10	00100	27	05	0000	0001	As needed	As needed	Add IP
0	10	00100	27	05	0000	0011	As needed	As needed	Read IP

Figure 6.17 Internet timestamps options layout.

- The timestamps and read IP addresses (read IP) flag collects 4-byte timestamps and reads the next 4-byte IP address of the interface that the datagram is to ask for its timestamp. We can collect four timestamps with this flag.

IP Sample Data Exchanges

Figure 6.18 shows some samples of IP data traffic between hosts on an Ethernet network. We provide them to give you data examples of what to expect in a good IP header.

```
Internet: 192.153.184.3 -> 192.153.184.1 hl: 5 ver: 4 tos: 0 len: 44
id: 0x0006 fragoff: 0 flags: 00 ttl: 60 prot: TCP(6) xsum: 0x908f

Internet: 192.153.184.1 -> 192.153.184.3 hl: 5 ver: 4 tos: 0 len: 44
id: 0xdc2a fragoff: 0 flags: 00 ttl: 30 prot: TCP(6) xsum: 0x6bd1

Internet: 192.153.184.3 -> 192.153.184.1 hl: 5 ver: 4 tos: 0 len: 40
id: 0x0007 fragoff: 0 flags: 00 ttl: 60 prot: TCP(6) xsum: 0x938f

Internet: 192.153.184.5 -> 192.153.184.1 hl: 5 ver: 4 tos: 0 len: 48
id: 0x034c fragoff: 0 flags: 00 ttl: 64 prot: UDP(17) xsum: 0x3988

Internet: 192.153.184.1 -> 192.153.184.255 hl: 5 ver: 4 tos: 0 len: 92
id: 0xdc32 fragoff: 0 flags: 00 ttl: 30 prot: UDP(17) xsum: 0x2cd0

Internet: 192.153.185.1 -> 192.153.185.5 hl: 5 ver: 4 tos: 0 len: 56
id: 0x0383 fragoff: 0 flags: 00 ttl: 64 prot: ICMP(1) xsum: 0x0a88

Internet: 192.153.185.5 -> 192.153.185.1 hl: 5 ver: 4 tos: 0 len: 56
id: 0x0385 fragoff: 0 flags: 00 ttl: 64 prot: ICMP(1) xsum: 0x0688

Internet: 192.153.186.1 -> 192.153.184.5 hl: 5 ver: 4 tos: 0 len: 1500
id: 0xdc83 fragoff: 0 flags: 0x01 ttl: 32 prot: ICMP(1) xsum: 0x63a9 Fragment

Internet: 192.153.186.1 -> 192.153.184.5 hl: 5 ver: 4 tos: 0 len: 538
id: 0xdc83 fragoff: 185 flags: 00 ttl: 32 prot: ICMP(1) xsum: 0x6ccc Fragment
```

Figure 6.18 Sample decoded IP datagrams.

Chapter

7

IP's Next Generation

Overview

No crystal ball could have predicted the popularity of the Internet and the need for an ever-increasing number of IP addresses. In 1987, estimates predicted a need for as many as 100,000 networks at some vague time in the future. We passed that mark in late 1996.

While the 32-bit addressing of IP's version 4 ideally can handle more than 4 billion interfaces on up to 16.7 million networks, true usable addressing is far less numerous than that. This was due, in part, to the use of Class A, B, and C addresses, which have expanded routing tables at an unmanageable rate.

The dilemma became one of choosing between limitations: On the one hand, there is the tremendous rate of growth of the Internet, and on the other hand, any major changes could seriously disrupt the network. The IETF formed the Routing and Addressing (ROAD) group in 1991 to explore this dilemma and guide the IETF in this area. The ROAD group's March 1992 recommendations ranged in kind from "immediate" to "long term." Their suggestions included the Classless Interdomain Routing (CIDR) proposal to reduce the rate of routing table growth and a call for "working groups to explore separate approaches for bigger Internet addresses."

In 1993, the IETF formed the Address Lifetime Expectations (ALE) working group "to develop an estimate for the remaining lifetime of the IPv4 [IP Version 4] address space." The group projected that the IPv4 address space would run dry sometime between 2005 and 2011.

One problem the ALE working group had was in determining the impact of Classless Interdomain Routing. In doing so, the InterNIC allocates IP addresses blocks of Class C addresses to regional Network Information Centers and access providers. The NICs and the providers then reallocate addresses to their customers. The ALE projections used the InterNIC batch-address assignments without counting the actual rate of assignment of individual network addresses to end-user organizations. These factors reduce the reliability of the ALE estimates but seem to indicate there is enough time remaining in the IPv4 address space to consider adding features in an IPng beyond expanding the address size.

The Contenders or Alphabet Soup

By November 1993, different groups in the Internet community had developed separate proposals for IPng. Those proposals had evolved and merged into other proposals and, in some cases, had simply changed names (see Fig. 7.1).

Simple ConnectionLess Network Protocol (Simple CLNP) evolved into TUBA. The TP/IX working group changed its name to Common Architecture for the Internet Protocol (CATNIP). IP Encaps had become IP Address Encapsulation (IPAE) before it merged with SIP; the new protocol kept the name SIP. This group then merged with PIP to become Simple Internet Protocol Plus (SIPP).

All the proposals would work to provide a way to overcome the obstacles presented by the Internet expansion. The IPng working group's job was to make sure that the IETF understood the proposals. The IETF would then provide a recommendation on which method best resolved the issues while providing the best future for the Internet.

In October 1994, RFC 1700 reported the Internet version numbers we see in Fig. 7.2. As we discussed in the last chapter, we currently use Version 4. Internet STream Protocol (STP) datagram mode is Version 5. Simple Internet Protocol (SIP) is Version 6. The TP/IX, which its author would change to Common Architecture for the Internet Protocol (CATNIP) is Version 7. The P Internet Protocol (PIP), whose working group later merged with the SIP to become Simple Internet Protocol Plus and Version 6, was Version 8. The TCP and UDP over Big Addresses (TUBA) is Version 9.

If all of this confuses you, you are not alone. The IAB issued a policy statement referring to IP Version 7 as the next version of IP. The statement was apparently the result of some incorrect information. The IETF then attempted to clear the confusion and called for proposals for "IP—The Next Generation." Like all TCP/IP terms, that quickly earned an acronym—IPng.

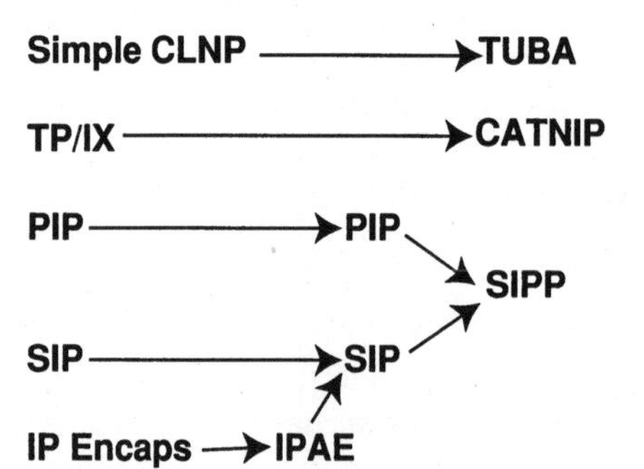

Figure 7.1 IP contenders combine.

VERSION NUMBERS

In the Internet Protocol (IP) [RFC791] there is a field to identify the version of the internetwork general protocol. This field is 4 bits in size.

Assigned Internet Version Numbers

Decimal	Keyword	Version	References
0		Reserved	[JBP]
1–3		Unassigned	[JBP]
4	IP	Internet Protocol	[RFC791,JBP]
5	ST	ST Datagram Mode	[RFC1190,JWF]
6	SIP	Simple Internet Protocol	[RH6]
7	TP/IX	TP/IX: The Next Internet	[RXU]
8	PIP	The P Internet Protocol	[PXF]
9	TUBA	TUBA	[RXC]
10–14		Unassigned	[JBP]
15		Reserved	[JBP]

Figure 7.2 Internet version numbers from RFC 1700.

The IPng working group evaluated the three IPng proposals: TUBA, CATNIP, and SIPP. We will examine each briefly to understand what each offers.

TUBA

The TUBA proposal attempts to reduce the migration risk to a new IP address space. It suggests that existing Internet protocols continue to run, by replacing the 32-bit IP addresses with larger addresses. The TUBA plan does not require moving completely to OSI. Instead, OSI's CLNP would replace IP while TCP, UDP, and the traditional TCP/IP applications would simply run on top of CLNP (see Fig. 7.3).

Part of TUBA's plan is also to expand the ability to route Internet datagrams by using addresses that support a more structured hierarchy than IP. It would require a transition from the use of IPv4 to ISO/IEC 8473 (CLNP) and OSI's matching large Network Service Access Point (NSAP) address space.

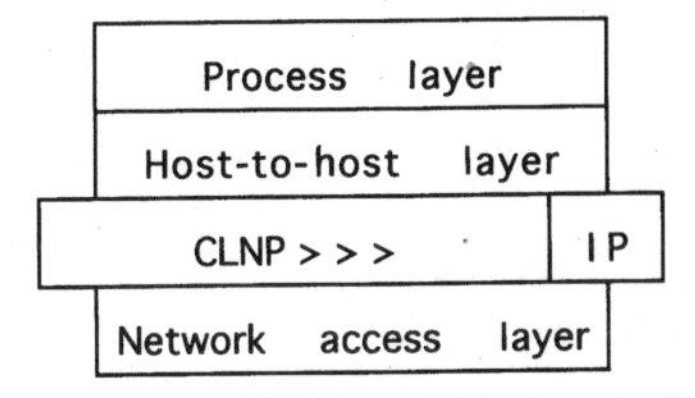

Figure 7.3 TUBA or CLNP replacing IP.

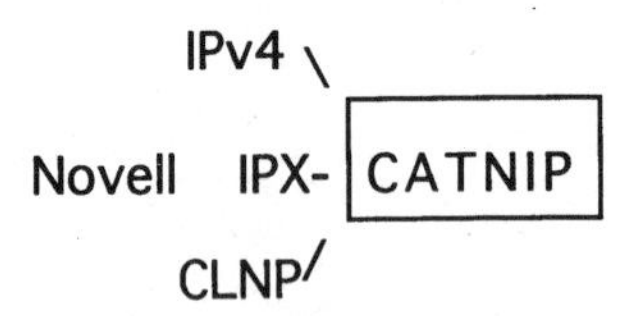

Figure 7.4 CATNIP, or pick a network layer.

The proposal uses a simple, long-term, gradual update of Internet hosts (to run Internet applications over CLNP) and DNS servers (to return larger addresses). This proposal requires that we update all routers to support forwarding of CLNP, along with IP and any other protocols. However, it does not require any datagram translation, encapsulation, or address mapping. We would assign, route, and use NSAP addresses independently throughout the transition period.

CATNIP

Common architecture for the Internet is a designed integration of CLNP, IPv4, and Novell's Internet Packet Exchange (IPX) (see Fig. 7.4). It allows any transport layer protocols in use to run over any of the network layer protocol formats. With some attention to detail, a TCP can run properly with one end system using IPv4 while the other end of the session uses another network protocol, such as CLNP. The plan is a common functionality between the Internet, OSI, and the Novell protocols. At the same time, we would be advancing Internet technology to the next generation.

Like TUBA, CATNIP supports OSI NSAP format addresses. It also offers the use of cache handles for high-performance routing as well as shorter network headers.

SIPP

Simple Internet Protocol Plus is a designed evolutionary step from IPv4. Functions that work in IPv4 stay in SIPP (see Fig. 7.5). Users may therefore install it as a normal upgrade in devices, since it is planned to interoperate with IPv4. The protocol will run well on both high-performance and low-bandwidth networks.

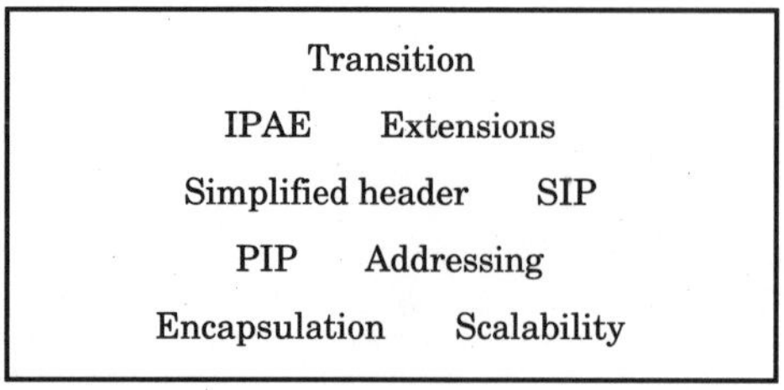

Transition
IPAE Extensions
Simplified header SIP
PIP Addressing
Encapsulation Scalability

Figure 7.5 SIPP: a blended solution.

The original version of SIPP calls for the extension of the IP address size from 32 bits to 64 bits. This allows a more structured addressing hierarchy and a much larger number of addressable interfaces. SIPP addressing was to allow extensions in 64-bit increments by using cluster addresses to identify regions instead of separate interfaces.

The use of SIPP increases routing efficiency, relaxes limits on the length of options, and increases the flexibility of adding new options just by changing the way the IP header encodes the options.

Proposal Reviews

The IPng directorate reviewed and discussed these three proposals during its biweekly teleconferences and through its mailing list. A two-day retreat, held near Chicago in May 1994, began with a roundtable airing of the views of each of the participants. This discussion also included area directors and guests invited by the chairs for each of the proposals.

The table in Fig. 7.6, reformatted from RFC 1752, summarizes each of the three proposals reviewed against the requirements in the IPng criteria document. The "yes" and "no" responses speak for themselves. A response of "mixed" means that the reviewers had mixed reviews with none dominating. An "unknown" response means the reviewers determined that the documentation did not clearly address the criterion.

Requirement	TUBA	CATNIP	SIPP
Access to standards	Mixed	Yes	Yes
Completely specified	Mostly	No	Yes
Configuration ease	Mixed	Unknown	Mixed
Control protocol	Mixed	Unknown	Yes
Datagram	Yes	Yes	Yes
Extensibility	Mixed	Unknown	Mixed
Media independent	Yes	Yes	Yes
Mobility	Mixed	Unknown	Mixed
Multicast	Mixed	Unknown	Yes
Performance	Mixed	Mixed	Mixed
Robust service	Yes	Mixed	Mixed
Scale	Yes	Yes	Yes
Security	Mixed	Unknown	Mixed
Service classes	Mixed	Unknown	Yes
Simplicity	No	No	No
Topology flexibility	Yes	Yes	Yes
Transition	Mixed	Mixed	No
Tunneling	Mixed	Unknown	Yes
Unique names	Mixed	Mixed	Mixed

Figure 7.6 Contenders compared to requirements.

A revised proposal

After the retreat, Steve Deering and Paul Francis of the SIPP working group sent a message to members on the SIPP mailing list proposing some changes in SIPP that would bring it closer to the requirements. These proposals include

- Changing to a fixed-length 16-byte address (128 bits)
- Specifying server-free autoconfiguration using the IEEE 802 address as the interface ID portion of the address
- Requiring that higher-layer protocols that use Internet-layer addresses as part of connection identifiers (such as TCP) use the entire 16-byte addresses

This revised proposal is a mix of multiple IETF processes. The basic protocol is from the SIPP proposal, the transition and autoconfiguration pieces came from the TUBA submission, and the newly extended addressing structure supports the implementation of CIDR. After much discussion and with the concurrence of the IPng directorate, the modified Simple Internet Protocol Plus specification (128-bit version) received the IPng working group's recommendation as the basis for the IPng protocol.

IPv6

As we noted earlier in this chapter, the IANA assigned version number 6 to SIPP. We can now call the next generation of the Internet protocol IPv6. IPv6 is an evolutionary step from IPv4. The functions that the SIPP group generally accepted as working in IPv4 are also included in IPv6, though in some different places or by different names. Functions that the SIPP group saw as getting infrequent use, or not working at all, they removed or made optional.

Feature	Benefit
128-bit, fixed length address	More levels of addressing hierarchy
Authentication extension	Data integrity
Autoconfiguration support	Ease of installation
Encryption extension	Data confidentiality
Extended option headers	More and longer options available
Multicast scope field	Improve multicast routing scalability
Option encoding	Incremental deployment with minimum disruption
Quality of service	Special router handling
Regional cluster addressing	Extra, source, path control
Simplified header format	Cut common processing and bandwidth
Source demand route support	Source directed routing
Transition plan	Incremental migration

Figure 7.7 IPv6 features and benefits.

They added a few new features to provide expanded functionality. Figure 7.7 provides a list of these features and their benefits.

The working group decided that some of the IPv4 header functions will remain while others will not carry over to the new version's header. The standard removes the header length field, the IP header checksum, and changes the time-to-live field to a hop count limit. Let's examine the new headers that will carry those features in the datagrams and packets. In a protocol analyzer, we will recognize IPv6 by the Ether type of 86 dd hex instead of the IPv4 value of 08 00 hex. Looking at the following headers will help us understand why that is so.

Header format

The IPv6 header, though longer than the IPv4 header as a result of the longer addresses, is greatly simplified, as can be seen in Fig. 7.8. Each of the IPv4 header functions is now in extension headers. Let's take a closer look at the fields of the IPv6 header to see what changed and what remained.

Version. The obvious value in this 4-bit field is 6.

Priority. This 4-bit field replaces the functions of the precedence field from IPv4. The lower the priority value, the more willing the source is to have a router discard the datagram.

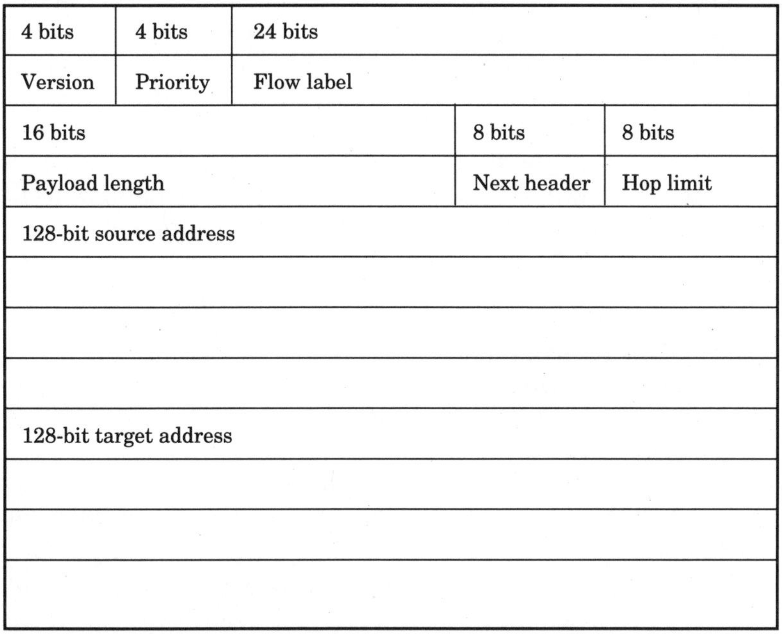

4 bits	4 bits	24 bits		
Version	Priority	Flow label		
16 bits			8 bits	8 bits
Payload length			Next header	Hop limit
128-bit source address				
128-bit target address				

Figure 7.8 IPv6 header layout.

Code	Priority
0	Uncharacterized
1	Background traffic
2	Unattended data transfer
3	Reserved, future use
4	Attended bulk data transfer
5	Reserved, future use
6	Interactive traffic
7	Control, management

Figure 7.9 Congestion-controlled priority values.

With 16 possible values, we can split this field into two components. A value of 0 indicates no specific priority. As Fig. 7.9 indicates, the next values (1 through 7) identify congestion-controlled traffic. The eight values (8 through 15) will set priorities for real-time datagrams. Those values have no standards as of this writing.

Flow label. This datagram's request for special handling by routers is the next 24-bit field. Many IPng designers consider this to be a key to running TCP/IP over the high-speed networks and high-performance routers and systems we foresee.

The IPng standard requires that the combination of source IP address and flow label uniquely define a flow. This way, the sequence of datagrams sent from a source to a target (whether unicast or multicast) on which the source wants special handling is the flow that the label defines. The routers that recognize the flow label can avoid any routing calculations and simply follow the previous calculations that were made in forwarding the datagram.

The flow label can also reserve resources on the target system for the application that the source user is requesting, such as voice or video transmission. By using the same flow label for all the datagrams during the session, the routers and systems reserve the network and system resources.

Any system, router, or host that does not support the flow label function must set the field to 0 for originating datagrams, pass on (route) datagrams without changing the field when forwarding datagrams, and ignore the field when receiving datagrams. The specification requires that all datagrams with the same (nonzero) flow label must have the same destination address, hop-by-hop options header, routing header, and source address contents.

By only looking up the flow label, the router can decide how to forward the datagram. It does not have to examine the rest of the header.

Payload length. This 16-bit unsigned integer field specifies the length, in octets (or 8-bit bytes), of the datagram after the IPv6 header. Since the specifications set the IPv6 header at a fixed length of 40 bytes, the IPv4 header length field becomes obsolete in IPv6.

Value	Next header
00	Hop-by-hop option header
04	Internet Protocol
06	Transmission Control Protocol (TCP)
17	User Datagram Protocol (UDP)
43	Routing header
44	Fragment
45	Interdomain Routing Protocol (IDRP)
46	Resource Reservation Protocol (RSVP)
50	Encapsulating security payload
51	Authentication header
58	Internet Control Message Protocol (ICMPv6)
59	No next header
60	Destination options header

Figure 7.10 IP next header values.

Next header. This field performs the same function as the IPv4 protocol field. To offer more flexibility for future additions, the standard identifies an 8-bit field. Figure 7.10 shows some of the more common values in the order that the specifications recommend. By following the recommendation's header order, the routers in the path can process the datagram much more efficiently.

Hop limit. This is the maximum number of nodes that may forward the datagram. As an 8-bit integer, that could be as many as 255. Though it replaces the IPv4 loop prevention of the time-to-live field, it may also support another capability.

The trace route application, as well as other similar service search implementations, needs to have a way to extend gradually a search for the target IP system. This hop limit field offers that support by placing a ceiling on the number of routers and gateways which can pass on the datagram.

Source address. This is the 128-bit, 16-byte address of the initial sender of the datagram.

Target address. This is the 16-byte address of the intended recipient (if a routing header is present, it may not be the ultimate recipient).

At this point, before looking at the IPv6 extensions, let's use Fig. 7.11 to see what has changed in the IP header from IPv4. Both start with a version number field and end with the source and target IP addresses, but the middle of the two headers is quite different. IPv4 uses a header length field that, when subtracted from the total IP length field value, can identify the payload length; IPv6 saves the subtraction step by providing the payload length later in its header.

IPv4	IPv6
Version	Version
Header length	40 bytes
Precedence	~Priority
Type of service	~Flow label
Total IP length	40 bytes + payload length
Datagram ID number	None
Fragmentation field	Fragmentation extension
Time to live	Hop limit
Protocol	Next header
Checksum	None
Source 4-byte IP address	Source 16-byte IP address
Target 4-byte IP address	Target 16-byte IP address

Figure 7.11 IPv4 and IPv6 header comparison.

Whereas IPv4 uses the first three bits of the type-of-service byte to identify the importance of the data it is carrying, IPv6 uses the priority field in a slightly different way. Instead of declaring the importance of the data being carried, IPv6 simply sets the type of traffic. The results are the same: the lower the priority, the more likely a router may discard the datagram. Also, IPv4 uses the remaining 5 bits of that byte, the type-of-service field, to tell routers how to handle this datagram. In IPv6 those instructions become part of the flow label that identifies a path the datagram should follow.

When IPv6 wants the total length of the IP datagram, it adds the 40-byte header to the payload length, whereas IPv4 has a separate field for the value. The value in IPv6 that comes closest to the IPv4 datagram ID number is the ID number in the fragment extension. There is no IPv6 header field to match that number. Since the source system will do all the fragmentation in IPv6, there is no fragmentation area in the basic IPv6 header. When necessary, the source adds a fragment extension.

The IPv4 time-to-live field has lost its deduction for time spent in a router's buffer in that IPv6 only tracks the maximum number of hops (network segments) available. The protocol identification field in IPv6 offers a variation on the IPv4 protocol identification field by identifying the next header. That next header, as we said earlier, may be an IPv6 extension or the next protocol after the IPv6 header.

The IPv4 header checksum is gone in IPv6. Most of the NAL protocols in use provide a checksum, a cyclic redundancy check, or a frame-check sequence. Since that error detection mechanism covers the IPv6 header and extensions, there is no need for duplicating that effort and taking up space in the new header.

Bit pattern	Purpose
00xx xxxx	Skip option and process the header
10xx xxxx	Discard and send an ICMP message
01xx xxxx	Discard the packet
11xx xxxx	Undefined

Figure 7.12 Unrecognized option-type response bits.

Extension headers

All IPv6 datagrams start with the same fixed length header. The IPv6 header can (usually) get the datagram to the appropriate destination. Of the many functions of IPv4 that remain, IPv6 encodes them in separate extension headers that follow the IPv6 header. To date, the standard has described only a small number of those extension headers as you can see in Fig. 7.10.

Each extension header has its own distinct next header value. Each of those extension headers (except 59, no next header) also has its own next header field to permit a string of header extensions.

The standard calls for encoding the option type identifiers so that their first two bits identify what an IPv6 node is to do if it does not recognize the option type. Figure 7.12 specifies those values.

Hop-by-hop options header. The hop-by-hop options header carries optional information that every router along a datagram's delivery path must examine. The hop-by-hop options header follows a next header value of 0 in the IPv6 header and uses the format shown in Fig. 7.13.

Next header. Identifies the type of header immediately following the hop-by-hop options header. It uses the same values as the IPv4 protocol (8-bit selector) field.

Header extension length. Length of the hop-by-hop options header in 8-octet units beginning with this (8-bit integer) field.

Options. One or more type-length-value-encoded options will use the layout in Fig. 7.14. Each is an integer multiple of 8 octets, as we see in Fig. 7.15.

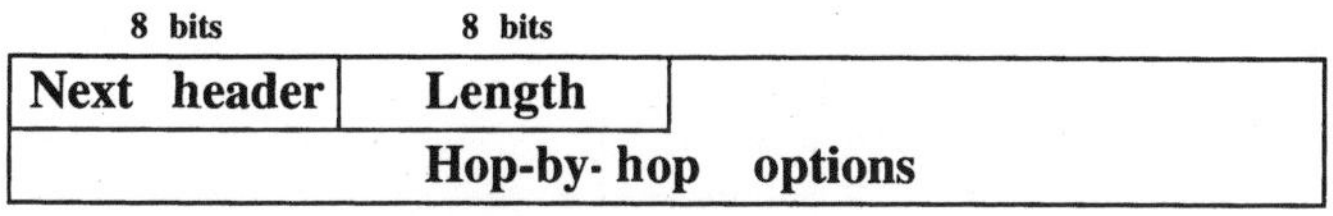

8 bits	8 bits	
Next header	Length	
	Hop-by- hop options	

Figure 7.13 Hop-by-hop header layout.

8 bits	8 bits	Variable length
Option type	Option length	Option value

Figure 7.14 Type-length-value options layout.

In the case of hop-by-hop options only, the third bit decides if IPv6 will include the data of this option in the integrity assurance calculation for an authentication header. Thus, IPv6 must exclude option data that changes en route from that computation.

The Pad1 option in Fig. 7.15 shifts the remaining options by 1 byte of padding to align those options for best processing by the receiving systems (hosts and routers). Invoking the PadN option, the source can shift the remaining option fields by 2 or more padding bytes to provide the same proper alignment.

The jumbo payload option replaces the IPv6 payload field, which IPv6 changes to 0. Since the basic IPv6 header limits the payload length to 65,535 bytes, we use this option when we need larger payloads. This lets us bypass the IPv4 datagram size limitations that new communications technologies have overcome.

The value of the option is the new 32-bit IPv6 payload length of up to 4,294,967,295 bytes. If future payloads threaten this size, we simply create another option with a bigger value. Like the basic header payload field, this value does not include the basic 40-byte header itself. It also does not include the hop-by-hop extension header that carries it.

Routing header. The routing header, which is portrayed in Fig. 7.16, provides support for the IPv6 equivalent of the IPv4 routing options: loose source routing and strict source routing. It lists one or more routers the datagram will "visit" on its way to the target system. The following list describes the header fields in more detail.

Next header. An 8-bit selector that identifies the type of header immediately following the routing header. It uses the same values as the IPv4 protocol field. We find these codes in the Assigned Numbers RFC.

Type	Option	Size
0	Pad1	1 byte
1	PadN	2 + N bytes
194	Jumbo payload	2 + 4 bytes

Figure 7.15 Hop-by-hop option types.

8 bits	8 bits	8 bits	8 bits
Next header	Header length	Routing type	Segments left
8 bits	**24 bits**		
Reserved	Strict/ loose bit map		
128-bit address(0)			
128-bit address(1)			
128-bit address(n)			

Figure 7.16 Routing option header layout.

Header extension length. An 8-bit unsigned integer that describes the length of the routing header in 8-octet (64-bit) words (not including the first 8 octets). For the Type 0 routing header, header extension length is equal to two times the number of addresses in the header and must be an even number less than or equal to 46.

Routing type. An 8-bit identification of a particular routing header type. Currently, Type 0 is the only one that the specifications identify.

Segments left. An 8-bit unsigned integer that indicates the number of route segments remaining. That is the number of intermediate addresses the datagram has left to visit before reaching the final target. The maximum legal value is 23 since the strict-loose bit map has only 24 bits. When the value reaches 0 the datagram is at the target address.

If, while processing a datagram, a router finds a routing header with an unrecognized routing type value, the router's response depends on the value of the segment's left field. If the segment's left field is 0, the router ignores the routing header and processes the next header in the datagram. It uses the value of the next header field in the routing header.

If the segment's left field is not 0, the router discards the datagram and sends an ICMPv6 Parameter Problem, Code 0, message to the datagram's source address. That ICMP message will point to the unrecognized routing type.

Reserved. An 8-bit reserved field that IPv6 sets to 0 for transmission and ignores upon receipt.

Strict-loose bit map. A 24-bit bit map (numbered 0–23, left-to-right). It indicates, for each segment of the route, the relationship of the next destination address to the preceding address. A value of 1 means strict source routing and requires that the next address must be a neighbor to the current address. A value of 0 means loose source routing and does not require the neighbor relationship.

If the first bit (bit number 0) of the strict-loose bit map has a value of 1, the IPv6 basic header's destination address identifies a neighbor of the originating node. If the first bit has the value of 0, the originator may use any legal, nonmulticast address as the first destination address.

The strict-loose bits beyond the *n*th, where *n* is the number of addresses in the routing header, must carry a value of 0 that the originator sets and the receivers ignore.

Address. A series of 128-bit (16-octet) addresses, numbered 0 to *n*. Multicast addresses cannot be values in a Type 0 routing header or in the IPv6 basic header destination address field of a datagram carrying a Type 0 routing header.

No node examines (or processes) a routing header until it reaches the node that the destination address field of the IPv6 header identifies. In that node, dispatching on the next header field of the immediately preceding header invokes the routing header functions.

Fragment header. The fragment header lets IPv6 sources send payloads larger than can be handled by the MTU's along the path to the target. (*Note*: To maintain faster router processing, source systems do all IPv6 fragmentation.) The header, depicted in Fig. 7.17, has the following fields.

Next header. An 8-bit selector that identifies the initial header type of the fragmentable part of the original datagram. It uses the same values as the IPv4 protocol field from the Assigned Numbers RFC.

Reserved. An 8-bit reserved field that the IPv6 source sets to 0 for transmission. The IPv6 target ignores this field.

Fragment offset. Like IPv4, IPv6 uses a 13-bit unsigned integer to measure the offset in 8-octet (64-bit) words. The offset is how far the data that follows this header is from the start of the fragmentable part of the original datagram.

Reserved. A 2-bit reserved field that the IPv6 source sets to 0 for sending. The IPv6 target ignores this field when it receives the datagram.

M flag. The equivalent of the IPv4 fragment status bit. A value of 1 tells the IPv6 target to expect more fragments; a value of 0 tells the IPv6 target that this is the last fragment.

Identification. A 32-bit value that the IPv6 assigns to the original datagram. It must set a unique value that is different from any other fragment-

8 bits	8 bits	13 bits	2	1
Next header	Reserved	Fragment offset	R	M
Identification				

Figure 7.17 Fragment option header layout.

ed payload that this device sent recently (within the maximum lifetime of the datagram) from the same IPv6 source address to the IPv6 target address. Like the IPv4 datagram ID number, this identification field is the same for all the original payload's fragments.

To send a datagram that is too large to fit in the MTU of the path to its destination, a source node may divide the datagram into fragments and send each fragment as a separate datagram that the receiver will reassemble (if possible).

We call the initial, large, unfragmented datagram the original datagram. It contains two parts. The unfragmentable part is the IPv6 header plus any extension headers that the nodes (routers) will process en route to the destination, that is, all headers up to and including the routing header. If no routing header is present, the unfragmentable part will include the hop-by-hop options header. If the hop-by-hop options header is absent, then no extension headers will be part of the unfragmentable part of the IPv6 header.

The fragmentable part contains the rest of the datagram. In other words, this includes any extension headers that the final target system(s) processes, plus the upper-layer header and data. Each fragment, except possibly the last, is a multiple of 8 octets or 64-bit words. The IPv6 source transmits the fragments in separate fragment datagrams.

Fragment reassembly. The target-addressed node reassembles the original datagram only from the fragment datagrams that have the same source address, destination address, and fragment identification.

The target system must complete reassembly of the original datagram within 60 seconds of receiving the first-arriving fragment of that datagram. If it cannot, it must abandon reassembly of that datagram. It also discards all the datagram fragments that it received. If it received the first fragment (i.e., the one with a fragment offset of 0), it will send an ICMP Fragment Reassembly Time Exceeded message to the source of that fragment.

The target must also discard any fragment whose length is not a multiple of 8 octets (if the M flag of that fragment is 1). The target will send an ICMP parameter problem message to the source of the original datagram, pointing to the payload length field of the fragment datagram.

The next header values in the fragment headers of different fragments of the same original datagram may differ. The receiver uses only the value from the offset zero fragment datagram for reassembly.

IPv6 Addressing

The IPv6 recognizes three types of addresses: unicast, multicast, and anycast. Just as we saw with the NIC or MAC address, a unicast address is unique to the interface it names. In IP Versions 4 and 6, a unicast address may be in the target or source field of an IP header.

Again, the similarity to NIL addressing shows in the multicast. A multicast IP address also identifies a set of interfaces, and each of these interfaces has a unicast address. We can use multicast addresses to contact groups of interfaces as in a videoconference or with router updates.

In IPv6 there is no broadcast IP address. Instead, we can designate a multicast address for every interface in a subnet or network. Since the IPv4 broadcast is technically the largest grouped multicast, the function remains under a different name. Logically, a multicast address can only work in the target IP address field of an IP header.

New to IPv6 is the anycast. The network takes the responsibility of getting the datagram with this target (only) address to any one in an anycast group (or set) of addresses. That one may be the DHCP server, a router with certain information, or an Internet Relay Chat (IRC) server.

Address formats

While we already know that IPv6 addresses are 128 bits, we also have a new way to write those addresses. In IPv4 we use a byte-dot-byte-dot-byte-dot-byte or dotted decimal notation. In IPv6 we use colons to separate the address into eight 16-bit pieces. Figure 7.18 shows an example of how that can look.

If any one of those pieces has a value of less than 1000 hex, we drop the leading 0s. If any of the pieces has a value of 0, we can write the address without anything between the colons. Figure 7.19 shows an IPv6 address sample with both of those situations. Note that IPv6 uses the double colon abbreviation only one time in an address. Thus, fedc::ba98:765 is the shortened version of the IPv6 address fedc:0:0:0:0:0:ba98:765, while 123::45::6789:abc::def is illegal since it uses more than one set of double colons.

Address hierarchy

To deal with the explosion of the number of routing table entries, IPv6 uses a hierarchy of bits, from the most significant to the least significant, in the address range. The left-most bits in the address signify a variable length address prefix. This variable length currently ranges between 3 and 10 bits. Figure 7.20 shows the prefixes that the RFCs have set so far. The rest of the possible addresses carry a designation of "unassigned."

```
1234:0567:0089:000A:0000:000B:00CD:0EF0
```

Figure 7.18 IPv6 written format.

1234:567:89:a::b:cd:ef0

Figure 7.19 A compact IPv6 address.

Some of the reserved addresses fall into an area called special addresses. These include the

- Unspecified address
- Loopback address
- Subnet-router address
- IPv4-compatible address
- IPv4-mapped address

The unspecified address (0:0:0:0:0:0:0:0) is available when we have no true address. It can come up when the node asks a server for an IP address. While it waits, it can use the unspecified address as the source address for any messages it must send. To prevent confusion, the unspecified address can never be in a target address field of any message.

The loopback address (0:0:0:0:0:0:0:1) never leaves the system that sends this message to itself. That message can be for diagnostic purposes as they can test the health of a TCP/IP stack without using a network. Typically, the system believes it sent and received two messages when both are the same loopback message.

The subnet-router address is an anycast address that IPv6 builds by using a nonzero subnet prefix followed by 0s. That subnet prefix identifies a particular subnet. As an anycast address, a system may send it in the target address field of a message to any single server (or router) when it wants only one answer and does not care which server (or router) answers.

Binary	Allocation
0000 0000	Reserved
0000 001	ISO network addresses
0000 010	Novell (IPX) network addresses
010	Provider-based unicast addresses
100	Geographic-based unicast addresses
1111 1110 10	Link local addresses
1111 1110 11	Site local addresses
1111 1111	Multicast addresses

Figure 7.20 IPv6 address allocations.

::bfff:2a0d G 191.255.42.13

IPv6 IPv4

Figure 7.21 IPv4-compatible addressing.

The IPv4-compatible and IPv4-mapped addresses support the transition from IPv4 to IPv6. These IPv6 systems can use them to communicate with each other over an IPv4 network. The IPv4-compatible addresses have a 96-bit zero field followed by the IPv4 address. This way the IPv6 address in Fig. 7.21, 0:0:0:0:0:0:bfff:2a0d (or ::bfff:2a0d in its short form), becomes IPv4 address 191.255.42.13.

These IPv4-compatible addressed systems rely on routers (or gateways) at the crossover point between the IPv4 network and the IPv6 network. That dual-IP router converts the IPv4-compatible address to a true IPv4 address for travel over the IPv4 network. At the entry point to the other IPv6 network, another dual-IP router reverses the process from an IPv4 to an IPv6 address.

The IPv4-mapped addresses identify systems that do not support IPv6. Routers may convert these addresses as well. However, IPv6-addressed systems use them to communicate with IPv4-only systems. An IPv4-mapped address has 80 bits of 0s, 16 bits of 1s (binary), and the 32-bit IPv4 address. Thus, the IPv4 address of 191.255.206.118 becomes the IPv6 address of 0:0:0:0:0:ffff:bfff:ce76 (see Fig. 7.22). As we learned earlier, we can shorten this to ::ffff:bfff:ce76.

Provider-based unicast address

While the IPv6 specifications are not cast in stone, some address planning is solid enough to discuss. The example in Fig. 7.23 shows the current plan for provider-based unicast addressing. The component parts are

- The provider-based prefix of 010 binary.
- A 5-bit registry identifier of 11010 binary (InterNIC administered) that combines with the provider-based prefix to read 5a in hex.
- A part of 56 bits of provider identification. The exact number of bits will be set by the registry assigning the address.
- The rest of the 56 bits carries the subscriber identification.
- Sixteen bits to set the subnet identifier.
- Forty-eight bits to identify the NIC (MAC) address of the interface. In Fig. 7.23 that is 0000 c0a0 5124.

Figure 7.22 IPv4-mapped addressing.

5a12:3456:789a:bcde:f012:0000:c0a0:5124

Figure 7.23 Provider-based unicast address formats.

So what does all this mean to us? First, it means that changes continue to unfold and in the time it took to write and publish this book some of the data in this and other chapters may have become obsolete. Second, it means we have some well-thought-out functions that IPv6 will bring.

Figure 7.24 shows some of the functions that seem to have a firmer foundation under them and should make it into the standard. Let's look at each of these to get a feel for what is to come.

- The streamlined IPv6 header may have twice as many bytes but it reduces the number of fields from 12 to 8 and so reduces unnecessary overhead from unused fragmentation and redundant error detection.
- Since the IPv6 source fragments IP datagrams, routers substantially reduce their processing delays and therefore speed an ever-growing volume through the network.
- The IPv6 header's flow field lets routers follow flow patterns and so streamlines their routing decision process and increases routing efficiency.
- By identifying IPv6 addresses as geographic or service provider–based addressing, datagrams can follow a faster, more direct pathway to their target-addressed system.
- Including the MAC address as part of the IPv6 address offers much easier IP configuration that can easily be automated by vendors or managers.
- With newer, faster networks, the payload can grow exponentially to offer far greater throughput by identifying jumbo payloads.

Function	Benefit
Streamlined IPv6 header	Reduces unnecessary overhead
Source fragments IP datagrams	Routers substantially reduce delays
Flow field	Streamlines routing decisions
Geographic or ISP-based addressing	Faster hierarchy-based routing
IPv6 address includes MAC address	Easier or automatic IP configuration
Jumbo payload identification	Eases migration to faster networks
IPv6 security	Integrates authentication and encryption
Header extensions	Eases adding new functionality
ICMPv6 neighbor discovery	Replaces ARP, detects duplicate addresses
ICMPv6 error identification	Streamlines reporting

Figure 7.24 Sample IPv6 changes.

- Integrating both authentication and encryption (while providing a default method for each) means IPv6 has much more security than IPv4 and can use it seamlessly without add-on or third-party software packages or external, additional protocols.
- Header extensions make it much easier to add IPv6 functions as technology changes offer those capabilities.
- By increasing the integration of ICMP with IPv6, functions such as neighbor discovery can replace ARP in identifying local addresses. An additional benefit is that neighbor discovery also detects duplicate addresses and identifies them as MAC or IP duplicates. This reduces troubleshooting headaches and time to locate the duplicates.
- ICMPv6 error identification works much like the current version of ICMP though it streamlines the possible reporting codes and so makes troubleshooting much easier.

IPv4 to IPv6 Transition

The transition plans move closer to completion with each day. The most current at this writing is RFC 1933 Transition Mechanisms for IPv6 Hosts and Routers by R. Gilligan and E. Nordmark of Sun Microsystems, Inc. To save rewriting a good explanation, we quote the abstract and introduction:

> **Abstract**
>
> This document specifies IPv4 compatibility mechanisms that can be implemented by IPv6 hosts and routers. These mechanisms include providing complete implementations of both versions of the Internet Protocol (IPv4 and IPv6), and tunneling IPv6 packets over IPv4 routing infrastructures. They are designed to allow IPv6 nodes to maintain complete compatibility with IPv4, which should greatly simplify the deployment of IPv6 in the Internet, and facilitate the eventual transition of the entire Internet to IPv6.
>
> **1. Introduction**
>
> The key to a successful IPv6 transition is compatibility with the large installed base of IPv4 hosts and routers. Maintaining compatibility with IPv4 while deploying IPv6 will streamline the task of transitioning the Internet to IPv6. This specification defines a set of mechanisms that IPv6 hosts and routers may implement in order to be compatible with IPv4 hosts and routers.
>
> The mechanisms in this document are designed to be employed by IPv6 hosts and routers that need to interoperate with IPv4 hosts and utilize IPv4 routing infrastructures. We expect that most nodes in the Internet will need such compatibility for a long time to come, and perhaps even indefinitely.
>
> However, IPv6 may be used in some environments where interoperability with IPv4 is not required. IPv6 nodes that are designed to be used in such environments need not use or even implement these mechanisms.

The mechanisms specified here include:

- *Dual IP layer.* Providing complete support for both IPv4 and IPv6 in hosts and routers.
- *IPv6 over IPv4 tunneling.* Encapsulating IPv6 packets within IPv4 headers to carry them over IPv4 routing infrastructures.

Two types of tunneling are employed: configured and automatic.

Additional transition and compatibility mechanisms may be developed in the future. These will be specified in other documents.

Figure 7.25 on page 148 provides the IANA's current information on IPv6 parameters and codes. We also recommend reading as many of the current IPv6 RFCs as possible, especially RFC 1933. Another excellent source of IPng information is available on the World Wide Web at:

http://playground.sun.com/pub/ipng/html/ipng-main.html

1. Version number = 6

2. Priority
   0 Uncharacterized
   1 Filler
   2 Unattended
   3 (Reserved)
   4 Attended bulk
   5 (Reserved)
   6 Interactive
   7 Internet control

3. Payload length

4. Flow label

5. Next header

5a. Header types
   00 = Hop-by-hop options
   43 = Routing
   44 = Fragment
   51 = Authentication
   60 = Destination options
   50 = Encapsulating security payload
   xx = Upper-layer header
   58 = Internet Control Message Protocol (ICMP)
   59 = No next header
   For the xx values see the file Protocol Numbers.

5b. Options

   The options have a type, length, value (TLV) structure.

   The type is a 2-bit action code, a 1-bit change code, and a 5-bit operation code.

   The action codes are:
   00 Skip
   01 Discard
   10 Discard and report
   11 Discard and report if not multicast

   The change codes are:
   0 No change en route
   1 Change allowed en route

   The operation codes are:
   0 Pad 1 (special case: this option is just this one octet)
   1 Pad N
   2 Jumbo payload length
   3 Endpoint identifier for Nimrod

5c. Routing types
   0 Source route
   1 Nimrod

6. Hop limit

7. Source address

8. Destination address

Figure 7.25 IPv6 parameters.

Chapter 8

User Datagram Protocol

Speed versus Reliability

As we work our way up the TCP/IP stack, the next layer above IP is the host-to-host or transport layer. The host-to-host layer provides multiplexing and demultiplexing between the IP layer and the applications at the process layer. It offers a choice of an error-free, flow-controlled (connection-oriented) datastream service or simply passing the connectionless IP services to the correct application. The protocols at this layer are Transmission Control Protocol (TCP) and User Datagram Protocol (UDP).

We will look at the reliability and connectivity that TCP provides in Chapter 11. In this chapter, we will see how UDP provides a speedier, connectionless transmission with reduced reliability. TCP is more reliable because it uses a 20-byte, 9-field header to support a connection-oriented session; UDP does not have those fields or the reliability they would provide. Instead, UDP reduces the overhead fields to four, to increase the speed.

TCP establishes a connection by sending three segments as a session handshake. Since UDP is connectionless and does not require the target system to acknowledge that it is ready to receive a message, it can carry data in its first packet. In other words, we can call the UDP datagram a "ready or not here I come" protocol. This offers speed to applications that frequently carry short or single messages between hosts on the same network. These short messages are often requests for time-critical information. We also see UDP in regularly repeated communications such as routing protocol updates.

Without the reliability of a connection, UDP may lose data in the transfer. UDP may have even less reliability since the checksum field is optional. That decision to use (or not use) the checksum, to verify the UDP header and its payload, is up to each vendor's software author.

If the network loses a message or generates an error condition, the application must provide error detection and recovery. The application may pass the retransmission decision to the user by indicating no response. Alternatively, the application may tell the user that there is a problem and ask the user to retry or cancel.

The connectionless nature of UDP also affects its message naming. Instead of the segments in TCP messages, we call UDP messages *datagrams.* Since UDP, like IP, is a best-effort protocol, it assumes the attitude that "if it is important, someone will send it again."

Header Fields

User Datagram Protocol provides a source port (logical service point) in its initial datagram so that a target (receiving system) has a return address for any messages it wants to send back to the other end of the conversation (see Fig. 8.1). It usually designates the client (or requester) side of the conversation. The target port in the initial datagram identifies the application service (server) that receives the data at the target host. These port numbers are not similar to physical port numbers. They are virtual numeric "nicknames" the communicating systems use for this session.

After that initial service request, the client and server ports do not correlate directly to the source and target port fields. The conversation typically sees the client and server ports change places (source and target port) with each consecutive UDP datagram in the conversation.

The message length field holds the value of the UDP datagram, including the UDP header and its payload, in octets or bytes. This is also the number of bytes that the checksum will cover (along with the pseudo header) when the vendor applies it.

As we said earlier, UDP's lower reliability shows in the fact that the checksum is optional. The checksum covers the UDP pseudo header (see Fig. 8.2) and the UDP payload to be sure that the correct IP addresses apply. Since it only supports the checksum, UDP does not include the pseudo header in the length field and does not send it with the datagram. It contains the source and target IP addresses, a byte of 0s to round out the 32-bit word, the IP protocol number field, and the UDP length field.

The UDP header includes its own error checking since IP provides no error checking beyond its own header. If the software author chooses to have UDP's checksum idle (unused), the value in the field will be 00 00 hex. Note that by not using the checksum, the vendor does not save any length. The UDP datagram is the same length either way.

When a system receives a UDP datagram with a checksum value of 00 00 hex, it does not attempt to calculate a verifying checksum. If the source's

2 bytes	2 bytes	2 bytes	2 bytes
Source port	Target port	Length	Checksum
05 c2	00 45	01 dc	00 00

Figure 8.1 UDP header layout.

Source IP address		
Target IP address		
Zero	Protocol	UDP length
8 bits	8 bits	16 bits

Figure 8.2 UDP pseudo header.

checksum calculation should derive a value of 00 00 hex, the checksum field will carry a value of ff ff to tell the receiving system to calculate a verifying value.

Port Basics

The process (or application) layer of the TCP/IP stack uses TCP or UDP to access the network. The host-to-host layer must identify the application and which user (especially on a multiuser system) is using that application. To identify these, it uses logical port numbers that fall into two basic types: one to identify the application and the other to identify the user.

The port numbers that identify the application may be either well-known ports or registered ports. They each have different points of assignment and control. Appendixes A and B at the end of this chapter show a listing of each. They fall into different numeric ranges for easy identification.

Well-known port numbers have a value of 0 through 1023. The Internet Assigned Numbers Authority assigns and controls these port numbers. The original plan for assigning these numbers was that only the system itself (or certain privileged users) could use many of the processes (or programs) these ports identify. As the protocols have evolved, that policy has changed. This can been seen in the variation of RFC 1700's well-known ports in App. A.

Registered port numbers range from 1024 through 65,535. The IANA lists the ports for convenience but does not control the assignment of these numbers. Evidence of that lack of control is shown in App. B. As we see there, ports 1525, 1898, and 1992 have duplicate assignments. This lack of standard control could present a problem if a network manager loaded the competing applications on the same server. On most systems, these port numbers identify ordinary user processes (applications).

The user application assigns a third type, random ports, with values of 1025 through 65,535 (see Fig. 8.3). These ports let the application match the data to the right client during each conversation. Each random port allows multiple sessions from a single IP-addressed interface to operate simultaneously. This way, individuals using terminals on a multiuser system can share one IP address to call for different sessions, running separate applications on diverse servers, because each user has a unique random port for each session.

Since they occupy the same numbering range, random ports can conflict

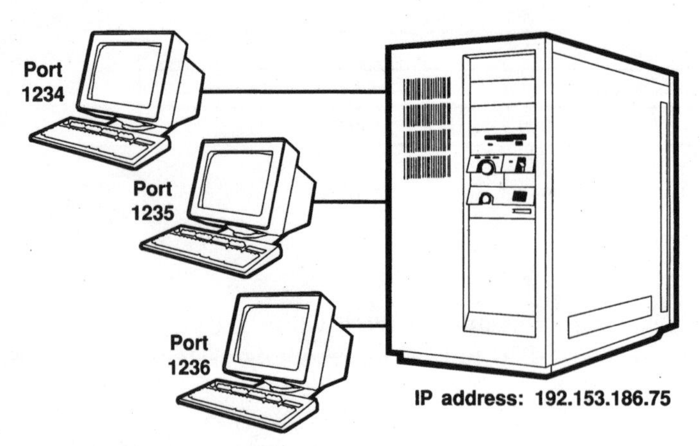

Figure 8.3 Random port example.

with registered ports. To prevent this, any system that provides services keeps a file that identifies the port numbers matching those services. Before assigning a random port number to the client side of a conversation, the host-to-host layer will check that file and skip any registered port numbers that the system may use.

Random ports are not truly random. The vendor's TCP/IP software may randomly select the first port that it will use after loading the TCP/IP stack. From then on, the random port number is usually the next sequential number in the allotted range. The software may increase or decrease in increments from the last number it uses to get to the next number. Most vendors choose to increase from a relatively low number, such as 1030.

If a client or user side system continues running long enough to use all of the available random ports, it will return to the same procedure it used to determine the first random port number and continue from there.

Ports and Sockets

When the use of a protocol grows faster than its standards can handle, we will coin new terms in an attempt to describe the work. Unfortunately, new terms in one discipline are often old terms from another discipline. We have that case here: The term *sockets* is one of those terms. It refers to a software Application Program Interface (API) between the host system and the TCP/IP applications in the programming arena. In our communication discipline, we use it for a pairing of the IP address and the port number. We also consider it to be the complete network address of one end of a UDP (or TCP) conversation.

In the language of UDP, the sockets in Fig. 8.4 are 192.136.118.30, 1427, and 192.153.185.62, 69. This indicates that the client on the left is asking the

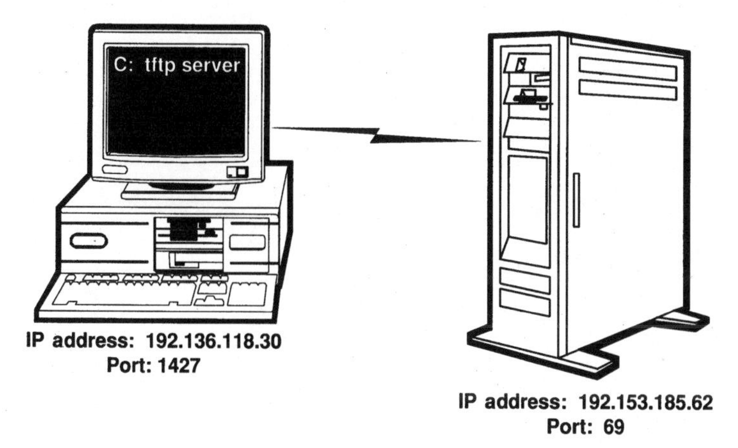

Figure 8.4 IP addresses, ports, and sockets.

system with an IP address of 192.153.185.62 (that it has identified as server) to start a TFTP session (well-known port 69). It goes on to request that anything the server wants to send should go to port 1427 on the system with the IP address of 192.136.118.30.

Applications

Two kinds of applications use UDP datagrams as their transport layer. What they have in common is the need to get the data to the target host as quickly as possible. In some of the cases, that is so important that the software authors are willing to take the chance of losing some datagrams and having to retransmit them to get them to their destination quickly.

One group includes the UDP applications that carry short messages that the sender can resend if they do not arrive as expected. These include such applications as Active Users, Quote of the Day, Whois, BootP, SNMP, or RIP. The other group of applications includes those in which the software author has embedded the reliability that is missing in UDP. That second type includes Domain Name Server, TFTP, and Sun's RPC. See Fig. 8.5 for a list of common UDP ports.

Sample Exchanges

As we have in previous chapters, we present in Fig. 8.6 some sample UDP datagrams that a protocol analyzer has decoded. For your review, we include a short explanation of each of them as follows:

1. The first datagram is the client's random port (5342) calling the TFTP well-known port (69) to request a TFTP file transfer.

Name	Dec	Hex
Active Users	11	00 0b
Netstat	15	00 0e
Quote of the Day	17	00 11
Whois	43	00 2b
Domain Name Server	53	00 35
BootP Server	67	00 43
BootP Client	68	00 44
TFTP	69	00 45
Sun's RPC	111	00 6f
SNMP Monitor	161	00 a1
SNMP Traps	162	00 a2
Ntalk	518	02 06
Routing Information Protocol	520	02 08

Figure 8.5 Common UDP named ports.

2. The next datagram is from an SNMP Network Management System to an SNMP agent. The NMS is requesting information from the agent. Note that the NMS vendor chose to not use the checksum.
3. The third UDP datagram is the SNMP agent's response to the NMS. As we can see, the agent's return message is larger than the NMS request and the UDP software is using the checksum.
4. UDP datagram four is a BootP client request for configuring information. It is calling from well-known port 68 to well-known port 67 (the BootP server port).
5. The following datagram is the BootP server's response to the client's request. Note that it is from well-known port 67 to well-known port 68 and does not use the checksum.
6. The sixth UDP datagram is also a communication between two well-known ports. In this case they are the same port. The protocol analyzer identifies them as route(520). If we looked them up in the well-known ports list in App. A, we would find local routing process. In fact, we are looking at a router's routing information protocol (RIP) broadcast to any other router that is using RIP to update its routing information. Once again, this vendor chose to not use the checksum.
7. The last datagram in Fig. 8.6 shows a request from random port 1042 to well-known port 518 to run the application ntalk. Ntalk is a keyboard-to-keyboard communication between two users. Each would see a dashed line across the middle of the screen with local characters on the bottom half of the screen and the remote characters appearing above the dashed line. The server running that application will respond from port 518 to port 1042 to confirm or deny the request.

```
UDP: 5342 -> TFTP(69) len: 12 xsum: 0xf7a7
UDP: 1369 -> SNMP-monitor(161) len: 49 xsum: ——
UDP: SNMP-monitor(161) -> 1369 len: 57 xsum: 0x83ab
UDP: BOOTPc(68) -> BOOTPs(67) len: 308 xsum: 0x5baf
UDP: BOOTPs(67) -> BOOTPc(68) len: 308 xsum: ——
UDP: route(520) -> route(520) len: 32 xsum: ——
UDP: 1042 -> ntalk(518) len: 84 xsum: 0xaf8d
```

Figure 8.6 UDP sample data exchanges.

Appendix A Well-Known Port Numbers from RFC 1700

0/tcp	0/udp	Reserved
1/tcp	1/udp	TCP Port Service Multiplexer
2/tcp	2/udp	Management Utility
3/tcp	3/udp	Compression Process
4/tcp	4/udp	Unassigned
5/tcp	5/udp	Remote Job Entry
6/tcp	6/udp	Unassigned
7/tcp	7/udp	Echo
8/tcp	8/udp	Unassigned
9/tcp	9/udp	Discard
10/tcp	10/udp	Unassigned
11/tcp	11/udp	Active Users
12/tcp	12/udp	Unassigned
13/tcp	13/udp	Daytime
14/tcp	14/udp	Unassigned
15/tcp	15/udp	Unassigned [was netstat]
16/tcp	16/udp	Unassigned
17/tcp	17/udp	Quote of the Day
18/tcp	18/udp	Message Send Protocol
19/tcp	19/udp	Character Generator
20/tcp		File Transfer [Data]
21/tcp		File Transfer [Control]
22/tcp	22/udp	Unassigned
23/tcp		Telnet
24/tcp	24/udp	Any private mail system
25/tcp		Simple Mail Transfer
26/tcp	26/udp	Unassigned
27/tcp	27/udp	NSW User System FE
28/tcp	28/udp	Unassigned
29/tcp	29/udp	MSG ICP
30/tcp	30/udp	Unassigned

31/tcp	31/udp	MSG Authentication
32/tcp	32/udp	Unassigned
33/tcp	33/udp	Display Support Protocol
34/tcp	34/udp	Unassigned
35/tcp	35/udp	Any private printer server
36/tcp	36/udp	Unassigned
37/tcp	37/udp	Time
38/tcp	38/udp	Route Access Protocol
39/tcp	39/udp	Resource Location Protocol
40/tcp	40/udp	Unassigned
41/tcp	41/udp	Graphics
42/tcp	42/udp	Host Name Server
43/tcp	43/udp	Who Is
44/tcp	44/udp	MPM FLAGS Protocol
45/tcp	45/udp	Message Processing Module [recv]
46/tcp	46/udp	MPM [default send]
47/tcp	47/udp	NI FTP
48/tcp	48/udp	Digital Audit Daemon
49/tcp	49/udp	Login Host Protocol
50/tcp	50/udp	Remote Mail Checking Protocol
51/tcp	51/udp	IMP Logical Address Maintenance
52/tcp	52/udp	XNS Time Protocol
53/tcp	53/udp	Domain Name Server
54/tcp	54/udp	XNS Clearinghouse
55/tcp	55/udp	ISI Graphics Language
56/tcp	56/udp	XNS Authentication
57/tcp	57/udp	Any private terminal access
58/tcp	58/udp	XNS Mail
59/tcp	59/udp	Any private file service
60/tcp	60/udp	Unassigned
61/tcp	61/udp	NI MAIL
62/tcp	62/udp	ACA Services
63/tcp	63/udp	Unassigned
64/tcp	64/udp	Communications Integrator (CI)
65/tcp	65/udp	TACACS-Database Service
66/tcp	66/udp	Oracle SQL*NET
67/udp	67/udp	Bootstrap Protocol Server
68/udp	68/udp	Bootstrap Protocol Client
69/udp	69/udp	Trivial File Transfer
70/tcp	70/udp	Gopher
71/tcp	71/udp	Remote Job Service netrjs-1
72/tcp	72/udp	Remote Job Service netrjs-2
73/tcp	73/udp	Remote Job Service netrjs-3
74/tcp	74/udp	Remote Job Service netrjs-4
75/tcp	75/udp	Any private dialout service
76/tcp	76/udp	Distributed External Object Store
77/tcp	77/udp	Any private RJE service
78/tcp	78/udp	vettcp
79/tcp	79/udp	Finger

80/tcp		World Wide Web HTTP
81/tcp	81/udp	HOSTS2 Name Server
82/tcp	82/udp	XFER Utility
83/tcp	83/udp	MIT ML Device
84/tcp	84/udp	Common Trace Facility
85/tcp	85/udp	MIT ML Device
86/tcp	86/udp	Micro Focus Cobol
87/tcp	87/udp	Any private terminal link
88/tcp	88/udp	Kerberos
89/tcp	89/udp	SU/MIT Telnet Gateway
90/tcp	90/udp	DNSIX Securit Attribute Token Map
91/tcp	91/udp	MIT Dover Spooler
92/tcp	92/udp	Network Printing Protocol
93/tcp	93/udp	Device Control Protocol
94/tcp	94/udp	Tivoli Object Dispatcher
95/tcp	95/udp	SUPDUP
96/tcp	96/udp	DIXIE Protocol Specification
97/tcp	97/udp	Swift Remote Vitural File Protocol
98/tcp	98/udp	TAC News
99/tcp	99/udp	Metagram Relay
100/tcp	100/udp	[unauthorized use]
101/tcp	101/udp	NIC Host Name Server
102/tcp	102/udp	ISO-TSAP
103/tcp	103/udp	Genesis Point-to-Point Trans Net
104/tcp	104/udp	ACR-NEMA Digital Imag. & Comm. 300
105/tcp	105/udp	Mailbox Name Nameserver
106/tcp	106/udp	3COM-TSMUX
107/tcp	107/udp	Remote Telnet Service
108/tcp	108/udp	SNA Gateway Access Server
109/tcp	109/udp	Post Office Protocol - Version 2
110/tcp	110/udp	Post Office Protocol - Version 3
111/tcp	111/udp	SUN Remote Procedure Call
112/tcp	112/udp	McIDAS Data Transmission Protocol
113/tcp	113/udp	Authentication Service
114/tcp	114/udp	Audio News Multicast
115/tcp	115/udp	Simple File Transfer Protocol
116/tcp	116/udp	ANSA REX Notify
117/tcp	117/udp	UUCP Path Service
118/tcp	118/udp	SQL Services
119/tcp	119/udp	Network News Transfer Protocol
120/tcp	120/udp	CFDPTKT
121/tcp	121/udp	Encore Expedited Remote Pro.Call
122/tcp	122/udp	SMAKYNET
123/tcp	123/udp	Network Time Protocol
124/tcp	124/udp	ANSA REX Trader
125/tcp	125/udp	Locus PC-Interface Net Map Ser
126/tcp	126/udp	Unisys Unitary Login
127/tcp	127/udp	Locus PC-Interface Conn Server
128/tcp	128/udp	GSS X License Verification

129/tcp	129/udp	Password Generator Protocol
130/tcp	130/udp	Cisco FNATIVE
131/tcp	131/udp	Cisco TNATIVE
132/tcp	132/udp	Cisco SYSMAINT
133/tcp	133/udp	Statistics Service
134/tcp	134/udp	INGRES-NET Service
135/tcp	135/udp	Location Service
136/tcp	136/udp	PROFILE Naming System
137/tcp	137/udp	NETBIOS Name Service
138/tcp	138/udp	NETBIOS Datagram Service
139/tcp	139/udp	NETBIOS Session Service
140/tcp	140/udp	EMFIS Data Service
141/tcp	141/udp	EMFIS Control Service
142/tcp	142/udp	Britton-Lee IDM
143/tcp	143/udp	Interim Mail Access Protocol v2
144/tcp	144/udp	NewS
145/tcp	145/udp	UAAC Protocol
146/tcp	146/udp	ISO-IP0
147/tcp	147/udp	ISO-IP
148/tcp	148/udp	CRONUS-SUPPORT
149/tcp	149/udp	AED 512 Emulation Service
150/tcp	150/udp	SQL-NET
151/tcp	151/udp	HEMS
152/tcp	152/udp	Background File Transfer Program
153/tcp	153/udp	SGMP
154/tcp	154/udp	NETSC
155/tcp	155/udp	NETSC
156/tcp	156/udp	SQL Service
157/tcp	157/udp	KNET/VM Command/Message Protocol
158/tcp	158/udp	PCMail Server
159/tcp	159/udp	NSS-Routing
160/tcp	160/udp	SGMP-TRAPS
	161/udp	SNMP
	162/udp	SNMPTRAP
163/tcp	163/udp	CMIP/TCP Manager
164/tcp	164/udp	CMIP/TCP Agent
165/tcp	165/udp	Xerox
166/tcp	166/udp	Sirius Systems
167/tcp	167/udp	NAMP
168/tcp	168/udp	RSVD
169/tcp	169/udp	SEND
170/tcp	170/udp	Network PostScript
171/tcp	171/udp	Network Innovations Multiplex
172/tcp	172/udp	Network Innovations CL/1
173/tcp	173/udp	Xyplex
174/tcp	174/udp	MAILQ
175/tcp	175/udp	VMNET
176/tcp	176/udp	GENRAD-MUX
177/tcp	177/udp	X Display Manager Control Protocol

178/tcp	178/udp	NextStep Window Server
179/tcp	179/udp	Border Gateway Protocol
180/tcp	180/udp	Intergraph
181/tcp	181/udp	Unify
182/tcp	182/udp	Unisys Audit SITP
183/tcp	183/udp	OCBinder
184/tcp	184/udp	OCServer
185/tcp	185/udp	Remote-KIS
186/tcp	186/udp	KIS Protocol
187/tcp	187/udp	Application Communication Interface
188/tcp	188/udp	Plus Five's MUMPS
189/tcp	189/udp	Queued File Transport
190/tcp	190/udp	Gateway Access Control Protocol
191/tcp	191/udp	Prospero Directory Service
192/tcp	192/udp	OSU Network Monitoring System
193/tcp	193/udp	Spider Remote Monitoring Protocol
194/tcp	194/udp	Internet Relay Chat Protocol
195/tcp	195/udp	DNSIX Network Level Module Audit
196/tcp	196/udp	DNSIX Session Mgt Module Audit Redir
197/tcp	197/udp	Directory Location Service
198/tcp	198/udp	Directory Location Service Monitor
199/tcp	199/udp	SMUX
200/tcp	200/udp	IBM System Resource Controller
201/tcp	201/udp	AppleTalk Routing Maintenance
202/tcp	202/udp	AppleTalk Name Binding
203/tcp	203/udp	AppleTalk Unused
204/tcp	204/udp	AppleTalk Echo
205/tcp	205/udp	AppleTalk Unused
206/tcp	206/udp	AppleTalk Zone Information
207/tcp	207/udp	AppleTalk Unused
208/tcp	208/udp	AppleTalk Unused
209/tcp	209/udp	Trivial Authenticated Mail Protocol
210/tcp	210/udp	ANSI Z39.50
211/tcp	211/udp	Texas Instruments 914C/G Terminal
212/tcp	212/udp	ATEXSSTR
213/tcp	213/udp	IPX
214/tcp	214/udp	VM PWSCS
215/tcp	215/udp	Insignia Solutions
216/tcp	216/udp	Access Technology License Server
217/tcp	217/udp	dBASE Unix
218/tcp	218/udp	Netix Message Posting Protocol
219/tcp	219/udp	Unisys ARPs
220/tcp	220/udp	Interactive Mail Access Protocol v3
221/tcp	221/udp	Berkeley rlogind with SPX auth
222/tcp	222/udp	Berkeley rshd with SPX auth
223/tcp	223/udp	Certificate Distribution Center
224——	——241	Reserved
242/tcp	242/udp	Unassigned
243/tcp	243/udp	Survey Measurement

244/tcp	244/udp	Unassigned
245/tcp	245/udp	LINK
246/tcp	246/udp	Display Systems Protocol
247——	——255	Reserved
256——	——343	Unassigned
344/tcp	344/udp	Prospero Data Access Protocol
345/tcp	345/udp	Perf Analysis Workbench
346/tcp	346/udp	Zebra server
347/tcp	347/udp	Fatmen Server
348/tcp	348/udp	Cabletron Management Protocol
371/tcp	371/udp	Clearcase
372/tcp	372/udp	Unix Listserv
373/tcp	373/udp	Legent Corporation
374/tcp	374/udp	Legent Corporation
375/tcp	375/udp	Hassle
376/tcp	376/udp	Amiga Envoy Network Inquiry Proto
377/tcp	377/udp	NEC Corporation
378/tcp	378/udp	NEC Corporation
379/tcp	379/udp	TIA/EIA/IS-99 modem client
380/tcp	380/udp	TIA/EIA/IS-99 modem server
381/tcp	381/udp	HP performance data collector
382/tcp	382/udp	HP performance data managed node
383/tcp	383/udp	HP performance data alarm manager
384/tcp	384/udp	A Remote Network Server System
385/tcp	385/udp	IBM Application
386/tcp	386/udp	ASA Message Router Object Def.
387/tcp	387/udp	Appletalk Update-Based Routing Pro.
388/tcp	388/udp	Unidata LDM Version 4
389/tcp	389/udp	Lightweight Directory Access Protocol
390/tcp	390/udp	UIS
391/tcp	391/udp	SynOptics SNMP Relay Port
392/tcp	392/udp	SynOptics Port Broker Port
393/tcp	393/udp	Data Interpretation System
394/tcp	394/udp	EMBL Nucleic Data Transfer
395/tcp	395/udp	NETscout Control Protocol
396/tcp	396/udp	Novell Netware over IP
397/tcp	397/udp	Multi Protocol Trans. Net.
398/tcp	398/udp	Kryptolan
399/tcp	399/udp	Unassigned
400/tcp	400/udp	Workstation Solutions
401/tcp	401/udp	Uninterruptible Power Supply
402/tcp	402/udp	Genie Protocol
403/tcp	403/udp	decap
404/tcp	404/udp	nced
405/tcp	405/udp	ncld
406/tcp	406/udp	Interactive Mail Support Protocol
407/tcp	407/udp	Timbuktu
408/tcp	408/udp	Prospero Resource Manager Sys. Man.
409/tcp	409/udp	Prospero Resource Manager Node Man.

410/tcp	410/udp	DECLadebug Remote Debug Protocol
411/tcp	411/udp	Remote MT Protocol
412/tcp	412/udp	Trap Convention Port
413/tcp	413/udp	SMSP
414/tcp	414/udp	InfoSeek
415/tcp	415/udp	BNet
416/tcp	416/udp	Silverplatter
417/tcp	417/udp	Onmux
418/tcp	418/udp	Hyper-G
419/tcp	419/udp	Ariel
420/tcp	420/udp	SMPTE
421/tcp	421/udp	Ariel
422/tcp	422/udp	Ariel
423/tcp	423/udp	IBM Operations Planning and Control Start
424/tcp	424/udp	IBM Operations Planning and Control Track
425/tcp	425/udp	ICAD
426/tcp	426/udp	smartsdp
427/tcp	427/udp	Server Location
428/tcp	428/udp	OCS_CMU
429/tcp	429/udp	OCS_AMU
430/tcp	430/udp	UTMPSD
431/tcp	431/udp	UTMPCD
432/tcp	432/udp	IASD
433/tcp	433/udp	NNSP
434/tcp	434/udp	MobileIP-Agent
435/tcp	435/udp	MobilIP-MN
436/tcp	436/tcp	DNA-CML
437/tcp	437/tcp	comscm
438/tcp	438/udp	dsfgw
439/tcp	439/udp	dasp
440/tcp	440/udp	sgcp
441/tcp	441/udp	decvms-sysmgt
442/tcp	442/udp	cvc_hostd
443/tcp	443/udp	https MCom
444/tcp	444/udp	Simple Network Paging Protocol
445/tcp	445/udp	Microsoft-DS
446/tcp	446/udp	DDM-RDB
447/tcp	447/udp	DDM-RFM
448/tcp	448/udp	DDM-BYTE
449/tcp	449/udp	AS Server Mapper
450/tcp	450/udp	TServer
451——	——511	Unassigned
512/tcp		Remote process execution
	512/udp	Used by mail system to notify users of new mail received
513/tcp		Remote login a la telnet
	513/udp	maintains data bases showing who's logged in to machines
514/tcp		cmd like exec, but with automatic authentication
	514/udp	syslog
515/tcp	515/udp	spooler

516/tcp	516/udp	Unassigned
517/tcp	517/udp	Like tenex link, but across machine
517/udp	517/udp	Like tenex link, but across machine
	518/udp	ntalk
519/tcp	519/udp	unixtime
520/tcp		Extended file name server
	520/udp	Local routing process (on site)
521——	——524	Unassigned
525/tcp	525/udp	timeserver
526/tcp	526/udp	newdate
527——	——529	Unassigned
530/tcp	530/udp	rpc
531/tcp	531/udp	chat
532/tcp	532/udp	readnews
533/tcp	533/udp	for emergency broadcasts
534——	——538	Unassigned
539/tcp	539/udp	Apertus Technologies Load Determination
540/tcp	540/udp	uucpd
541/tcp	541/udp	uucp-rlogin
542/tcp	542/udp	Unassigned
543/tcp	543/udp	
544/tcp	544/udp	krcmd
545——	——549	Unassigned
550/tcp	550/udp	new-who
551——	——555	Unassigned
555/tcp	555/udp	
556/tcp	556/udp	rfs server
557——	——559	Unassigned
560/tcp	560/udp	rmonitord
561/tcp	561/udp	
562/tcp	562/udp	chcmd
563/tcp	563/udp	Unassigned
564/tcp	564/udp	plan 9 file service
565/tcp	565/udp	whoami
566——	——569	Unassigned
570/tcp	570/udp	demon
571/tcp	571/udp	udemon
572——	——599	Unassigned
600/tcp	600/udp	Sun IPC server
607/tcp	607/udp	nqs
606/tcp	606/udp	Cray Unified Resource Manager
608/tcp	608/udp	Sender-Initiated/Unsolicited File Transfer
609/tcp	609/udp	npmp-trap
610/tcp	610/udp	npmp-local
611/tcp	611/udp	npmp-gui
634/tcp	634/udp	ginad
666/tcp	666/udp	doom Id Software
704/tcp	704/udp	errlog copy/server daemon
709/tcp	709/udp	EntrustManager

729/tcp	729/udp	IBM NetView DM/6000 Server/Client
730/tcp	730/udp	IBM NetView DM/6000 send/tcp
731/tcp	731/udp	IBM NetView DM/6000 receive/tcp
741/tcp	741/udp	netGW
742/tcp	742/udp	Network based Rev. Cont. Sys.
744/tcp	744/udp	Flexible License Manager
747/tcp	747/udp	Fujitsu Device Control
748/tcp	748/udp	Russell Info Sci Calendar Manager
749/tcp	749/tcp	kerberos administration
750/tcp		rfile
	750/udp	loadav
751/tcp	751/udp	pump
752/tcp	752/udp	qrh
753/tcp	753/udp	rrh
754/tcp	754/udp	tell
758/tcp	758/udp	nlogin
759/tcp	759/udp	con
760/tcp	760/udp	ns
761/tcp	761/udp	rxe
762/tcp	762/udp	quotad
763/tcp	763/udp	cycleserv
764/tcp	764/udp	omserv
765/tcp	765/udp	webster
767/tcp	767/udp	phonebook
769/tcp	769/udp	vid
770/tcp	770/udp	cadlock
771/tcp	771/udp	rtip
772/tcp	772/udp	cycleserv2
773/tcp		Submit
	773/udp	Notify
774/tcp		rpasswd
	774/udp	acmaint_dbd
775/tcp		entomb
	775/udp	acmaint_transd
776/tcp	776/udp	wpages
780/tcp	780/udp	wpgs
786/tcp	786/udp	concert
800/tcp	800/udp	mdbs_daemon
801/tcp	801/udp	device
996/tcp	996/udp	xtreelic
997/tcp	997/udp	maitrd
998/tcp		busboy
	998/udp	puparp
999/tcp		garcon
	999/udp	applix
999/tcp	999/udp	puprouter
1000/tcp		cadlock
	1000/udp	ock
1023/tcp	1023/udp	Reserved

Appendix B Registered Port Numbers (Range: 1024–65535) from RFC 1700

1024/tcp	1024/udp	Reserved
1025/tcp	1025/udp	Network blackjack
1030/tcp	1030/udp	BBN IAD 1
1031/tcp	1031/udp	BBN IAD 2
1032/tcp	1032/udp	BBN IAD 3
1067/tcp	1067/udp	Installation Bootstrap Proto. Server
1068/tcp	1068/udp	Installation Bootstrap Proto. Client
1080/tcp	1080/udp	Socks
1083/tcp	1083/udp	Anasoft License Manager 1
1084/tcp	1084/udp	Anasoft License Manager 2
1155/tcp	1155/udp	Network File Access
1222/tcp	1222/udp	SNI R&D network
1248/tcp	1248/udp	hermes
1346/tcp	1346/udp	Alta Analytics License Manager
1347/tcp	1347/udp	Multimedia conferencing
1348/tcp	1348/udp	Multimedia conferencing
1349/tcp	1349/udp	Registration Network Protocol
1350/tcp	1350/udp	Registration Network Protocol
1351/tcp	1351/udp	Digital Tool Works (MIT)
1352/tcp	1352/udp	Lotus Notes
1353/tcp	1353/udp	Relief Consulting
1354/tcp	1354/udp	RightBrain Software
1355/tcp	1355/udp	Intuitive Edge
1356/tcp	1356/udp	CuillaMartin Company
1357/tcp	1357/udp	Electronic PegBoard
1358/tcp	1358/udp	CONNLCLI
1359/tcp	1359/udp	FTSRV
1360/tcp	1360/udp	MIMER
1361/tcp	1361/udp	LinX
1362/tcp	1362/udp	TimeFlies
1363/tcp	1363/udp	Network DataMover Requester
1364/tcp	1364/udp	Network DataMover Server
1365/tcp	1365/udp	Network Software Associates
1366/tcp	1366/udp	Novell NetWare Comm Service Platform
1367/tcp	1367/udp	DCS
1368/tcp	1368/udp	ScreenCast
1369/tcp	1369/udp	GlobalView to Unix Shell
1370/tcp	1370/udp	Unix Shell to GlobalView
1371/tcp	1371/udp	Fujitsu Config Protocol
1372/tcp	1372/udp	Fujitsu Config Protocol
1373/tcp	1373/udp	Chromagrafx
1374/tcp	1374/udp	EPI Software Systems
1375/tcp	1375/udp	Bytex
1376/tcp	1376/udp	IBM Person-to-Person Software
1377/tcp	1377/udp	Cichlid License Manager
1378/tcp	1378/udp	Elan License Manager
1379/tcp	1379/udp	Integrity Solutions

1380/tcp	1380/udp	Telesis Network License Manager
1381/tcp	1381/udp	Apple Network License Manager
1382/tcp	1382/udp	udt_os
1383/tcp	1383/udp	GW Hannaway Network License Manager
1384/tcp	1384/udp	Objective Solutions License Manager
1385/tcp	1385/udp	Atex Publishing License Manager
1386/tcp	1386/udp	CheckSum License Manager
1387/tcp	1387/udp	Computer Aided Design Software Inc LM
1388/tcp	1388/udp	Objective Solutions DataBase Cache
1389/tcp	1389/udp	Document Manager
1390/tcp	1390/udp	Storage Controller
1391/tcp	1391/udp	Storage Access Server
1392/tcp	1392/udp	Print Manager
1393/tcp	1393/udp	Network Log Server
1394/tcp	1394/udp	Network Log Client
1395/tcp	1395/udp	PC Workstation Manager software
1396/tcp	1396/udp	DVL Active Mail
1397/tcp	1397/udp	Audio Active Mail
1398/tcp	1398/udp	Video Active Mail
1399/tcp	1399/udp	Cadkey License Manager
1400/tcp	1400/udp	Cadkey Tablet Daemon
1401/tcp	1401/udp	Goldleaf License Manager
1402/tcp	1402/udp	Prospero Resource Manager
1403/tcp	1403/udp	Prospero Resource Manager
1404/tcp	1404/udp	Infinite Graphics License Manager
1405/tcp	1405/udp	IBM Remote Execution Starter
1406/tcp	1406/udp	NetLabs License Manager
1407/tcp	1407/udp	DBSA License Manager
1408/tcp	1408/udp	Sophia License Manager
1409/tcp	1409/udp	Here License Manager
1410/tcp	1410/udp	HiQ License Manager
1411/tcp	1411/udp	AudioFile
1412/tcp	1412/udp	InnoSys
1413/tcp	1413/udp	Innosys-ACL
1414/tcp	1414/udp	IBM MQSeries
1415/tcp	1415/udp	DBStar
1416/tcp	1416/udp	Novell LU6.2
1417/tcp	1417/udp	Timbuktu Service 1 Port
1418/tcp	1418/udp	Timbuktu Service 2 Port
1419/tcp	1419/udp	Timbuktu Service 3 Port
1420/tcp	1420/udp	Timbuktu Service 4 Port
1421/tcp	1421/udp	Gandalf License Manager
1422/tcp	1422/udp	Autodesk License Manager
1423/tcp	1423/udp	Essbase Arbor Software
1424/tcp	1424/udp	Hybrid Encryption Protocol
1425/tcp	1425/udp	Zion Software License Manager
1426/tcp	1426/udp	Satellite-data Acquisition System 1
1427/tcp	1427/udp	mloadd monitoring tool
1428/tcp	1428/udp	Informatik License Manager

1429/tcp	1429/udp	Hypercom NMS
1430/tcp	1430/udp	Hypercom TPDU
1431/tcp	1431/udp	Reverse Gosip Transport
1432/tcp	1432/udp	Blueberry Software License Manager
1433/tcp	1433/udp	Microsoft-SQL-Server
1434/tcp	1434/udp	Microsoft-SQL-Monitor
1435/tcp	1435/udp	IBM CISC
1436/tcp	1436/udp	Satellite-data Acquisition System 2
1437/tcp	1437/udp	Tabula
1438/tcp	1438/udp	Eicon Security Agent/Server
1439/tcp	1439/udp	Eicon X25/SNA Gateway
1440/tcp	1440/udp	Eicon Service Location Protocol
1441/tcp	1441/udp	Cadis License Management
1442/tcp	1442/udp	Cadis License Management
1443/tcp	1443/udp	Integrated Engineering Software
1444/tcp	1444/udp	Marcam License Management
1445/tcp	1445/udp	Proxima License Manager
1446/tcp	1446/udp	Optical Research Associates License Mgr
1447/tcp	1447/udp	Applied Parallel Research LM
1448/tcp	1448/udp	OpenConnect License Manager
1449/tcp	1449/udp	PEport
1450/tcp	1450/udp	Tandem Distributed Workbench Facility
1451/tcp	1451/udp	IBM Information Management
1452/tcp	1452/udp	GTE Government Systems License Man
1453/tcp	1453/udp	Genie License Manager
1454/tcp	1454/udp	interHDL License Manager
1455/tcp	1455/udp	ESL License Manager
1456/tcp	1456/udp	DCA
1457/tcp	1457/udp	Valisys License Manager
1458/tcp	1458/udp	Nichols Research Corp.
1459/tcp	1459/udp	Proshare Notebook Application
1460/tcp	1460/udp	Proshare Notebook Application
1461/tcp	1461/udp	IBM Wireless LAN
1462/tcp	1462/udp	World License Manager
1463/tcp	1463/udp	Nucleus
1464/tcp	1464/udp	MSL License Manager
1465/tcp	1465/udp	Pipes Platform
1466/tcp	1466/udp	Ocean Software License Manager
1467/tcp	1467/udp	CSDMBASE
1468/tcp	1468/udp	CSDM
1469/tcp	1469/udp	Active Analysis Limited License Manager
1470/tcp	1470/udp	Universal Analytics
1471/tcp	1471/udp	csdmbase
1472/tcp	1472/udp	csdm
1473/tcp	1473/udp	OpenMath
1474/tcp	1474/udp	Telefinder
1475/tcp	1475/udp	Taligent License Manager
1476/tcp	1476/udp	clvm-cfg
1477/tcp	1477/udp	ms-sna-server

1478/tcp	1478/udp	ms-sna-base
1479/tcp	1479/udp	dberegister
1480/tcp	1480/udp	PacerForum
1481/tcp	1481/udp	AIRS
1482/tcp	1482/udp	Miteksys License Manager
1483/tcp	1483/udp	AFS License Manager
1484/tcp	1484/udp	Confluent License Manager
1485/tcp	1485/udp	LANSource
1486/tcp	1486/udp	nms_topo_serv
1487/tcp	1487/udp	LocalInfoSrvr
1488/tcp	1488/udp	DocStor
1489/tcp	1489/udp	dmdocbroker
1490/tcp	1490/udp	insitu-conf
1491/tcp	1491/udp	anynetgateway
1492/tcp	1492/udp	stone-design-1
1493/tcp	1493/udp	netmap_lm
1494/tcp	1494/udp	ica
1495/tcp	1495/udp	cvc
1496/tcp	1496/udp	liberty-lm
1497/tcp	1497/udp	rfx-lm
1498/tcp	1498/udp	Watcom-SQL
1499/tcp	1499/udp	Federico Heinz Consultora
1500/tcp	1500/udp	VLSI License Manager
1501/tcp	1501/udp	Satellite-data Acquisition System 3
1502/tcp	1502/udp	Shiva
1503/tcp	1503/udp	Databeam
1504/tcp	1504/udp	EVB Software Engineering License Manager
1505/tcp	1505/udp	Funk Software, Inc.
1506——	——1523	Unassigned
1524/tcp	1524/udp	Ingres
1525/tcp	1525/udp	Oracle (duplicate port)
1525/tcp	1525/udp	Prospero Dir. Svc. non-priv (duplicate port)
1526/tcp	1526/udp	Prospero Data Access Prot non-priv
1527/tcp	1527/udp	Oracle
1529/tcp	1529/udp	Oracle
1600/tcp	1600/udp	issd
1650/tcp	1650/udp	nkd
1651/tcp	1651/udp	Proshare conf audio
1652/tcp	1652/udp	Proshare conf video
1653/tcp	1653/udp	Proshare conf data
1654/tcp	1654/udp	Proshare conf request
1655/tcp	1655/udp	Proshare conf notify
1661/tcp	1661/udp	Netview-aix-1
1662/tcp	1662/udp	Netview-aix-2
1663/tcp	1663/udp	Netview-aix-3
1664/tcp	1664/udp	Netview-aix-4
1665/tcp	1665/udp	Netview-aix-5
1666/tcp	1666/udp	Netview-aix-6
1986/tcp	1986/udp	Cisco license management

1987/tcp	1987/udp	Cisco RSRB Priority 1 port
1988/tcp	1988/udp	Cisco RSRB Priority 2 port
1989/tcp	1989/udp	MHSnet system (duplicate port)
1989/udp	1989/udp	Cisco RSRB Priority 3 port (duplicate port)
1990/tcp	1990/udp	Cisco STUN Priority 1 port
1991/tcp	1991/udp	Cisco STUN Priority 2 port
1992/tcp	1992/udp	Cisco STUN Priority 3 port (duplicate port)
1992/tcp	1992/udp	IPsendmsg (duplicate port)
1993/tcp	1993/udp	Cisco SNMP TCP port
1994/tcp	1994/udp	Cisco serial tunnel port
1995/tcp	1995/udp	Cisco perf port
1996/tcp	1996/udp	Cisco Remote SRB port
1997/tcp	1997/udp	Cisco Gateway Discovery Protocol
1998/tcp	1998/udp	Cisco X.25 service (XOT)
1999/tcp	1999/udp	Cisco identification port
2000/tcp	2000/udp	callbook
2001/tcp		dc
	2001/udp	wizard curry
2002/tcp	2002/udp	globe
2004/tcp		mailbox
	2004/udp	emce CCWS mm conf
2005/tcp		berknet
	2005/udp	Oracle
2006/tcp		invokator
	2006/udp	raid-cc
2007/tcp		DECtalk
	2007/udp	raid-am
2008/tcp		conf
	2008/udp	terminaldb
2009/tcp		news
	2009/udp	whosockami
2010/tcp		search
	2010/udp	pipe_server
2011/tcp		raid-cc
	2011/udp	servserv
2012/tcp		ttyinfo
	2012/udp	raid-ac
2013/tcp		raid-am
	2013/udp	raid-cd
2014/tcp		troff
	2014/udp	raid-sf
2015/tcp		cypress
	2015/udp	raid-cs
2016/tcp	2016/udp	bootserver
2017/tcp		cypress-stat
	2017/udp	bootclient
2018/tcp		terminaldb
	2018/udp	rellpack
2019/tcp		whosockami

	2019/udp	about
2020/tcp	2020/udp	xinupageserver
2021/tcp		servexec
	2021/udp	xinuexpansion1
2022/tcp		down
	2022/udp	xinuexpansion2
2023/tcp	2023/udp	xinuexpansion3
2024/tcp	2024/udp	xinuexpansion4
2025/tcp		ellpack
	2025/udp	xribs
2026/tcp	2026/udp	scrabble
2027/tcp	2027/udp	shadowserver
2028/tcp	2028/udp	submitserver
2030/tcp	2030/udp	device 2
2032/tcp	2032/udp	blackboard
2033/tcp	2033/udp	glogger
2034/tcp	2034/udp	scoremgr
2035/tcp	2035/udp	imsldoc
2038/tcp	2038/udp	objectmanager
2040/tcp	2040/udp	lam
2041/tcp	2041/udp	interbase
2042/tcp	2042/udp	isis
2043/tcp	2043/udp	isis-bcast
2044/tcp	2044/udp	rimsl
2045/tcp	2045/udp	cdfunc
2046/tcp	2046/udp	sdfunc
2047/tcp	2047/udp	dls
2048/tcp	2048/udp	dls-monitor
2049/tcp	2049/udp	shilp
2065/tcp	2065/udp	Data Link Switch Read Port Number
2067/tcp	2067/udp	Data Link Switch Write Port Number
2201/tcp	2201/udp	Advanced Training System Program
2500/tcp	2500/udp	Resource Tracking system server
2501/tcp	2501/udp	Resource Tracking system client
2564/tcp		HP 3000 NS/VT block mode telnet
2784/tcp	2784/udp	World Wide Web - development
3049/tcp	3049/udp	NSWS
3264/tcp	3264/udp	cc:mail/lotus
3333/tcp	3333/udp	DEC Notes
3421/tcp	3421/udp	Bull Apprise portmapper
3900/tcp	3900/udp	Unidata UDT OS
3984/tcp	3984/udp	MAPPER network node manager
3985/tcp	3985/udp	MAPPER TCP/IP server
3986/tcp	3986/udp	MAPPER workstation server
4132/tcp	4132/udp	NUTS Daemon
4133/tcp	4133/udp	NUTS Bootp Server
4343/tcp	4343/udp	UNICALL
4444/tcp	4444/udp	KRB524
4672/tcp	4672/udp	Remote file access server

5000/tcp	5000/udp	commplex-main
5001/tcp	5001/udp	commplex-link
5002/tcp	5002/udp	Radio-free Ethernet
5010/tcp	5010/udp	TelepathStart
5011/tcp	5011/udp	TelepathAttack
5050/tcp	5050/udp	Multimedia conference control tool
5145/tcp	5145/udp	rmonitor_secure
5190/tcp	5190/udp	America-Online
5236/tcp	5236/udp	padl2sim
5300/tcp	5300/udp	# HA cluster heartbeat
5301/tcp	5301/udp	# HA cluster general services
5302/tcp	5302/udp	# HA cluster configuration
5303/tcp	5303/udp	# HA cluster probing
5304/tcp	5304/udp	hacl-local
5305/tcp	5305/udp	hacl-test
6000-6063/tcp	6000-6063/udp	X Window System
6111/tcp	6111/udp	HP SoftBench Sub-Process Control
6141/tcp	6141/udp	Meta Corporation License Manager
6142/tcp	6142/udp	Aspen Technology License Manager
6143/tcp	6143/udp	Watershed License Manager
6144/tcp	6144/udp	StatSci License Manager - 1
6145/tcp	6145/udp	StatSci License Manager - 2
6146/tcp	6146/udp	Lone Wolf Systems License Manager
6147/tcp	6147/udp	Montage License Manager
6558/udp	6558/udp	xdsxdm
7000/tcp	7000/udp	File server itself
7001/tcp	7001/udp	Callbacks to cache managers
7002/tcp	7002/udp	Users & groups database
7003/tcp	7003/udp	Volume location database
7004/tcp	7004/udp	AFS/Kerberos authentication service
7005/tcp	7005/udp	Volume management server
7006/tcp	7006/udp	Error interpretation service
7007/tcp	7007/udp	Basic overseer process
7008/tcp	7008/udp	Server-to-server updater
7009/tcp	7009/udp	Remote cache manager service
7010/tcp	7010/udp	onlinet uninterruptable power supplies
7100/tcp	7100/udp	X Font Service
7121/tcp	7121/udp	Virtual Prototypes License Manager
7200/tcp	7200/udp	FODMS FLIP
7201/tcp	7201/udp	DLIP
7491/tcp	7491/udp	telops-lmd
7511/tcp	7511/udp	pafec-lm
7777/tcp	7777/udp	cbt
8450/tcp	8450/udp	npmp
9000/tcp	9000/udp	CSlistener
9535/tcp	9535/udp	man
9876/tcp	9876/udp	Session Director
9998/tcp	9998/udp	Distinct32
9999/tcp	9999/udp	distinct

17007/tcp	17007/udp	isode-dua
18000/tcp	18000/udp	Beckman Instruments, Inc.
21845/tcp	21845/udp	webphone
21846/tcp	21846/udp	NetSpeak Corp. Directory Services
21847/tcp	21847/udp	NetSpeak Corp. Connection Services
21848/tcp	21848/udp	NetSpeak Corp. Automatic Call Distribution
21849/tcp	21849/udp	NetSpeak Corp. Credit Processing System
25000/tcp	25000/udp	icl-twobase1
25001/tcp	25001/udp	icl-twobase2
25002/tcp	25002/udp	icl-twobase3
25003/tcp	25003/udp	icl-twobase4
25004/tcp	25004/udp	icl-twobase5
25005/tcp	25005/udp	icl-twobase6
25006/tcp	25006/udp	icl-twobase7
25007/tcp	25007/udp	icl-twobase8
25008/tcp	25008/udp	icl-twobase9
25009/tcp	25009/udp	icl-twobase10
47557/tcp	47557/udp	Databeam Corporation
47808/tcp	47808/udp	Building Automation and Control Networks

Chapter

9

IP Routing

Bridging or Routing

Beyond the obvious though overly simplified statement that bridges bridge and routers route, there is an easy way to define these functions. Both connect separate network segments and pass traffic between them, but bridges (and the newer variation: switches) operate at the network interface layer of the TCP/IP stack while routers work at the Internetwork (IP) layer of the TCP/IP stack. Another difference is that bridges will pass almost all data until we tell them what to filter out, while a router will not pass anything until we (or another router) tell it what to pass.

A bridge typically has two separate sides (or interfaces) to different segments of media (pieces of cabling) that both carry the same media-based protocol (like Ethernet or IEEE 802.3 or token ring). The two segments may use different cables, (twisted pair, fiber, ThinNet, etc.). In other words, a bridge focuses on passing data between two segments of cable with the same protocol that may have different cable types. Routers focus on passing data between two segments of media that carry different protocols.

A bridge often passes packets from one side of itself to the other if the destination MAC address is on the remote side from the source NIC-addressed side. Bridges separate traffic by not passing data packets whose target address is on the same side of the bridge as the source. In this way, bridges reduce traffic on adjacent network segments by filtering out (not forwarding) traffic destined for a system on the same network segment it came from.

When a segment shows sustained traffic peaks of over 30 percent with Ethernet or over 50 percent with token ring, it is time to consider the traffic separation that a bridge will provide. Long-term statistics that protocol analyzers and remote monitoring probes offer can be a great help in assessing network traffic trends to support the decision to bridge.

Routing or Bridging

When there are more than three servers and/or more than (about) 100 users, it is time to look at the network (IP) layer separation that a router can provide. This way, the IP traffic in one subnet does not contribute to congestion in another, because the router helps ensure traffic is only sent on to networks that help it reach its destination, instead of appearing on all networks.

If we add a new segment running a different NIL protocol (like Ethernet, token ring or FDDI) to a network, it is time to consider using a router to pass the traffic to the next segment of media. When the topology of the network is subnetted and a segment needs direct connectivity to several other segments, the answer is a router.

A router can pass only the traffic it knows to pass, based on how the network administrator configured it or by what other routers have told it. It also has (typically) multiple interfaces, one to each segment or subnet, instead of the two that most bridges have.

Subnetting any network without using routers is a fancy numbering system that may help locate IP-addressed interfaces for troubleshooting. A numbering system alone does not separate the traffic (like broadcasts) that could flood a bridged network. A router can accomplish all of that and offer connectivity to multiple segments and/or other networks.

Routers and Gateways

The exponential growth of the Internet and the availability of constantly evolving technology means we are creating new terminology (and redefining older terminology), which can be very confusing. At one time, we called any translating device a *gateway*. For example, an SNA gateway or an OSI gateway could translate from the local network operating system to a remote network or device's language.

Now the term applies to a router that acts as the gate for traffic from one subnet or network to another. As you look at the routers (labeled *R*) in Fig. 9.1, any one of them could qualify as a gateway that provides external access from a subnet, or one of them (label *G* in the figure and on the edge of the network) could be the router/gateway that provides access beyond the network (circle) boundary to other networks. In the strictest sense, that router is a translating system from an interior routing protocol to an exterior routing protocol.

In the following pages, we will examine the primary open routing protocols that administrators use for the interior of TCP/IP-based enterprise networks. We will also look at the exterior protocols from those networks to the Internet.

Direct versus Indirect Routing

When an IP-addressed interface sends a datagram onto a network, routing is involved—either direct routing or indirect routing.

Figure 9.1 Router or gateway?

Direct routing. If the two IP-addressed interfaces are on the same subnet (like 192.136.118.1 and 192.136.118.2 in Fig. 9.2), no router is necessary for communication from one system to the other, and the protocols can provide the data needed to deliver the datagram to the correct network interface physical address.

Indirect routing. If the user on a host (192.136.118.1 in Fig. 9.2) wants to make use of the power of the IP (to connect to any host interface anywhere), things become a little more complicated. We know that the data goes to an IP router and picks the correct route and keeps picking routers and routes closer to the target, proceeding from network to network until the IP datagram reaches the target host (192.153.184.3 in Fig. 9.2) or IP's time-to-live field reaches 0.

How do the IP routers along the way know which port to use to reach the target that the sending user wants? Indirect routing uses a third party (an IP router) to pass the datagram from the source to the target on another subnet

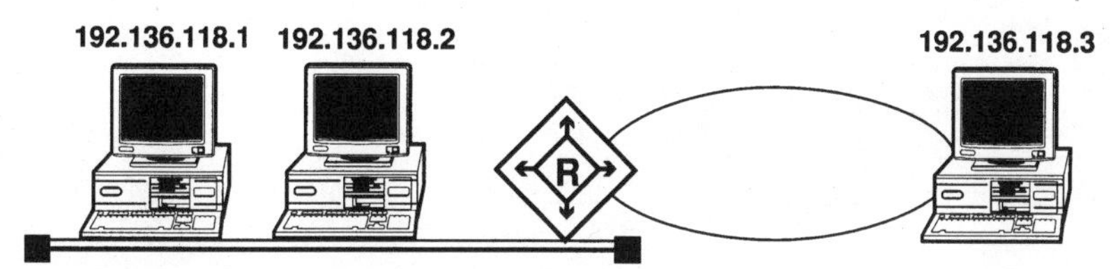

Figure 9.2 Direct or indirect routing?

or network. The indirect delivery uses the store-and-forward capability of an IP router.

The IP router stores the datagram while it checks its routing table to determine where it will forward the datagram next (to another router or the target IP-addressed system). When it knows where to forward the datagram, it sends the datagram down to the network interface for sending out toward the target system, subnet, or network.

Let's explore three ways that an IP router can find a route for the datagram: locally in the same network; connecting to neighbor networks; and reaching out across networks such as the Internet. The router's decision calls for a routing table that the router consults to determine which interface is on the IP datagram's path to its target.

Table-Driven Routing

Local

As we saw in Chap. 4, when an IP-addressed host begins the communicating process with another IP-addressed host, it first checks to determine that it is in the same network and then ensures that the host is in the same subnet. If the subnet is the same, direct routing is the easiest method to get to the target host interface.

If the network is the same and the subnet is different, the local router (supporting the source's subnet) receives the datagram. It must consult its routing table (see Fig. 9.3 for a subset sample) to determine which way to send

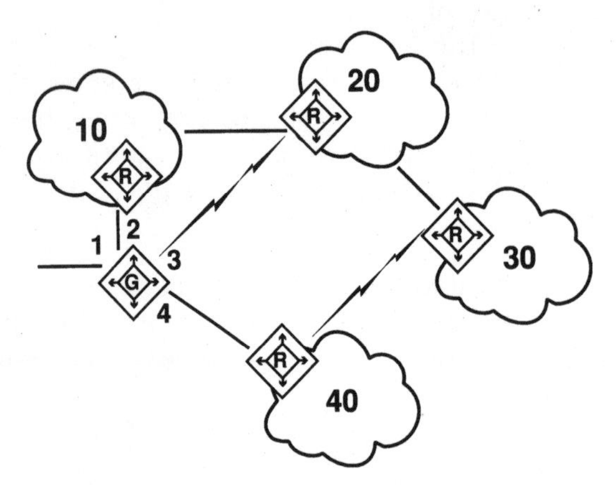

Subnet	Interface	Next Hop
190.236.10.0	2	190.236.10.251
190.236.20.0	3	190.236.20.251
190.236.30.0	4	190.236.40.251
190.236.40.0	4	190.236.40.251

Figure 9.3 A local routing table sample.

the datagram toward its specified target IP-addressed interface. If the next stop is not the datagram's target IP address (located in the same subnet), the router sends it toward that address by forwarding it to a network or subnet it knows is en route to the target.

The example in Fig. 9.3 demonstrates indirect routing when the gateway sends the datagram from the gateway (G) through interface 4 to 190.236.40.251 to reach the target IP address in subnet 190.236.30.0. The gateway chose that interim router over subnet 20's router. Those decisions may be a result of the first router to identify a route or the router that has the smallest distance (hop count) to the target IP-addressed network (or subnet).

This works fine inside an organization's network, but what happens when the destination is outside that network? The following section shows us how a router may use a table to handle this situation.

Extended

Routers, especially those in use as gateways, typically use a "fall through" method of looking at their routing table. In this method the subnets appear at the top of the list and the neighboring networks follow them lower on the list. This listing method ensures the routing software will find the highest volume lookups (local addresses) early in the search. When the gateway, shown in Fig. 9.4, consults its routing table for ways to reach external networks, its search passes through its own autonomous system first.

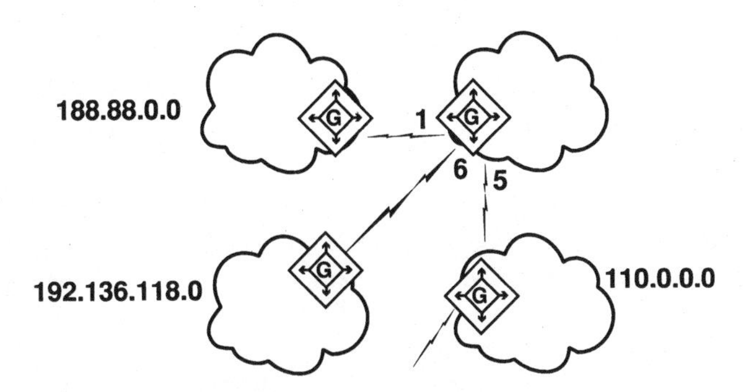

Group	Interface	Next Hop
190.236.10.0	2	190.236.10.251
190.236.20.0	3	190.236.20.251
190.236.30.0	4	190.236.40.251
190.236.40.0	4	190.236.40.251
110.0.0.0	5	110.2.0.1
188.88.0.0	1	188.88.1.1
192.136.118.0	6	192.136.118.10
0.0.0.0	5	110.2.0.1

Figure 9.4 Expanded routing table sample.

An autonomous system is a convenient collection of interfaces, hosts, and networks that fall under a single administrative (organizational) authority. That way, if an organization, company, or agency has a Class B IP address and a couple of Class C addresses, the enterprise network falls under a single administrative authority (organization). We consider that enterprise network an autonomous system. An autonomous system is a convenient reference to a virtual collection of networks—as many or as few as the organization in question has.

Default

What happens if, after working its way down through the routing table, the router cannot find the target IP address requested by the sending host? That is, when the routing table relies on a last entry ("none of the above") or the default route. This default route may not be an entry in the routing table. The router vendor can use a configuration file to identify where to send a datagram without a routing table entry.

The router forwards all datagrams containing an IP address for which the router has no reference in the routing table to this default IP gateway for further distribution. We choose a gateway as default because it has more routing information than the router sending the datagram to it. With that in mind, the default gateway should be one that is closer to the core of the Internet in both private and public systems.

Automatic versus Manual Routing

If you have the time and a small network that only requires a few changes, you could manually load and maintain a routing table in each router (including the one acting as the network gateway). If, on the other hand, you have a medium to large network or a network that has to be flexible for many moves, adds, and changes, then you need an automated way to create and maintain the Internet routing tables.

Let's take a look at the standard (current and proposed) ways we can automate these tables. As in any automated methods, there are certain rules. Routers implement these rules in standardized protocols, like Exterior Gateway Protocol (EGP), Routing Information Protocol (RIP), Open Shortest Path First (OSPF), and Border Gateway Protocol (BGP).

Exterior or interior protocol

Gateway to Gateway Protocol (GGP) was the first automated routing protocol in ARPAnet. It formally divided what is now the Internet into two parts: exterior gateways and interior gateways. The division of routing tables was to reduce the number of gateways sending routing table updates and the size of those updates, and to keep the size of routing tables manageable.

Routers often build their routing tables from the static entries the network administrator configured and the updates or advisories periodically sent from other routers. We call the manual entries static routes because they change only when replaced by a network administrator. The automatic method of routers sharing dynamic routing information in updates is more common. Most network managers or administrators have little time to devote to the manual maintenance of router tables. By configuring the routers to talk with each other, network managers have more time to focus on more important network issues.

When we categorize these devices into exterior and interior gateways, the interior gateways are routers that act as subnet gateways. A subnet (or interior) gateway is an IP node that is part of an autonomous system whose routing table is only of concern to members of that IP network (including the network's router that is acting as the gateway to the rest of the Internet). Each autonomous system is free to choose its own interior routing protocol(s).

An exterior gateway is one that takes traffic from an autonomous system's network(s) out onto the Internet. The entire Internet cares about this gateway's availability and Internet routing tables.

External gateways have to provide local routing information to one another, and the Internet is currently migrating away from EGP. As we see in Fig. 9.5, the current *Internet Official Protocol Standards* RFC (RFC 1920 as of this writing) identifies GGP and EGP as historic protocols. The BGP (whose fourth version, BGP4, we will examine later in this chapter) is the draft standard that is replacing EGP. The IPv6 RFCs specify the Interdomain Routing Protocol (IDRP) as the eventual BGP replacement.

The Interior Gateway Protocols (IGPs) in use in IP networks include Cisco's proprietary Interior Gateway Routing Protocol (IGRP) and its enhanced version, EIGRP; OSPF in its second version; and RIP. Though the standards RFC declares RIP to be an historic protocol and sets RIPv2 as the draft standard to replace it, many network managers continue to use the original version.

Interior	Status	Exterior	Status
RIP	Historic in use	GGP	Historic not used
RIPv2	Draft standard in use	EGP	Historic in use
OSPFv2	Draft standard in use	BGP4	Draft standard in use
IGRP	Cisco proprietary	IDRP	Proposed standard
EIGRP	Cisco proprietary		

Figure 9.5 Interior and exterior routing protocols.

Routing information protocol

One of the simplest and most common of the interior gateway protocols is RIP. It comes from the University of California at Berkeley's UNIX program routed (pronounced *route D*). It was in multiple implementations before the first RFC was written to standardize the specifications. Though vendor implementations vary, the differences are small enough, and vector distance routing simple enough, to cause no major problems.

Vector distance routing is a fancy way of saying that RIP bases its routing decisions on the number of hops (networks and matching routers) that it will take to get to the target IP-addressed network. RIP sends messages along the path (vector) that takes the fewest hops to reach the datagram's destination. Though the designers intended it for use with multiple protocols, RIP has only supported IP.

RIP functions at two levels: active and passive. In the passive mode, the system (usually a host) listens to RIP updates that come along and simply makes note of the information. In active mode, the system (usually a router or a server doubling as a router) broadcasts the routing information it has to the networks and subnets it serves.

Routing information basics. RIP sends its routing updates by listing (in a RIP message) each network or subnet and the number of hops to reach it. A hop count of 1 says the router needs no other IP gateways to reach the destination subnet or network, while a hop count of 16 identifies the network or subnet as unreachable. If a new route arrives with an equal value, the router does not publish the new route as the new route provides no improvement over the current route.

The hop count only specifies the number of routers in the path. It does not consider the speed or quality of the circuits used. That way, a route with a hop count of 4 could take less time to travel than a route with a metric of 2. The Routing Information Protocol faces other challenges as well.

Route loops. By calling RIP a vector-distance routing protocol, we said it cares only about the number of hops and not the quality, speed, or reliability of the circuits. There is potential for creating RIP route loops that can cause datagram storms (too many datagrams swamping the network so it cannot function properly). For example, this may happen when the route between a gateway and the Internet fails, as Fig. 9.6 shows.

If the route between the gateway (G) and the router (R) fails after the gateway updated the router, the router may announce in its next RIP message that it is 2 hops from the Internet. The gateway accepts this and updates its table so that it can advertise 3 hops in its next update.

Since RIP sends its updates every 30 seconds, it takes time for one of the devices to find that the Internet is unreachable (hop count reaches 16). We call this delayed recognition of an unreachable target "slow convergence."

Figure 9.6 RIP route loops.

This means the devices slowly become aware (converge on) that the route has failed.

While this loop (of the router and gateway increasing each other's hop count by an increment of 1) continues, the gateway sends the datagrams that are trying to reach the Internet back to the router. This means datagrams flow back and forth between the two RIP devices, clogging the network and not reaching their targets.

In the solution, known as *split horizon* (named for dividing the view of the network), the router does not send updates (about the networks or subnets) that it received from the gateway back to the gateway. That means the gateway does not get an alternate route (through the router to the Internet) and so quickly converges on the unreachability of the Internet. Most router vendors extend this function with an additional function called *poison reverse*.

Poison reverse goes beyond split horizon's not repeating back to a link what it learned from that link. The extra step means that the router will report the groups it learned from a link as unreachable to that link. Poison reverse does this by setting hop count values to 16.

In other words, the router (on the right in Fig. 9.6) advertises the groups it heard on the gateway link as 16 hops away. This way the gateway discovers quickly (fast convergence) that it cannot route to the Internet using the path to (and through) the router.

Message format. In the following sections we will look at the fields in a RIP message (both versions 1 and 2) to see how RIP passes information to other RIP-based systems (see Fig. 9.7). Before we look at those sections, let's see how we work our way through a packet to the RIP portion and why. In the

```
ff ff ff ff ff ff 00 c0 - 93 21 88 a7 08 00 45 00
00 48 09 00 00 00 3c 11 - 84 0d c0 99 b8 64 c0 99
b8 ff 02 08 02 08 00 34 - 14 a2 02 01 00 00 00 02
00 00 c0 99 ba 00 00 00 - 00 00 00 00 00 00 00 00
00 01 00 02 00 00 c0 99 - b9 00 00 00 00 00 00 00
00 00 00 00 00 01
```

Figure 9.7 Hex sample of a RIP message.

Ethernet Header															IP	
Target Ethernet Address						Source Ethernet Address							Protocol Type		ver hln	tos
ff	ff	ff	ff	ff	ff	00	c0	-	93	21	88	a7	08	00	45	00
IP Header																
IP Len		ID No.		Frag.		ttl	pro		Cksum		Source IP Address				Target	
00	48	09	00	00	00	3c	11	-	84	0d	c0	99	b8	64	c0	99
IP		UDP Header									RIP Message					
Target Address		Source Port		Target Port		UDP Length			Cksum		Cd	Ver	reserved		Family 1	
b8	ff	02	08	02	08	00	34	-	14	a2	02	01	00	00	00	02
reserved		Advertised Group 1													Group 1	
00	00	c0	99	ba	00	00	00	-	00	00	00	00	00	00	00	00
Metric		Family 2		reserved		Advertised Group 2										
00	01	00	02	00	00	c0	99	-	b9	00	00	00	00	00	00	00
		Group 2 Metric														
00	00	00	00	00	01											

Figure 9.8 RIP message layout.

following examples, we have identified some key pieces we will see in a protocol analyzer.

The Ethernet header specifies the broadcast address (hex ff ff ff ff ff ff) as the target Ethernet address, and IP (hex 08 00) as the Ether type. The IP header identifies the next protocol as UDP (hex 11) and the target IP address (hex c0 99 b8 ff) as every IP interface on network 192.153.184.255.

In the UDP header, the source and target ports (both values of hex 02 08) indicate that the source and target applications are the same (local routing process or RIP). In other words, it is being sent by RIP to any RIP application on the target network. The layout of the same RIP message (see Fig. 9.8) begins with the third byte, to the right of the hyphen, on the third line of hex, right after the UDP header.

Command. The first field in the RIP message is the 1-byte RIP command field, the possible values of which are shown in Fig. 9.9. Figure 9.6 shows us

Code	Command
1	Request for routing information
2	Response with routing information
5	Reserved for Sun's internal use

Figure 9.9 RIP command codes.

02 as its value. Figure 9.7 shows the three current functional RIP command codes and the 02 value in the sample hex header identifies this as a RIP response. Though other commands are available, RIP messages with the response code are the vast majority. We will rarely see the other codes.

Version. The second field (the fourth byte, to the right of the hyphen, on the third row of the hex in Fig. 9.6) identifies the version of RIP (hex 01 equals Version 1) that the sending router is using in this message. The current draft standard version is 2. Many routers and network administrators, as we see in the example in Fig. 9.6, still use Version 1. Fortunately, the field placement is the same in both versions, though some fields that the developers marked as reserved in Version 1 contain data in Version 2.

In Version 1, the first reserved field is the 2 bytes that immediately follow the version field (at the third and fourth bytes from end of the third row of hex, in Fig. 9.7). As with reserved fields in other TCP/IP protocols, the router fills this field with hex 0s. Version 2 of RIP specifies the same 2 bytes as the routing domain field. This, along with the next hop field that follows later in the message, lets multiple autonomous systems share a single cabling system without misrouting.

Network family of Group 1. The last two bytes (on the third line of the hex in Fig. 9.6) identify the Network Family of Group 1. To understand this we need to define both *family* and *group.*

Family, as we use it here, specifies the network protocol that will use the routing information, and thus the protocol whose network or subnet addresses appear in this message. *Group* indicates the collection of nodes, interfaces, or hosts that is being advertised in this portion of the RIP message.

Since IP is the only protocol or family to use RIP, you can expect the same hex value of 00 02 (indicating IP) as we see here. Apparently the RIP developers did not worry about consistency: They could have had that consistency by using the Ether type of 08 00 hex instead.

Like the version field, the router follows the Network Family of Group 1 with another 2-byte reserved field. It fills this field with 0s in Version 1. Version 2 specifies these 2 bytes as the route tag. The route tag flags external routes for use by BGP. Version 2 adds a family identification for authentication data. Only the first family in the message can indicate that authentication data follows by the value of ff ff hex. That first RIP entry does not have a route tag. Instead, that 2-byte field will carry the authentication type. Currently, the only type is "simple password" (a value of 02 hex). The rest of the RIP entry (16 bytes) is the left-justified (and 0-filled if necessary) password.

Network address of Group 1. The next field is the longest in the RIP message (12 bytes). It contains the network address of the group being advertised in this portion of the message. In this case the decimal value of the hex entry of c0 99 ba 00 is 192.153.186.0, Group 1. The specification calls for justifying the entry left and filling all unused bytes with the hex value of 00.

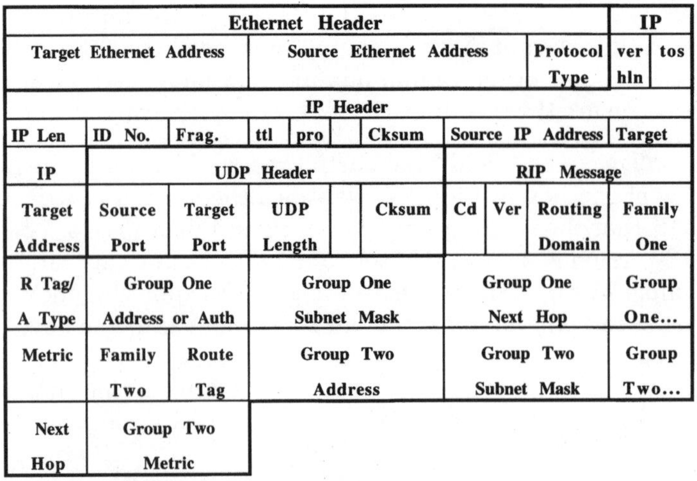

Ethernet Header									IP
Target Ethernet Address	Source Ethernet Address	Protocol Type	ver hln	tos					
IP Header									
IP Len	ID No.	Frag.	ttl	pro		Cksum	Source IP Address	Target	
IP	UDP Header						RIP Message		
Target Address	Source Port	Target Port	UDP Length		Cksum	Cd	Ver	Routing Domain	Family One
R Tag/ A Type	Group One Address or Auth	Group One Subnet Mask	Group One Next Hop	Group One...					
Metric	Family Two	Route Tag	Group Two Address	Group Two Subnet Mask	Group Two...				
Next Hop	Group Two Metric								

Figure 9.10 RIP Version 2 message layout.

The original plan was to leave enough space so RIP could serve multiple protocols or families. Since an OSI address needs 21 bytes of address field, no one ever considered using RIP to route CLNP because of its 12-byte address limitation. With IPv6 working on a 16-byte IP address and only 12 bytes available in RIP, there may be more changes coming. Since the last 6 bytes of the IPv6 address are to be the NIC address, there is room to use RIP to identify networks and subnets in that protocol.

RIP Version 2 variations. Figure 9.10 shows us that Version 2 splits the 12 bytes that Version 1 reserves for the IP address into three fields. The first 4 bytes will carry the IP address of the group. The second 4-byte field contains the subnet mask. The third field is also 4 bytes, which indicate the IP address of the next hop (router), on the way to the target group (the first field).

Group 1 cost. The last 4-byte field in the RIP entry (the last two bytes on the fourth row of hex and the first two bytes on the fifth row of hex) in Figs. 9.7 and 9.8 identifies the number of hops required to reach the network specified in the previous field. The challenge is that the field carries so many different names by so many different vendors. Some call it *cost,* while others call it *hop count, distance,* or *metric.*

The value found in the Fig. 9.7 example message (hex 00 00 00 01) tells us that the router sending this RIP message has a direct connection (1 hop) to the identified network (192.153.186.0). If a value of 16 is in this field, the router sending the message indicates that the group is unreachable. That may translate to out of range (too many hops) or a failure in the network path between the router and the group in this entry.

It is confusing that there is a 4-byte (32-bit) field being used to carry a value that cannot exceed 16. The only explanation that we have found is to keep the important fields aligned to a 32-bit or 4-octet arrangement.

Group 2, etc. That Group 1 metric field completed the advertising of Group 1. What happens if there is more than one group to advertise? How do we send the contents of an entire routing table in an efficient manner?

As you can see from the example in Fig. 9.6, for every Group from 2 (shown) through 25 (possible) in a message, RIP Version 1 repeats the three fields of the RIP entry. That entry includes another network family of group n, the group address, and the hop count to reach that group address from this router. Overflow advertisements appear in another RIP message that follows the initial message containing 25 groups.

Version 2 repeats the six fields: family of group n, route tag, IP address, subnet mask, next hop, and cost. Like Version 1, Version 2 repeats the six-field sequence up to 24 times in one message. That means 25 RIP entries or an authentication entry and 24 RIP entries.

While it is very rare, a router could advertise an individual host address as a group address. Much more typical is the presence of a network or subnet address.

Open Shortest Path First

We probably should shorten "open shortest path first" to "O,SPF" since the word *open* implies that this is a nonproprietary protocol known as "shortest path first." It is different from RIP in many ways, including different terminology. Let's review some of these terms:

Area. A group of routers working with a range of IP addresses. We call this a *subnet* in other protocols.

Area border router. The default router that controls access to an area.

Backbone. A circuit that connects all areas together via their area border routers.

Boundary router. A router that controls access from its autonomous system to other autonomous systems.

Virtual route. A path for data to an area that does not connect directly to the backbone.

Multiaccess areas. An OSPF area that contains more than one router from or to other areas.

Designated router. The router responsible for routing traffic to a multiaccess area.

OSPF versus RIP

The Open Shortest Path First protocol differs from RIP (see Fig. 9.11) in that it is a link state protocol and RIP (as we said) is a hop counter. In other words, OSPF cares about all the aspects of the link (reliability, speed, cost,

OSPF	RIP
Link state	Vector distance
TOS routing	Hop count routing
Multiple paths	Single path
Load sharing	No choice
Multicast advertising	Broadcast advertising
Authentication	Risk
Variable masking	Common masking

Figure 9.11 OSPF-RIP contrast.

and so on) and not just the number of hops. The OSPF routers may make use of the value in the type-of-service field in the IP header. That way OSPF routers may choose a path based on those bits (delay, reliability, throughput, and cost).

The OSPF protocol can route onto multiple paths where RIP can only use one path to the target IP-addressed subnet or network. With multiple routes available, OSPF lets routers share the traffic load down these multiple routes (if the routes being shared provide the type of service the datagram requests or if the datagram makes no type-of-service request). This is not load balancing because the router does not check for the size of each datagram when making its routing decision.

Another difference is that OSPF directs its route advertisements to other OSPF routers by using multicasting; RIP, on the other hand, broadcasts its information to all devices on a network or subnet. This protects nonrouting systems from having to handle the traffic that the routing updates can cause.

A protection provided by OSPF is its ability to include authentication that the routing information is from a known source. Without this protection (which RIP does not offer), anyone with a little programming ability can send phony routing updates that corrupt the network's routing ability. This can seriously affect network traffic as the bogus information is propagated across the network through router updates.

Variable subnet masking

When network managers and administrators look at OSPF, one of the first advantages they see is the ability to adjust the network to meet the needs of the organization. This is an improvement over making the organization change to meet the network's structure as RIP requires. Since OSPF passes the local subnet mask as part of its routing updates, variable subnet masking is a key to meeting the needs of the organization and helps fulfill the original flexibility-driven goals of TCP/IP.

This aspect of OSPF lets each area have a different subnet mask instead of the common mask requirement that is part of the restriction of RIP. With this flexibility, we need to monitor closely assignments of area IP addresses so that addressing does not overlap. This is especially important in moving from one mask (subnet) to another. We can better understand this flexibility by looking at how an organization can enjoy the benefits of variable subnet masking.

An example of OSPF subnetting

By applying the table from Fig. 9.12, the organization in Fig. 9.13 can make the most of its Class B IP network. As you can see, each area has a different mask, and the autonomous system presents no mask to the rest of the world. This way the boundary router assumes the responsibility of sending the inbound IP datagram to the appropriate area border router for delivery to the target-addressed system.

The headquarters, according to the mask and chart in the last row of Fig. 9.12, could have 24,574 separately addressed interfaces (without separate subnets) in its multiaccess area. The regional office can support up to 4094 different IP addresses, while the smaller district office uses a subnet mask that supports up to 254 IP addresses (which is appropriate for its size). Using OSPF removes the RIP-based need to use a common subnet mask (such as 255.255.240). That common subnet mask would provide many more IP addresses than the district office will ever need.

Note that OSPF's variable subnet masking will work quite well for Class A and B networks with their range of possible subnets, but is not flexible enough for regular use with Class C networks. Due to size constraints, there are only so many ways to subnet a single Class C network. However, OSPF

Area portion IP address	Subnet mask	Usable hosts
191.252.16.0	255.255.240.0	4,094
191.252.32.0	255.255.240.0	4,094
191.252.48.0	255.255.240.0	4,094
191.252.64.0	255.255.240.0	4,094
191.252.80.0	255.255.240.0	4,094
191.252.96.0	255.255.240.0	4,094
191.252.128.0	255.255.240.0	4,094
191.252.144.0	255.255.240.0	4,094
191.252.160.0	255.255.240.0	4,094
191.252.176.0 through 191.252.191.0	255.255.255.0	254
191.252.192.0 through 191.252.255.254	255.255.0.0	24,574

Figure 9.12 Variable subnet mask example.

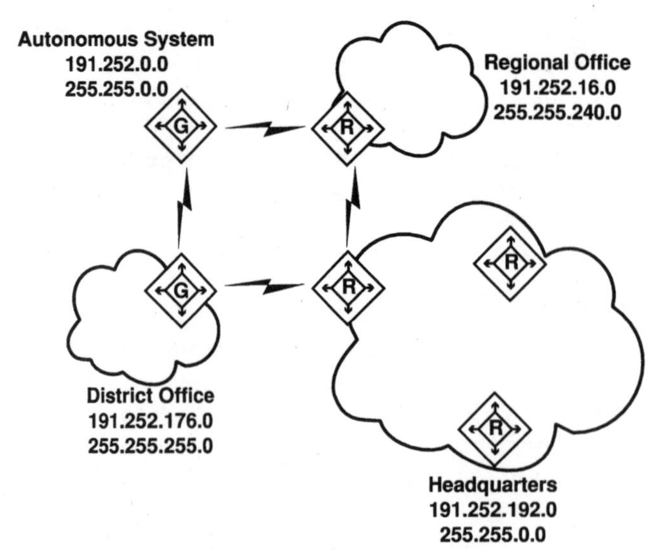

Figure 9.13 OSPF subnet example.

can summarize multiple Class C networks into one supernet route between areas.

Border Gateway Protocol

Border Gateway Protocol is the successor of EGP. Now in its fourth version, BGP is an interautonomous system routing protocol. That means it advertises routes between autonomous systems (AS) instead of between networks. Using ASs cuts down on the number of entries that each EGP router must advertise and maintain.

Like EGP, it is also a vector distance-routing protocol, but it differs by sending the sequence of autonomous system numbers in the path instead of the cost (or metric or distance or hop count) that EGP sends. In other words, each BGP gateway calculates the preferred route that it uses to a target IP-addressed network and passes that route to its BGP-based neighbors instead of identifying the network and its distance.

The BGP also uses the more reliable TCP to send its routing updates while EGP simply passes its update through the best efforts of IP. With EGP, the network administrator must take steps to make sure there are no loops in the network; BGP, on the other hand, eliminates those loops by identifying the actual route to the AS rather than a hop-by-hop move closer to the target. BGP can support more complexity in the routing process by using routing information bases to store its data rather than the routing table that EGP uses.

Version 4 introduced mechanisms that offer aggregation of routes by combining AS paths in preparation for supporting CIDR supernetting.

Routing table size and CIDR

A major issue in expanding the number of Internet addresses is the increasing size of the routing tables that the backbone routers require. Imagine each Internet router storing a routing entry for every Class A, B, and C network address that both RS.InterNIC.Net and NIC.DDN.MIL have assigned. So many organizations wanted to participate in the Internet that the number of address assignments has grown exponentially. The backbone routing table expanded about 1½ times as fast as memory technology.

The CIDR recommendation offers relief by assigning blocks of contiguous (Class C) addresses to the same organization, particularly ISP. Assigning blocks of Class C addresses brings up another new concept: supernetting.

Like subnetting, supernetting uses a mask to identify the network significant bits. The difference is that a Class C supernet mask contains a smaller value in the third byte than the default mask. For example, Fig. 9.14 shows the mask of the addresses assigned to Pat's Internet Service and Bait Shoppe.

Since the block of 32 Class C addresses is from 223.255.0.0 through 223.255.31.0, the supernet mask is 255.255.224.0 to account for as many of the common bits as possible. When BGP4 passes that information to its neighbor BGP routers, it will send 223.255.0.0/19 to show that the first 19 bits of the block are common to all the addresses. As you can see, this is a much smaller update than would occur if each of the 32 Class C addresses were identified separately.

The RS.InterNIC.Net began this aggregation process to relieve the depletion of Class B addresses. Aggregating block address assignments and aggregating routes reduced the size of the tables for a while, but they are now growing again.

Even though the IPng working group designed the IPv6 addresses to make the most of aggregation, switching to IPv6 will not solve the routing-table size problem by itself. This efficient route advertising can work well with

	Binary	Decimal
Supernet mask	11111111 11111111 11100000 00000000	255.255.224.0
Network 1	11011111 11111111 00000000 00000000	223.255. 0.0
Network 2	11011111 11111111 00000001 00000000	223.255. 1.0
.	.	.
.	.	.
.	.	.
Network 32	11011111 11111111 00011111 00000000	223.255. 31.0

Figure 9.14 Supernetting example.

BGP Path Attributes	
Mandatory	Discretionary
Well-known	Optional
Transitive	Nontransitive
Partial	

BGP Attribute Categories
Well-known mandatory
Well-known discretionary
Optional transitive
Optional nontransitive

Figure 9.15 BGP path attributes and categories.

IPng, however, since it includes address autoconfiguration capability for easy renumbering if a customer decides to switch providers.

Path attributes

In BGP language, we call a field (part of the route description) an *attribute*. Though some find these terms confusing, we will attempt to explain them in a way that helps you understand them (see Fig. 9.15).

All vendor implementations of BGP must recognize well-known attributes. We call the well-known attributes that all vendor implementations of BGP do not have to recognize *optional attributes*. Mandatory attributes must appear in each route description. Instead of the mandatory attributes being referred to as optional, in BGP we call them *discretionary*. We also consider the mandatory attributes (Origin, As_Path, and Next_Hop) to be well known.

We categorize discretionary attributes as transitive (for use by all BGP gateways that support the attribute) or nontransitive (for use by this BGP gateway). A BGP gateway that has not implemented the attribute can pass along a transitive attribute without modifying it. If this happens, the gateway marks the attribute as partial. If the attribute is nontransitive, any BGP gateway that has not implemented the attribute deletes it.

Routing information base

Instead of a routing table, BGP stores its routes in the three separate (though they may reside in one file) parts of the routing information base (RIB): the Adj-RIBs-In, the Loc-RIB, and the Adj-RIBs-Out (see Fig. 9.16).

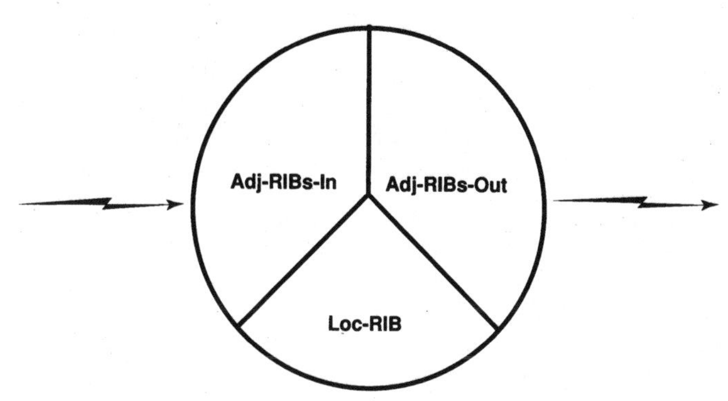

Figure 9.16 BGP RIBs.

- The Adj-RIBs-In portion stores routing information that comes from other (adjacent) BGP speakers (or peers). Their contents represent routes from which the router can choose when making its routing decision.
- The Loc-RIB contains information on routes that the local BGP speaker uses after applying its local routing policies to the routing information contained in its Adj-RIBs-In. Routing policies can include special Internet access points through alternate routes that the network administrator assigns.
- The Adj-RIBs-Out portion stores the routing information that the BGP router advertises to other BGP speakers (or peers).

Although the model in Fig. 9.13 distinguishes between Adj-RIBs-In, Loc-RIB, and Adj-RIBs-Out, it does not imply or require that any implementation must have three separate copies of the routing information. The choice: three copies of the information versus one copy with pointers, which is the vendor's or implementing manager's decision.

Messages

BGP uses four message types: open, update, keep alive, and notification. Since BGP runs over TCP with no periodic update sent, each BGP speaker must maintain the current version of the entire RIB for all of its BGP peers as long as it holds a connection to them. Functions of the four message types are as follows:

- The open message goes out to confirm the connection parameters.
- An update message follows the open message and carries the entire BGP Adj-RIBs-Out from the sender, which becomes the Adj-RIBs-In for the receiver.

- The keep alive message reassures the BGP peer that the sender is alive and well. The router sends that message when the peer's hold time is approaching its limit so that the time will not expire. The hold time (set during the opening BGP exchange) is the maximum delay the peer system should have to wait from one message to the next.
- In keeping with the alternate BGP terms, error messages are notifications. They respond to an error or a special condition. If the connection finds an error, the notification goes out and the host closes the connection.

Like many other interior and exterior routing protocols, the BGP gateway may be a router or a host (usually a server) system that is doubling as a router to conserve equipment costs.

Chapter

10

Internet Control Message Protocol

Overview

The Internet Control Message Protocol lets TCP/IP internetworks handle logical errors that may arise from the time the IP datagram leaves the source host to the time it reaches the target host. For example, if a router cannot deliver an IP datagram because the target host does not respond to an ARP, it will send an ICMP message back to the source of that IP datagram to report the problem.

Note that ICMP does not discover problems, it only reports them to the source of the IP datagram that made the problem show itself. ICMP also offers the ability to transmit messages to help diagnose existing or potential connectivity problems. The best known of that kind of ICMP message is the Echo Request and Response. At the screen we call the ping command and the system sends an echo request.

Any network interface with a full IP address can send or receive and process ICMP messages. This list usually includes hosts, routers, gateways, and intelligent hubs. An IP datagram carries the ICMP message with its protocol number field set to the hex value 01.

ICMP messages, riding inside IP datagrams, have no special priority. Worse than that, many TCP/IP stack vendors treat ICMP messages as interruptions. In an IP network, the ICMP messages will probably be the first that systems will lose or discard. Since ICMP is an internal TCP/IP reporting mechanism, applications tend to get the first use of limited CPU cycles.

Message Destinations

The ICMP message goes where the category of the message (variation reporting or diagnostic) designates (see Fig. 10.1). The variation (or error) reporting messages identify network problems. They can only go back to the problem IP datagram's source IP address, which is most often a client host interface. If

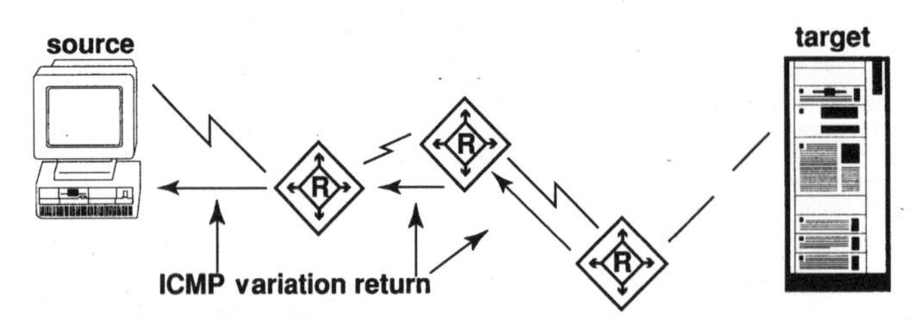

Figure 10.1 ICMP variation reporting message destination.

the variation relates to something that the host cannot control or change, the host does not fix the real cause of the problem.

This limitation is the result of TCP/IP's flexibility. Since any datagram can follow any route that is open, they contain only source and target addresses. There is no way to route the ICMP message to the problematic intermediate IP router without its IP address. That address is not in the IP datagram unless it is the source or target interface.

There is an additional restriction on ICMP variation reporting messages. If one of these messages encounters an error in returning to the source address of the original IP datagram, the system discovering that error discards the datagram and will not send a report. So, unlike financing with its interest on interest, there are no error reports on error reports.

That is not the case with ICMP diagnostic messages, however. These messages go where the originating source IP-addressed node sends them. That may also include a designated route to the target. If an error occurs here, the device that discovers the error can send an ICMP variation reporting message to the source of the diagnostic message to report the error.

Messages

All ICMP messages begin with the same format (in Fig. 10.2). The first byte of the message indicates the type of this ICMP message. As you can see from Fig. 10.3, RFCs specify 11 types of messages. Each variation reporting message has one type number. Diagnostic messages have different type numbers for the requests and responses.

Beyond the type, the next ICMP field (ICMP code) is also a single byte that gives further information about each type. As you can see in Fig. 10.3, only the variation reporting messages use a code other than 0. We will discover

1 byte	1 byte	2 bytes
TYPE	CODE	CHECKSUM

Figure 10.2 ICMP common header.

	Type	Codes	Function
	3	0–12	Destination unreachable
Variation	4	0	Source quench
reporting	5	0–3	Redirect
	11	0–1	Time exceeded
	12	0	Parameter fault
	8/0	0	Echo request/response
Diagnostic	13/14	0	Timestamp request/response
	17/18	0	Subnet mask request/response

Figure 10.3 ICMP types.

the further meanings of those codes as we look at each of the messages in more detail.

The final 2-byte field that is common to all ICMP messages is the ICMP checksum. Since IP covers only its own header with its checksum, ICMP must cover itself and any data it contains with a separate checksum. The checksum confirms that the source sent a good message and the target received it in the same condition.

Beyond those first four bytes, the message formats vary from type to type. Let's take a look at the different types.

Echo request (Type 8) and response (Type 0)

These two messages form the major part of TCP/IP managers' troubleshooting ability. When users or technicians want to test the connections on the network, they issue the ping command. That system sends an ICMP echo request message (see Figs. 10.4 and 10.5) to the designated IP address. If the IP address is valid and the receiving vendor supports the requested functions, an echo response will return.

In its basic form, this message can verify connectivity. It ensures that systems can send and receive packets correctly. Most implementations can support multiple flags that expand functionality to include the IP header options. For example, ping-r can use an IP option to record the route while ping-n can set the number of echo requests the user wants sent to the target. These flags are vendor specific.

The ID number (when implemented per the RFC) changes for each new ping command issued. If the ping command calls for multiple echo requests to

```
        08  00  02  cf  0a  3c  -  01  10  c9  11  4e  90  06  b2
1f  00  08  09  0a  0b  0c  0d  -  0e  0f  10  11  12  13  14  15
16  17  18  19  1a  1b  1c  1d  -  1e  1f  20  21  22  23  24  25
```

Figure 10.4 Echo request hex sample.

1 byte	1 byte	2 bytes
TYPE	CODE	CHECKSUM
ID number		Sequence number
Data to echo or be echoed		

Figure 10.5 ICMP echo request (Type 8/0) and response message layout.

the same target, the sequence number changes for each request and the ID number stays the same. The data field is optional and can contain any data the sender wishes. Some vendors compress their own information into the first 7 bytes and mark the end of that with an eighth byte of 00 hex. That can include the data size and any other information that the user entered when issuing the ping command.

Only IP's limitation of a total IP length of 65,535 bytes restricts the total number of bytes that the source system wants the target system to echo. The target system's memory, software, and other capabilities can restrict what it can echo in response. To show an example of this variable length, we have left the end of the field open and did not display all the data in the hex sample in Fig. 10.4.

A troubleshooting guideline. In troubleshooting an IP network, one school suggests setting up batch files in which network technicians ping the key network resources (like routers, hubs, and servers) 12 times each. By doing this from different locations around the network, we can get a good picture of how well the network is working. (See Fig. 10.6.)

If we get 9 to 12 positive responses, then we know the network is in good (though not the best) shape. Most network managers prefer 12 out of 12 responses and any fewer than 12 gives them reason to investigate the problems this effort uncovered. If we receive four to eight good answers, there must be some congestion out in the network. Obviously, four responses indicate much more congestion than eight. Many technicians would follow such a response with other, more thorough diagnostics such as trace route.

Successes	Indicates
9 to 12	Good network
4 to 8	Congestion
0 to 3	Major problems

Figure 10.6 A 12-ping guideline.

		03	03	04	47	00	00	-	00	00	45	00	00	68	dc	7d
00	00	1e	11	00	00	c0	99	-	b7	01	c0	99	b7	05	04	12
02	06	00	54	00	00			-								

Figure 10.7 ICMP Type 3 hex sample.

If there are three or less successful responses to 12 pings, the network is not working sufficiently to rely on it. In *Star Trek* terms, "It's dead, Jim." If we consider that ICMP's purpose is to discover and report network problems, the ping (and the 12-ping process) offers that capability.

Destination unreachable example (Figs. 10.7 to 10.9)

As you can see in Fig. 10.9, the Type 3 message has 13 codes (0 through 12) to specify the reason for the inability to reach the target. As with other variation-reporting messages, the specification reserves the second 4 bytes and follows them with the IP header that discovered the error.

That failed IP header leads into the first 8 bytes of the IP header's data field or payload. That means it could include the first 8 bytes of the TCP header, the entire UDP header, the first 8 bytes of the ICMP (diagnostic) header, and so on. The reporting system adds these 8 bytes to help the source that receives the ICMP destination unreachable message to find the reason for the failure and designate exactly which datagram failed.

In Fig. 10.7 the Type 3 Code 3 is a destination unreachable, port unreachable message. The tenth byte of the failed IP header is 11 hex (17 decimal), which indicates the failed port is part of the failed UDP header. The UDP

1 byte	1 byte	2 bytes
Type	Code	Checksum
Reserved		
Failed		
IP		
header		
that caused		
the error.		
The first 8 bytes of		
the datagram's data field.		

Figure 10.8 ICMP destination unreachable layout.

Code	Indicates	Purpose
0	Network unreachable	Cannot reach requested network
1	Host unreachable	Cannot reach requested host
2	Protocol unreachable	Target host cannot find protocol
3	Port unreachable	Target host's port is not available
4	Fragmentation blocked	DF bit is blocking fragmentation
5	Source route failed	Source route unavailable
6	Target network unknown	Network not in routing table
7	Target host unknown	Host not in routing table
8	Source host isolated	Source can't talk to Internet
9	Target network prohibited	Target network blocking access
10	Target host prohibited	Target host blocking access
11	Network TOS problem	Requested TOS prevents access
12	Host TOS problem	Requested TOS prevents access

Figure 10.9 Destination unreachable codes.

source port is 04 12 hex (1042 decimal), a random port. The UDP target port is 02 06 hex (518 decimal), a well-known port. According to the excerpt of RFC 1700 in App. A of Chap. 8, the server denied access to the application ntalk.

Destination unreachable codes. The details (listed in Fig. 10.9) explain the various reasons that a network node might send a Type 3 (destination unreachable) message back to the source IP address. The first four codes (0 through 3) work in a hierarchical arrangement with each other indicating just how far the datagram got before a node discovered the problem, sent a portion back to the source IP-addressed node in the ICMP header, and discarded the datagram. Codes 6 and 7, 9 and 10, and 11 and 12 work together to identify the network or host that has the problem. In some cases, the problem is a reaction to an Internet event. That type of reaction can cause more problems than it fixes.

For example, when two lawyers spammed (sent unwanted advertisements to) a lot of news groups, some organizations reacted by blocking access from the ISP the lawyers used by sending ICMP Type 3, Code 9 messages. The ISP stopped providing the lawyers with access, but some of these organizations continued to block any communications from the other subscribers. The solution became worse than the original problem.

Source quench message

Routers, gateways, and servers use the ICMP source quench (message type 4) (see Fig. 10.10) to indicate that they are near their buffer capacity, which means they cannot continue to process all the messages they are receiving. Its format (see Fig. 10.11) is the same as the destination unreachable mes-

```
        04 00 0a 7b 00 00 - 00 00 45 00 04 18 dc 7d
00 00 1e 06 00 00 c0 a0 - f0 01 c0 a0 f0 3c 05 b3
00 14 0c a4 03 f0       -
```

Figure 10.10 ICMP source quench hex sample.

sage. The message goes to the IP address whose datagram helped the router recognize the problem.

When a system receives this message, it should slow down its transmission of datagrams until the source quench messages stop coming. Then it may slowly increase its transmission rate back to normal. This will give the router or server time to handle the data in the buffer and get the volume back down to a manageable amount.

Redirecting traffic with an ICMP Type 5

Let's work through the example in Fig. 10.12 to understand what this Type 5 message does. The source computer frequently talks to network 1 and so has its default router set to A. If it wants to send traffic to network 2, it still sends that traffic (along with its usual traffic) to its default router.

When router A gets the datagram and realizes that its routing table directs it to forward the datagram back onto the same originating network, it also sends an ICMP redirect message to the source to tell it to send this session's datagrams to router B's IP address instead. It includes that IP address in the ICMP message field that was previously (in other variation reporting messages) reserved.

1 byte	1 byte	2 bytes
Type 04	Code 00	Checksum
Reserved		
Failed		
IP		
header		
that caused		
response.		
The first 8 bytes of		
the IP datagram's data field.		

Figure 10.11 ICMP Type 4 message layout.

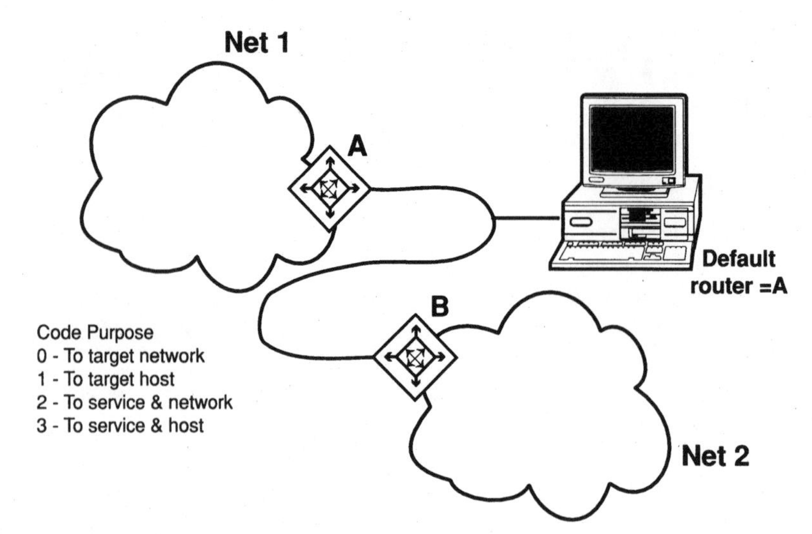

Figure 10.12 Redirecting failed messages.

The rest of the message format is the same as the other variation reporting messages. The effect is different. The redirecting router forwards the identified datagram to the correct receiving router. Let's look at the layout of the message and a sample in Fig. 10.13.

Redirect message (Type 5). In this message, as we saw in the previous section, a router or gateway returns instructions to the originating IP address. It suggests that the rest of this session be conducted through the address it provides in bytes 5 through 8. In the sample in Fig. 10.13, that gateway address is c0 a0 f0 64 or 192.160.240.100.

Other variation-reporting ICMP messages return the failed IP header and the first 8 bytes after that header. This returns a copy of the message that caused the ICMP message. In other words, the redirecting router forwarded the offending IP datagram to its destination and copied only part of it here as the instruction to redirect future IP datagrams in this session (see Fig. 10.14).

The router does not send a Type 5 message if the original IP datagram contains a loose (or strict) source routing option. It assumes that the operator who sent that routing instruction had a good reason for it.

```
            05 00 1a 79 c0 a0 - f0 64 45 00 00 68 dc 7d
00 00 3f 06 00 00 c0 a0 - f0 31 c0 99 b9 03 04 9b
00 19 01 7c 00 00       -
```

Figure 10.13 ICMP Type 5 hex sample.

1 byte	1 byte	2 bytes
Type 05	Code 00	Checksum
Correct router address		
Failed		
IP		
header		
that found		
the error.		
The first 8 bytes of		
the IP datagram's data field.		

Figure 10.14 ICMP redirect message layout.

Time-exceeded message

There are two circumstances that will cause a node to send this ICMP message. As with the other variation reporting messages, it returns the failed IP header and the first 8 bytes of its data field to the source address in the original datagram.

1. A router or gateway sends this message when the time-to-live field in an IP header falls to a value of 0. In that case, the router sets the code field to 0. See Chap. 6 for more information on the time-to-live field.

2. A host (usually a server) sends this message when the target host's timer expires before the target-addressed station can reassemble the fragments of a datagram. This usually means one or more of the fragments has not arrived at the target system. In this case, the host sets the code field to 1 (as in Fig. 10.15). See Chap. 6 for more information on fragmentation.

If the final fragment in the series does not arrive, the target-addressed node will not send a time-exceeded message since that last fragment may still be in the last router's outbound buffer. See Fig. 10.16 for a time-exceeded message layout.

```
                  0b 01 04 47 00 00 - 00 00 45 00 00 68 dc 7d
00 00 3f 06 00 00 c0 a0 - f0 31 c0 99 b9 03 04 9b
00 45 01 7c 00 00       -
```

Figure 10.15 ICMP Type 11 hex sample.

Type 0b	Code 00	Checksum
Reserved		
Failed		
IP		
header		
that found		
the error.		
The first 8 bytes of		
the IP datagram's data field.		

Figure 10.16 ICMP time-exceeded message layout.

Parameter problem message

If a router or host discovers a parameter problem in a datagram, it may return a Type 12 message to the source host. A parameter problem is a violation of the RFC-specified protocol. The format of the message follows that of the other variation reporting messages for the most part. The exception is that the first byte of the reserved field is a pointer that tells us which byte of the failed IP header and its 8 bytes of data is a problem.

In the example in Fig. 10.17, the ICMP pointer field points to the first byte as a problem. As you can see, the pointer field indicates the first byte of the failed IP header that is the problem. The code on a parameter problem ICMP message is always 0. See Fig. 10.18 for a layout of a parameter problem.

Timestamp request and response message

This message is very useful in troubleshooting or evaluating the effectiveness of a circuit. The ICMP timestamp request (Type 13) and response (Type 14) helps in searching for response time problems in the network. It lets the technician (or user) send a message and receive a timestamp of the receipt at the other end and the time that the datagram began its return.

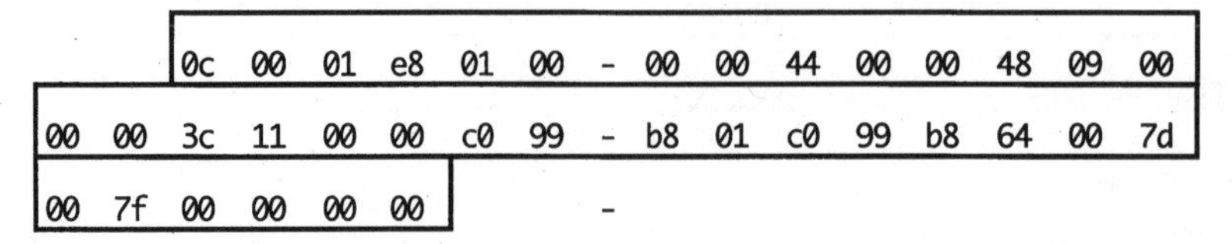

0c 00 01 e8 01 00 - 00 00 44 00 00 48 09 00
00 00 3c 11 00 00 c0 99 - b8 01 c0 99 b8 64 00 7d
00 7f 00 00 00 00 -

Figure 10.17 ICMP Type 12 hex sample.

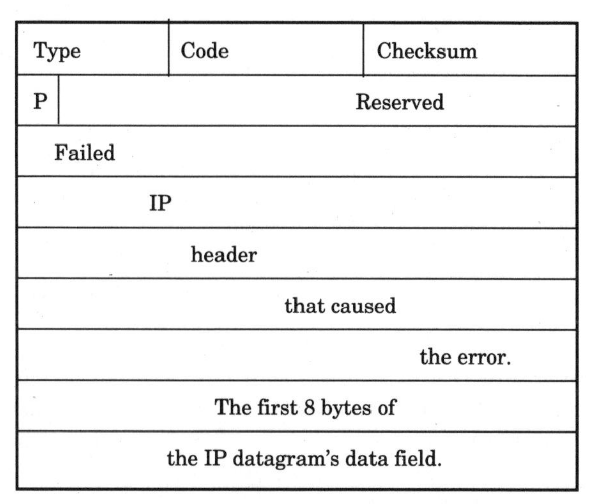

Figure 10.18 ICMP parameter problem layout.

In Fig. 10.19, you can see the advantage that this timestamp offers over the similar IP option (Collect Internet Timestamps). Like that option, this message is usually part of a ping. It gives us origin, receive, and transmit times. By adding an arrival time, the CPU marks when the response arrives at the original sending station.

With these times we can create simple equations to calculate realistic times without being concerned about operating in a time-locked network. A time-locked network has all the nodes' CPU clocks ticking to the same beat of the atomic clock. Since most network administrators are happy to do without another protocol (like Network Time Protocol) to manage, this ICMP message can provide useful times.

By subtracting the arrival time from the origin time (since they are both in the same system and use the same CPU clock), we calculate the round-trip time. With the same CPU clock providing the receive and transmit times, we can subtract to find the buffer time or the extent of the congestion in that device's buffer.

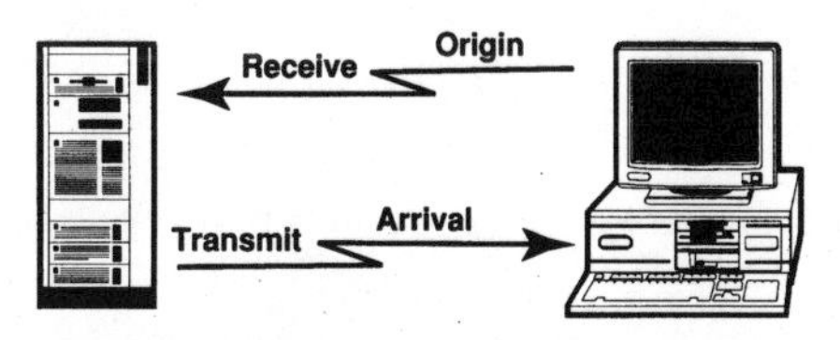

Type	Code	Checksum
ID number		Sequence number
Origin timestamp		
Receive timestamp		
Transmit timestamp		

Figure 10.19 ICMP timestamp request layout.

Type	Code	Checksum	
ID number		Sequence number	
Address mask			

Figure 10.20 ICMP subnet mask request or response layout.

Since we now have two times, in hundredths of a second, we can subtract the buffer time from the round-trip time to determine the amount of time the datagram took to travel on the media (cable). That could include the time it took to cross any bridges or pass through any routers. Locally, it can identify the time it took for "nothing but net."

Subnet mask request or response message

The subnet mask message uses the same request (17) and response (18) format as the other diagnostic ICMP messages. The similarity does not end there. The message also includes an ID number and sequence number to match the right response to the request (see Fig. 10.20).

If a network node needs to know the subnet address mask (i.e., the number of bits in the IP address used for subnets versus hosts), it sends this message to a router. It can do this by using the router's address (if it knows it) or the broadcast address (if it does not). The response normally comes from the default IP router. The format for the subnet bits is up to each site, as no standard applies at this time. The only valid code is 0.

ICMP Samples

In Fig. 10.21 we provide some samples of decoded ICMP messages. This translation of live data from a network shows us the way ICMP should work. In our samples we provided

- A Type 3 destination unreachable with the Code 3 to indicate a port unreachable message
- Three attempts and one success at sending Type 8 echo requests to receive Type 0 echo responses
- A Type 11 time-exceeded message with a code to identify it as a fragment reassembly timeout message

The destination unreachable message shows that a node (192.153.185.5) was trying to access the UDP port 517 on the server (192.153.185.1) to run

```
ICMP: Type: port_unreachable(0x0303) cksum: 0x0447
Packet that caused error:

Internet: 192.153.185.1 -> 192.153.185.5 hl: 5 ver: 4 tos: 0 len: 104
id: 0xdc7d fragoff: 0 flags: 00 ttl: 30 prot: UDP(17) xsum: 0x0000
UDP: 1042 -> talk(517) len: 84 xsum: 0x0000

ICMP: Type: echo_request(0x0800) cksum: 0x6180 id: 0x2e01 seq: 0000

ICMP: Type: echo_request(0x0800) cksum: 0xb057 id: 0x2e01 seq: 0x0100

ICMP: Type: echo_request(0x0800) cksum: 0x6235 id: 0x2e01 seq: 0x0200

ICMP: Type: echo_request(0x0800) cksum: 0x5132 id: 0x4470 seq: 0x0001

ICMP: Type: echo_reply(0000) cksum: 0x513a id: 0x4470 seq: 0x0001

ICMP: Type: frag_reassembly_time_exceeded(0x0b01) cksum: 0x2d86
Packet that caused error:
Internet: 192.153.183.1 -> 192.153.183.5 hl: 5 ver: 4 tos: 0 len: 1500 id:
0xdc84 fragoff: 32 flags: 00 ttl: 32 prot: ICMP(1) xsum: 0x62a9 Fragment
```

Figure 10.21 Decoded ICMP sample data exchanges.

the talk program. Either the server did not support that protocol or the requester did not meet one or more of the application requirements. In this case, the server found no invitation to support the user's request.

Notice that the server had verified both the UDP and IP checksums and cleared them on the way up the stack. When it returned the IP and UDP headers, the ICMP checksum provided the error detection.

The three attempted echo requests show us how the ID number can stay the same for repeated pings to the same target while the sequence number changes to show that these are different pings that the source sent from one user command. The successful echo request and echo reply both show the same ID number and sequence number, indicating a match.

The time-exceeded message is carrying a fragment. Its fragment status flag of 0 and the fragment offset at thirty-two 64-bit words indicate that it is the last fragment in a series. The router discarded the other fragments before sending this fragment back to the original source IP address.

Chapter

11

Transmission Control Protocol

Reliable Transport Services

Although IP does its best to ensure the arrival of datagrams carrying messages, few of the data-link layer protocols provide data integrity. Therefore, TCP must accept that responsibility at the host-to-host layer. If an application author wants to be sure that all the data arrive, and do so in order, there are two parallel choices:

- Provide each application with its own error-handling software for the necessary reliability
- Use a single piece of software, like TCP, to provide these services for all applications

Most TCP/IP software authors choose the second option because it is the most efficient and flexible. As an added benefit, standardized application methods ensure interoperability among a variety of vendor systems and software.

TCP Header

As in previous chapters, we start by determining that it is TCP we have in Fig. 11.1. The Ethernet header carries the protocol type of 08 00 hex, which indicates IP is to follow. The IP header that follows Ethernet has a protocol number of 06 hex, which identifies TCP as the protocol that follows IP. This means that TCP should begin with the third byte on the third line of hex.

The first two fields in the TCP header (Fig. 11.2) are the source and target ports. As with other protocols in the TCP/IP protocol suite, the source precedes the target. Note that the target precedes the source only in the NIL.

```
00 00 0c 4c 7c 2b 02 07 - 01 2b 31 a5 08 00 45 00
00 2c 00 01 00 00 40 06 - 8b 3a c0 99 b8 2c c0 99
b8 21 34 2b 00 15 0a 31 - 1b 81 00 00 00 00 60 02
20 00 f4 28 00 00 02 04 - 05 b4 00 00
```

Figure 11.1 Ethernet, IP, and TCP hex sample.

When starting a TCP session, as we are doing in Fig. 11.1, the source port is the random port (third and fourth bytes on the third line of hex) of the device requesting the service (we may also call it the client system). This becomes clearer when we see that the decimal value of hex 34 2b is 13,355, a number well into the random port range (1025 through 65,535).

Converting a four-character hex value to a single decimal value is basic multiplication. From the right, multiply the values by 1, 16, 256, and 4096. In Fig. 11.1 that gives us $1 \times$ b (b hex is 11 decimal), which equals 11, 2×16 which is 32, 4×256 for a value of 1024, and 3×4096, which equals 12,288. Adding those results ($11 + 32 + 1024 + 12{,}288$) equals 13,355.

So we can determine that this is the client end of a TCP session (with the source port well into the 1024 through 65,535 random port number range). Figure 11.1 shows it making a request to talk with the well-known target port (fifth and sixth bytes on the third line of hex) of 00 15 hex (21 decimal). By looking up that port number in the Assigned Numbers RFC (or in Fig. 8.7 at the end of the UDP chapter) we find that this is the beginning of an FTP session.

Each TCP session starts with a three-step handshake to establish the connection that the TCP applications uses for reliability. TCP also uses sequence numbers to increase that reliability.

Source sequence number

To keep track of each byte of data (and so provide the reliability that we expect), TCP uses separate sequence numbers from each end of the session. If TCP sent 1 byte with each TCP header, each byte would have its own sequence number.

The first step of the three-step handshake must include the source sequence number so the destination system can use it to acknowledge the

2 bytes	2 bytes	4 bytes	4 bytes	4 bits	4 bits	1 byte	2 bytes	2 bytes	2 bytes
Source port	Target port	Source seq number	Ack seq number	Header length	Reserved	Session bits	Window size	Check sum	Urgent data pointer
34 2b	00 15	0a 31 1b 81	00000000	6	0	02	20 00	f4 28	00 00

Figure 11.2 TCP header layout.

receipt of data bytes. In Fig. 11.1, that 4-byte field is the 2 bytes on both sides of the hyphen on the third line of hex.

The source sequence number indicates that a TCP segment is being sent and where the segment fits in relation to other segments in this session. This lets the receiving host properly resequence data if necessary but does not identify how many bytes of user data may be in each segment.

The way to discover the number of user data bytes in a datagram is to look at a datagram and subtract the TCP and IP header lengths from the IP total length field. In Fig. 11.1 the total IP length field is the first 2 bytes on the second line of hex.

The implementation of transmission control protocol running on UNIX systems usually calculates the initial sequence number from the CPU clock time and makes it a 32-bit, nonzero number as the RFC 793 specification states. Other implementations simply use a random number or start with the same number for each session. Starting each session with the same number can cause problems with multiuser, multitasking systems by not using unique numbers.

As we will see next, the acknowledgment sequence number performs a similar function and initially comes from the same calculation in the server.

Acknowledgment sequence number

This number shows the number of the next byte that the receiver expects from the sending host. In this first step of the startup handshake, we have not heard from the server to get its sequence number. That means TCP must fill that field (the 4 bytes beginning with the third byte to the right of the hyphen on the third line of hex) for correct field spacing.

As we will discover later in the chapter, the acknowledgement (Ack) sequence number also provides proof that the sender has completed certain tasks. Since most vendors fill the field with hex 0s when not using it (as in Fig. 11.1), this is a good indicator that we are looking at the first step in the handshake to start a TCP session. After the first step of the handshake, the field should contain valid data.

TCP header length

The next byte contains two fields. The first 4-bit field, the TCP header length, tells the target machine the size of the current TCP header in 32-bit words. Like IP's header length field, if the value here is greater than the default of 5 (translated to 20 bytes), the header contains options.

Normally the value in this field is 5, denoting that the TCP header contains five 32-bit words (20 bytes) of data. When a value greater than 5 appears, the destination host can then know that the optional extra bytes are present. TCP reserves the remaining 4 bits in this byte (octet) for future use and so fills that space with 0s.

Reserved	Reserved	Urgent	Valid	Push	Reset	Synch.	Final
Bit	Bit	Data	Ack.	Request	Session	Sequence number	Data sent
8	4	2	1	8	4	2	1

Figure 11.3 Session byte flag layout.

Session bit flags

As you can see in Fig. 11.3, TCP reserves the first two bits in this byte. The 6 remaining bits have separate functions. Some of those functions can occur together in the same header. In other words, they are not all mutually exclusive like the IP type of service bits.

The urgent data flag dictates that the receiving host must process some of the data in this segment before all other data. The vital data usually makes changes in the state of the session. While all stations must process urgent data first, this bit (like the push flag described below) is not on during the start-up or shutdown procedures.

If the data in the acknowledgment sequence number field is a valid number, TCP sets the valid acknowledgment bit to 1. Since the acknowledgment sequence number in this first handshake step (Fig. 11.1) segment contains hex 0s, the valid acknowledgement bit carries a value of 0.

The push request bit requests the nonurgent data in this segment be processed ASAP. In this case ASAP really means "as soon as possible after any urgent data that may be present." When a rapid response time is important, as in Telnet or FTP, TCP may set this bit to 1. Like the urgent pointer bit, the push flag is off during the start-up handshake and the shutdown procedure. The reset session flag is part of an abnormal end to a TCP session. Later in this chapter we will discuss it and ways to end a TCP session.

TCP sets the next flag (synchronize sequence number) to 1 at the start of a TCP session to indicate a request that the destination host prove synchronization with the sender's starting sequence number. Since both ends of the session (client and server) need this function, this bit will carry a value of 1 for the first TCP message in each direction. After those first two steps, TCP sets the bit to 0.

The final data sent flag occurs in the normal four-step, TCP session–termination procedure. We will discuss it with the reset session flag later in this chapter.

Sender window size

This is the number of bytes that the sender will accept from the other end of the session without requiring the other end to wait for an acknowledgment. In essence, it is the amount of receive buffer that the sender has available for storage of TCP data. Since this field (the first 2 bytes on the fourth line of hex in Fig. 11.1) is part of each TCP header, it gives a constant update of space availability.

In our example, the hex value of 10 00 is 4096 bytes (octets) decimal. If the value in this field reaches 0, the sending host indicates that it cannot accept any more TCP data. When space is available again, the value in this field indicates the amount. As we will see later in this chapter, the maximum segment size field influences the window size.

When the architectural difference between the two hosts in a session is significant, the sender window size on the less-capable system can send (throughout the session) many TCP segments with this field set at 0. The sending system is indicating congestion. A large difference in processing power often causes inefficient use of the network.

TCP checksum

The third and fourth byte on the fourth line of the Fig. 11.1 hex is the error check for the TCP header fields and any data that the header is carrying. Since TCP checks these fields, the IP header checksum need only concern itself with its own fields. Like the UDP checksum, TCP bases its checksum on a pseudo-header that it creates for this purpose but does not send on the network.

As Fig. 11.4 shows, the TCP pseudo-header includes the source and target IP addresses, the protocol number from the IP header and the total TCP segment length (including the TCP header).

Urgent pointer

If TCP sets the urgent data flag (in the session flags) to 1, this field (fifth and sixth byte on the fourth line of the Fig. 11.1 hex) contains an offset pointing to the first byte after the urgent data in the message. Otherwise, the urgent data bit is 0 and TCP keeps this field at a hex value of 00 00 as a second indicator that no urgent data is present. When the target device has congestion or other data throughput problems, it will clear the correct amount of buffer space to receive and process the urgent data message.

Option fields (type, length, and option)

When the TCP header length field has a value of 6 hex (as does the left character in the next-to-the-last byte on the third line of Fig. 11.1), it signifies

4 bytes	4 bytes	1 byte	1 byte	2 bytes
Source IP address	Target IP address	Zero	Protocol	TCP length
bf ff 0d 2a	bf ff 6d c9	00	06	14

Figure 11.4 TCP pseudo-header layout.

1 byte	1 byte	2 bytes
Type	Length	Option
02	04	05 b4

Figure 11.5 TCP option field layout.

that there are 24 bytes of TCP header. Like IPv4, any value larger than 20 bytes of header means there is an option field at the end of the TCP header. (See Fig. 11. 5 for a layout of a TCP option field.)

Currently, the only time TCP uses this field is during the start of a TCP session. While RFC 1323 has proposed two new TCP options (window scale and round-trip time measurement), the only standard TCP option is to set the maximum segment size. If TCP does not use the field, the maximum segment size (MSS) defaults to 536 bytes. That is the IPv4 maximum datagram size of 576 bytes, less the standard 20-byte IP header and the standard 20-byte TCP header.

The number of the MSS option is 02 and the option length is always 04 (1 byte each for the option number and option length and 2 bytes for the MSS itself). The remaining 2 bytes are the message portion of the option. They hold the value of the largest segment size the sender is able to receive during this session.

In this case, TCP sets the MSS (the 2 bytes to the right of the hyphen on the fourth line of the Fig. 11.1 hex) to 05 b4 hex (1460 bytes decimal). That leaves room for the 40 additional bytes of IPv4 and TCP headers to reach the Ethernet (in this example) MTU of 1500.

Maximum segment size and window sizing. There are two camps in the vendor community of what value setting this field should contain. One says that it should be as close to the local network's MTU as possible, which is why we saw 1460 used in the option field in Fig. 11.1. The other camp says that the value carried here should be an evenly divisible part of the original sender window size that does not exceed the local network's MTU. In that way a sender window size of 4096 would probably have an MSS of 1024.

The relationship between the MSS and sender window size does not end there. The two sides of the TCP session constantly update each other by providing the current sender window size. TCP sets the MSS once for the entire TCP session. Given those facts, the largest piece of data that TCP sends is always the smaller of the two values.

For example, consider A's MSS of 1024 and a current window size of 192 (in Fig. 11.6). With that limit, we can send no more than 192 TCP data bytes from B to A. If someone mistakenly configured A's maximum sender window size to

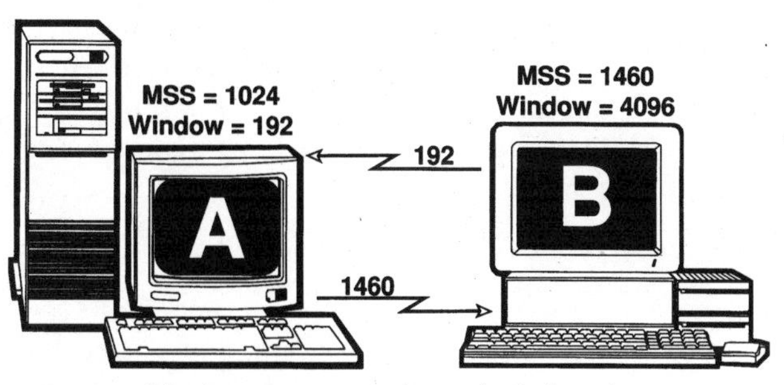

Figure 11.6 Maximum segment size and window size.

192 bytes, we could take a long time to send a large file to A, at never more than 192 bytes per segment. That same result occurs for another reason.

The silly window syndrome. The same problem of slow file transmission has been plaguing TCP users for years. The silly window syndrome typically shows up just after one of the systems fills its receive buffer. That system correctly sends a window size of 0 as does System D in Fig. 11.7. System C in step two of our example (Fig. 11.7) complies with the zero window size and sends nothing but a plain TCP header to let System D know it is still in the session. The silliness shows in step three when System D announces a window of 00 10 hex (16 bytes decimal).

System C perpetuates the silliness by sending 16 more bytes of data to fill System D's receive buffer again. As you can see in Fig. 11.7, if this continues very long, System C's throughput to System D will be so small as to make it silly to continue. Since the problem resides in both systems, both sides of the problem need to be fixed in both systems. The first fix is to have a minimum advertised window size configuration. For example, do not give System C a window size advisory smaller than one-half of System C's MSS.

	System C		System D
1		<——	Window = 00 00
2	Keep alive, 0 data	——>	
3		<——	Window = 00 10
4	16 bytes of data	——>	
5		<——	Window = 00 00
6	Keep alive, 0 data	——>	
7		<——	Window = 00 10
8	16 bytes of data	——>	

Figure 11.7 Silly window example.

The second fix is to configure each system with a low transmit window threshold. In that way, System C does not send any data until System D advertises a window size equal to at least half of System D's MSS. This way, either fix can restrict the incidence of silly window syndrome, and both would virtually eliminate the problem.

The Three-Step Handshake

We described the hex in Fig. 11.1 as the first stage of a three-step handshake. The TCP begins a session by creating a virtual connection using a three-step handshake procedure (see Figs. 11.8 to 11.10). The term *virtual connection* refers to the fact that TCP considers the session to be maintained over a single connection.

Even though each TCP segment follows a different IP route, neither end of the session knows nor cares. This flexibility lets each IP datagram use whatever path is available, while maintaining session continuity. This handshake exchange takes place when a user at a client host wants to start a session with another server host using TCP, which is outlined as follows:

1. The client (originator) sends its
   - Random port number
   - Source sequence number
   - Window size
   - Synchronization request (the synchronize request flag set to one)
   - (Usually) maximum segment size (with the TCP header length set to 6)

   to the well-known port on the server system.
2. The server synchronizes with the client sequence number and sends
   - Its own source sequence number
   - The acknowledgment sequence number (the client's source sequence number increased in increments of 1 to prove synchronization)
   - The synchronize sequence number flag set to 1
   - The valid acknowledgment flag set to 1
   - The window size
   - (Usually) a maximum segment size
   - With the TCP header length set to 6

   to the random port the client provided.

```
b8 21 34 2b 00 15 0d 6c - 00 00 00 00 00 00 60 02
10 00 56 f8 00 00 02 04 - 05 b4 00 00
```

Figure 11.8 TCP handshake, step 1 hex.

b8	2c	00	15	34	2b	3b	ae	-	0a	d0	0d	6c	00	01	60	12
05	38	1e	4a	00	00	02	04	-	02	9c	54	72				

Figure 11.9 TCP handshake, step 2 hex.

3. The client acknowledges the server's sequence number by
   - Increasing the Ack sequence number by increments of 1
   - Setting the valid acknowledgment flag to 1
   - Setting the synchronize sequence number flag to 0
   - Setting the TCP header length to 5

   and sending all these to the server's well-known port.

After this three-step process is complete, the session remains open until each end sends and acknowledges the other's final data sent flags, or one or both systems make a reset session request as a result of a failure in one of the hosts or in the underlying internetwork.

Congestion and TCP

Congestion is a state that results when too much data arrives for the amount of available buffer space and/or bandwidth. TCP cannot see the congestion directly. All it knows is that the receiver's response time is declining. If the receiver shows problems with its buffers (by setting its window size too small), TCP can identify the problem in the target host.

This dwindling window size happens when the application does not clear the TCP buffer (the sender window size) fast enough. This can happen if an application tries to support more sessions than it can handle with available memory. The system may be swapping the application, or a subroutine, in and out of virtual memory. In essence, some problem is slowing the system's ability to process the data that TCP queued in the buffer.

Congestion can also happen in a network when too many sessions use a router at the same time, there is a high error rate on a circuit, or there is some other problem along the path being used. If a router or gateway runs out of buffer space it begins to drop IP datagrams, and any services that do not use TCP's reliable transport will start to see data vanish.

b8	21	34	2b	00	15	0d	6c	-	00	01	3b	ae	0a	d1	50	10
10	00	28	27	00	00	00	00	-	00	00	00	00				

Figure 11.10 TCP handshake, step 3 hex.

b8	2c	00	15	34	2b	3b	ae	-	0c	cf	0d	6c	00	53	50	11
05	38	c8	e5	00	00	00	00	-	00	00	00	00				
b8	21	34	2b	00	15	0d	6c	-	00	53	3b	ae	0c	d0	50	10
10	00	d8	1e	00	00	00	00	-	00	00	00	00				

Figure 11.11 Final data steps 1 and 2.

One symptom is when TCP segments appear with a large sender window size but no change in the acknowledgment sequence number. This indicates that the remote end of the session has not acknowledged the sender's recent data.

Ending a TCP Virtual Connection Normally

When the application or process on one host (usually the server) has completed its need for the use of a TCP port, it has TCP send a header (with the final data sent bit set to 1) to the other end of the session (see the first hex code in Fig. 11.11). This tells the destination system that the application program at the sending system has completed its use of the connection.

The destination's TCP acknowledges receipt and processing of the final data sent bit by increasing the acknowledgment sequence number by increments of 1 (see the bottom hex code in Fig. 11.11). If the remote system is not increasing the acknowledgment sequence number, most TCP/IP vendors' software will retransmit the step 1 segment.

If the remote system is increasing the acknowledgment sequence number and the acknowledging system's application has finished its use of the TCP port, it sends a TCP header with the final data sent bit turned on (see the top hex code in Fig. 11.12). When the sender receives the final data sent bit request from the other end of the session, an acknowledgment sequence number (increasing the source sequence number by 1 as in the bottom hex code in Fig. 11.12) goes to the target and both systems close the session.

Reset Session

Another (less normal) method of closing down a session is to send a TCP header with the reset connection bit turned on. (See Fig. 11.13 for an example.) It can happen when the system tells a sending application that the system is going down because it detects an internal failure, or it may be simply the lack of communication with the other end of the session. When TCP sees a header with a reset session bit arrive, it immediately notifies the application and returns an acknowledgment to the sender. The system closes the session and the application notifies the user of the failed session's TCP port number.

b8	21	34	2b	00	15	0d	6c	-	00	53	3b	ae	0c	d0	50	11
10	00	d9	10	00	00	00	00	-	00	00	00	00				
b8	2c	00	15	34	2b	3b	ae	-	0c	d0	0d	6c	00	54	50	10
00	00	bf	d8	00	00	00	00	-	00	00	00	00				

Figure 11.12 Final data steps 3 and 4.

If the other end of the session receives no response, the session takes a time-out by sending the reset session bit three (or more) times at 30-second intervals. If TCP finishes this retransmit procedure without hearing a response from the target, it closes the session and notifies the local application.

TCP Sample Session

As in previous chapters, we provide a sample decoded TCP session (Fig. 11.14). We added the numbers, on the left, for easier identification of each segment in our discussion. Segments 1, 2, and 3 are the three-step handshake

Server		Client
Resend unacknowledged data	\	
	/	Resend last acknowledgment
Resend unacknowledged data	\	
	/	Resend last acknowledgment
Resend unacknowledged data	\	
	/	Retransmit timer expires
Retransmit timer expires	\	
	/	Transmit reset flag
Transmit reset flag	\	
	/	Transmit reset flag
Transmit reset flag	>	Receive reset flag
	/	Send acknowledgment
Transmit reset flag	>	End session locally
No response		
End session locally		

Figure 11.13 Abnormal session end.

```
1   TCP: 16515 -> smtp(25) seq: 38b33c80 ack: ——
    win: 1336 hl: 6 xsum: 0x00f1 urg: 0 flags: <SYN> mss: 668

2   TCP: smtp(25) -> 16515 seq: 0589f801 ack: 38b33c81
    win: 4096 hl: 6 xsum: 0x38e7 urg: 0 flags: <ACK><SYN> mss: 1024

3   TCP: 16515 -> smtp(25) seq: 38b33c81 ack: 0589f802
    win: 1336 hl: 5 xsum: 0x0a08 urg: 0 flags: <ACK>

4   TCP: smtp(25) -> 16515 seq: 0589f802 ack: 38b33c81
    win: 4096 hl: 5 xsum: 0x3cf4 urg: 0 flags: <ACK><PUSH>

5   TCP: 16515 -> smtp(25) seq: 38b33c81 ack: 0589f811
    win: 1321 hl: 5 xsum: 0x0a08 urg: 0 flags: <ACK>

6   TCP: 16515 -> smtp(25) seq: 38b33c81 ack: 0589f811
    win: 1336 hl: 5 xsum: 0xfa07 urg: 0 flags: <ACK><FIN>

7   TCP: smtp(25) -> 16515 seq: 0589f811 ack: 38b33c82
    win: 4096 hl: 5 xsum: 0x31fd urg: 0 flags: <ACK>

8   TCP: smtp(25) -> 16515 seq: 0589f811 ack: 38b33c82
    win: 4096 hl: 5 xsum: 0x30fd urg: 0 flags: <ACK><FIN>

9   TCP: 16515 -> smtp(25) seq: 38b33c82 ack: 0589f812
    win: 0 hl: 5 xsum: 0x310d urg: 0 flags: <ACK>
```

Figure 11.14 TCP sample data exchange.

to start the session. From the well-known port (25), we know this is a Simple Mail Transfer Protocol (e-mail) session. Segments 6, 7, 8, and 9 are the normal four-step session shutdown procedure. Notice the acknowledgment sequence number increasing without data passing from host to host. You can now see the extra overhead that reliability adds to the number of packets.

That leaves us with segments 4 and 5. By examining these two segments, we can see that the mail server (identified by the well-known port) sends data (note the PUSH flag) to the client system. We can now determine how many bytes the server sent. Simply subtract the source sequence number (0589f802) in segment 4 from the acknowledgment sequence number (0589f811) in segment 5: The answer is 15 bytes.

Chapter

12

Telnet

Overview

Telnet is both a protocol and a program. The program uses the protocol with TCP to make a virtual session with a program on a server. Telnet lets the remote user act as a dumb terminal that the network has directly attached to the server. In that way, the user can execute the applications that the target server can provide. This lets the user log into a remote host and use it almost as if it were local.

Its name comes from its normal use over dial-up TELephone NETwork connections. It offers a range of features to control the details of data displayed on the terminal. These details include

- Local or remote character echo
- Request terminal types and functions
- Full or half duplex operation

Client, Server, and Network Virtual Terminal

The central idea behind TCP/IP is to connect as many different computers and terminals as possible. Its authors designed it so that different systems can make sense to each other when they exchange data. This is also the purpose of the Telnet client and server software. Its job is to translate host commands into a standard format at the client site and must also translate those commands back to the format used by the Server.

Let's imagine we want to delete a character while in a Telnet session. From the user's perspective it is very easy. As we have seen behind the scenes, however, things are often much more complicated. Some systems use the backspace key, others use the delete key, and so on. Telnet network virtual terminal (NVT) commands step up to that challenge, translating smoothly and invisibly.

Like a universal translator, the NVT commands offer a validated method for the client and the server to send information about various terminal functions. It also ensures the receiving system will not misunderstand the instructions. To make sure there is no confusion between user data and NVT commands, the hex characters ff, called the Interpret as Command (IAC) instruction, tells the destination application that the next octet in the data message is an NVT command.

Telnet Option Negotiations

Suppose two systems want to communicate through Telnet, but one is running a version of Telnet NVT that does not provide all the options the other offers. How do we make sure that commands do not get lost, or worse yet, are incorrectly interpreted? The answer lies in the special IAC NVT commands that clients and servers use to check with each other before using an option. We have listed some of the most common commands in Fig. 12.1. A list of the negotiable Telnet options, taken from RFC 1700's Telnet Options section, is shown in Fig. 12.2.

We will see some of these options negotiated during our examination of a sample Telnet session in the sections to follow.

Purpose	Command	Hex code
The next byte is a Telnet command.	IAC	ff
Requester demands the target stop using the option.	Don't	fe
Responder confirms the option will no longer be used.	Don't	fe
Requester asks for the option	Do	fd
Responder expects the option from the target	Do	fd
Requester refuses the provide the option.	Won't	fc
Responder will no longer support the option.	Won't	fc
Requester wants to provide the option.	Will	fb
Responder confirms that it will provide the option.	Will	fb
Beginning of a multibyte subnegotiation field.	SB	fa
End of a multibyte subnegotiation field	SE	f0
Go ahead; a line turnaround in half-duplex sessions.	GA	f9

Figure 12.1 Telnet common IAC commands.

TELNET OPTIONS

The Telnet Protocol has a number of options that may be negotiated. These options are listed here. "Internet Official Protocol Standards" (STD 1) provides more detailed information.

Options	Name	References
0	Binary Transmission	[RFC856,JBP]
1	Echo	[RFC857,JBP]
2	Reconnection	[NIC50005,JBP]
3	Suppress Go Ahead	[RFC858,JBP]
4	Approx Message Size Negotiation	[ETHERNET,JBP]
5	Status	[RFC859,JBP]
6	Timing Mark	[RFC860,JBP]
7	Remote Controlled Trans and Echo	[RFC726,JBP]
8	Output Line Width	[NIC50005,JBP]
9	Output Page Size	[NIC50005,JBP]
10	Output Carriage-Return Disposition	[RFC652,JBP]
11	Output Horizontal Tab Stops	[RFC653,JBP]
12	Output Horizontal Tab Disposition	[RFC654,JBP]
13	Output Formfeed Disposition	[RFC655,JBP]
14	Output Vertical Tabstops	[RFC656,JBP]
15	Output Vertical Tab Disposition	[RFC657,JBP]
16	Output Linefeed Disposition	[RFC657,JBP]
17	Extended ASCII	[RFC698,JBP]
18	Logout	[RFC727,MRC]
19	Byte Macro	[RFC735,JBP]
20	Data Entry Terminal	[RFC1043,RFC732,JBP]
22	SUPDUP	[RFC736,RFC734,MRC]
22	SUPDUP Output	[RFC749,MRC]
23	Send Location	[RFC779,EAK1]
24	Terminal Type	[RFC1091,MS56]
25	End of Record	[RFC885,JBP]
26	TACACS User Identification	[RFC927,BA4]
27	Output Marking	[RFC933,SXS]
28	Terminal Location Number	[RFC946,RN6]
29	Telnet 3270 Regime	[RFC1041,JXR]
30	X.3 PAD	[RFC1053,SL70]
31	Negotiate About Window Size	[RFC1073,DW183]
32	Terminal Speed	[RFC1079,CLH3]
33	Remote Flow Control	[RFC1372,CLH3]
34	Linemode	[RFC1184,DB14]
35	X Display Location	[RFC1096,GM23]
36	Environment Option	[RFC1408,DB14]
37	Authentication Option	[RFC1409,DB14]
38	Encryption Option	[DB14]
39	New Environment Option	[RFC1572,DB14]
40	TN3270E	[RFC1647]
255	Extended-Options-List	[RFC861,JBP]

URL = ftp://ftp.isi.edu/in-notes/iana/assignments/telnet-options

Figure 12.2 RFC 1700 Telnet options.

```
TCP: 1363 -> telnet(23) seq: 308003c0 ack: ——
win: 2048 hl: 6 xsum: 0xb5fe urg: 0 flags: <SYN> mss: 1024

TCP: telnet(23) -> 1363 seq: 0b248a01 ack: 308003c1
win: 4096 hl: 6 xsum: 0x535d urg: 0 flags: <ACK><SYN> mss: 1460

TCP: 1363 -> telnet(23) seq: 308003c1 ack: 0b248a02
win: 2048 hl: 5 xsum: 0x247e urg: 0 flags: <ACK>
```

Figure 12.3 Telnet session, TCP startup.

Telnet sample session

In Fig. 12.3 we see a sample of a decoded Telnet session. Let's step through the entire session to understand how Telnet works.

By now you can recognize the three-step, TCP startup sequence. Looking carefully, we find the key pieces of information:

Client port: 1363	Server port: 23
Client sequence number: 308003c1	Server sequence number: 0b248a02
Client window size: 2048	Server window size: 4096
Client maximum segment: 1024	Server maximum segment: 1460

Suppress go-ahead request. The first negotiation in our sample session (see Fig. 12.4) is the client's request that the server agree to suppress go ahead (SGA). As you can see, it takes only 3 bytes of commands to accomplish this: hex ff fd 03. Notice that the client turned on the PUSH flag so the server processes the data quickly. The following sequence then takes place.

- The hex ff tells the server to interpret the following byte as a command (IAC).
- The hex fd translates to *do* (the sender, the client in this case, wants to have the server provide the following option).

```
TCP: 1363 -> telnet(23) seq: 308003c1 ack: 0b248a02
win: 2048 hl: 5 xsum: 0x1b7b urg: 0 flags: <ACK><PUSH>
Telnet: Do SGA
 data (3/3): FFFD03

TCP: telnet(23) -> 1363 seq: 0b248a02 ack: 308003c4
win: 4093 hl: 5 xsum: 0x5c73 urg: 0 flags: <ACK>
```

Figure 12.4 Telnet session, SGA negotiation.

```
TCP: telnet(23) -> 1363 seq: 0b248a02 ack: 308003c4
win: 4096 hl: 5 xsum: 0x505b urg: 0 flags: <ACK><PUSH>
Telnet: Do TermType
 data (3/3):FFFD18

TCP: 1363 -> telnet(23) seq: 308003c4 ack: 0b248a05
win: 2045 hl: 5 xsum: 0x217e urg: 0 flags: <ACK>
```

Figure 12.5 Telnet session, terminal type request.

- The hex 03 identifies the option the client is requesting as SGA. The Telnet options are listed by their decimal value in Fig. 12.2 (Telnet Options section of the current Assigned Numbers RFC 1700).

Note the PUSH flag is on in the TCP header to indicate data is present. If the server agrees, this will mean that this session is full duplex. In other words, either end will not have to tell the other to go ahead when it is ready to listen for the other end's data; they will simply send a steady stream of information to each other.

The second TCP segment here is not an agreement to do so: It is simply an acknowledgment that the server has received the 3 bytes. Notice that the 3 bytes are in the server's buffer and the server's window size is 3 bytes smaller.

Terminal type request. The server suggests the next negotiation in Fig. 12.5. It wants the client to request a terminal type for use during the session. Remember, the session is using NVT for now. The negotiation is as follows:

- The ff tells the client to interpret the following byte IAC.
- The following fd (we call *do*), means the sender (the server in this case) wants to have the client provide an option.
- The requested option is 18 hex (24 decimal), or request terminal type.

If the client agrees, it requests the terminal type that it wants to use. This lets the client request the terminal type but does not guarantee the server will support the client-requested type. The server may give no indication about that support until after the user enters a login name and password. While there is no specified default terminal type, many vendors can use VT100 or TTY (the NVT default) until further negotiations.

The second TCP segment here is the client's acknowledgment of the 3 bytes that it received from the server. Notice that the 3 bytes are in the client's buffer and the client's window size is 3 bytes smaller (2045).

Suppress go-ahead response. In the exchange shown in Fig. 12.6, the server responds to the previous client's request to suppress go ahead. By answering "will SGA," the server is saying it will provide the suppress go ahead option.

```
TCP: telnet(23) -> 1363 seq: 0b248a05 ack: 308003c4
win: 4096 hl: 5 xsum: 0x4f70 urg: 0 flags: <ACK><PUSH>
Telnet: Will SGA
 data (3/3): FFFB03

TCP: 1363 -> telnet(23) seq: 308003c4 ack: 0b248a08
win: 2042 hl: 5 xsum: 0x217e urg: 0 flags: <ACK>
```

Figure 12.6 Telnet session, suppress go-ahead response.

The rest of the session (unless one end or the other reopens this negotiation) will be at full duplex.

- The ff tells the client to interpret the following byte IAC.
- The next byte, fb (we call *will*), means the sender (the server in this case) agrees to provide the client-requested option.
- The 03 identifies the agreed option as SGA.

The second TCP segment acknowledges receipt of the 3 bytes in the option negotiation and reduces the window size by 3 more bytes.

Terminal type responses. Figure 12.7 shows the client's answer to the server about the terminal type option. The client has agreed to request a terminal type for the session, which proceeds as follows:

- The ff tells the server to interpret the following IAC.
- The following byte, fb (we call *will*), says the sender agrees to provide the option.
- The agreed option is 18 hex (24 decimal) or request terminal type.

```
TCP: 1363 -> telnet(23) seq: 308003c4 ack: 0b248a08
win: 2042 hl: 5 xsum: 0x1a66 urg: 0 flags: <ACK><PUSH>
Telnet: Will TermType
 data (3/3): FFFB18

TCP: telnet(23) -> 1363 seq: 0b248a08 ack: 308003c7
win: 4096 hl: 5 xsum: 0x555b urg: 0 flags: <ACK><PUSH>
Telnet: Sub begin: TermType Sub end
 data (6/6): FFFA1801FFF0

TCP: 1363 -> telnet(23) seq: 308003c7 ack: 0b248a0e
win: 2036 hl: 5 xsum: 0x1e7e urg: 0 flags: <ACK>
```

Figure 12.7 Telnet session, terminal type responses.

The second segment is the server's instruction on the format for the terminal type request. The system knows the terminal type will take more than 1 byte in ASCII characters. To adjust, the server specifies brackets around the option information and the ASCII characters. Then,

- The hex ff says IAC.
- The hex fa tells the client to begin the multibyte subnegotiation field.
- The option is 18 hex or request terminal type.
- The 01 is fill to make this a multibyte field.
- The hex ff says IAC.
- The hex f0 is the end bracket of a multibyte subnegotiation field.

The third TCP segment is the client's acknowledgment of the 6 bytes from the server. As in previous segments, the client reduces the window size by the 6-byte data size.

Requesting vt220. In Fig. 12.8 the client continues with the terminal type request to use vt220 for this session with the server. Remember that this is only a request. The following sequence then takes place.

- The ff tells the server to interpret the following byte IAC.
- The fa tells the server to begin a multibyte subnegotiation field.
- The option is 18 hex or request terminal type.
- The 01 is spacing between the option ID and the option value.
- The option value (in ASCII characters) is vt220.
- Again, the ff means IAC.
- The f0 finishes this data as the end bracket of a multibyte subnegotiation field.

The second TCP segment is the server's acknowledgment of the 11 bytes that it received from the client. Notice that the server has processed all the bytes from its buffer and sends its current window size as 4096.

```
TCP: 1363 -> telnet(23) seq: 308003c7 ack: 0b248a0e
win: 2036 hl: 5 xsum: 0x1766 urg: 0 flags: <ACK><PUSH>
Telnet: Sub begin: TermType Sub end
 data (11/11): FFFA1801vt220FFF0

TCP: telnet(23) -> 1363 seq: 0b248a0e ack: 308003d2
win: 4096 hl: 5 xsum: 0x3f73 urg: 0 flags: <ACK>
```

Figure 12.8 Telnet session, requesting vt220.

```
TCP: telnet(23) -> 1363 seq: 0b248a0e ack: 308003d2
win: 4096 hl: 5 xsum: 0xda33 urg: 0 flags: <ACK><PUSH>
Telnet: Will Echo Do Echo
 data (46/53): FFFB01FFFD01130A130AUNIX(r) System V Release 4.0
(unix)13

TCP: 1363 -> telnet(23) seq: 308003d2 ack: 0b248a43
win: 1983 hl: 5 xsum: 0x077e urg: 0 flags: <ACK>

TCP: 1363 -> telnet(23) seq: 308003d2 ack: 0b248a43
win: 1983 hl: 5 xsum: 0xfd7c urg: 0 flags: <ACK><PUSH>
Telnet: Won't Echo Do Echo
 data (6/6): FFFC01FFFD01

TCP: telnet(23) -> 1363 seq: 0b248a43 ack: 308003d8
win: 4096 hl: 5 xsum: 0xff71 urg: 0 flags: <ACK><PUSH>
Telnet: Don't Echo
 data (3/3): FFFE01

TCP: 1363 -> telnet(23) seq: 308003d8 ack: 0b248a46
win: 1980 hl: 5 xsum: 0x047e urg: 0 flags: <ACK>
```

Figure 12.9 Telnet session, echo negotiations.

Echo negotiations. In previous negotiations, we have seen requests and responses separated by other network traffic (as is typical in a normal network). In Fig. 12.9 we see remote echo negotiated by TCP segments. In the first segment, the server says it

- Will echo (ff fb 01) the client's data
- Asks the client to do echo (ff fd 01) for it as well
- Contains screen control characters in the 13 0A entry that position the UNIX System V Release 4.0 and other data on the client's screen.

The second TCP segment acknowledges those 53 bytes of data and cuts the window size to show that they are in the buffer along with 12 previously received bytes.

In the third segment, the client says it will not echo the server (ff fc 01) though it expects the server to echo its data (ff fd 01). Segment 4 shows the server agreeing that it will no longer expect this option. The last segment is a plain acknowledgment of the 3 bytes in segment 4.

Login. The two systems have completed the setup negotiations for this session. All that we discussed in the last few sections frequently occurs in a few seconds. Now we can get down to the working part of the Telnet session. In Fig. 12.10 notice that the server's window size remains at its original 4096.

```
TCP: telnet(23) -> 1363 seq: 0b248a46 ack: 308003d8
win: 4096 hl: 5 xsum: 0xe210 urg: 0 flags: <ACK><PUSH>
 data (7/7): login:

TCP: 1363 -> telnet(23) seq: 308003d8 ack: 0b248a4d
win: 2041 hl: 5 xsum: 0xcc7d urg: 0 flags: <ACK>
```

Figure 12.10 Telnet session, login.

In the first segment, the server sends the client's user screen the instruction to provide a login or user ID. Does this mean that we, simply by using a protocol analyzer, can see the user IDs as they pass on the network segment we are watching? Absolutely!

Now we learn the bad news: Network packets are open to anyone with protocol analysis capability unless the participants encrypt the data. The good news: protocol analysis software and hardware can only capture packets when they have a connection to the same media (cable).

Segment 2 acknowledges receipt of those 7 bytes and reduces the window size (that had returned to its original size of 2048) to 2041.

User ID. In response to the server's request for a user ID, designated by the login: prompt we just saw, the user enters his ID in Fig. 12.11. It may seem, at first glance, that Telnet's user interface is character based. In reality, most vendor systems base the default on a periodic sweep of the keyboard buffer to collect and transmit the character entered since the last sweep.

We can configure most Telnet clients to wait for a line feed before sending the typed characters. This is a more efficient use of the network and lets the

```
TCP: 1363 -> telnet(23) seq: 308003d8 ack: 0b248a4d
win: 2048 hl: 5 xsum: 0xbc17 urg: 0 flags: <ACK><PUSH>
 data (1/1): f

TCP: 1363 -> telnet(23) seq: 308003d9 ack: 0b248a4d
win: 2048 hl: 5 xsum: 0xbb0b urg: 0 flags: <ACK><PUSH>
 data (2/2): re

TCP: 1363 -> telnet(23) seq: 308003db ack: 0b248a4d
win: 2048 hl: 5 xsum: 0xb919 urg: 0 flags: <ACK><PUSH>
 data (1/1): d

TCP: telnet(23) -> 1363 seq: 0b248a4d ack: 308003dc
win: 4096 hl: 5 xsum: 0xf972 urg: 0 flags: <ACK>
```

Figure 12.11 Telnet session, user ID.

```
TCP: 1363 -> telnet(23) seq: 308003dc ack: 0b248a4d
win: 2048 hl: 5 xsum: 0xb770 urg: 0 flags: <ACK><PUSH>
 data (2/2): 1301

TCP: telnet(23) -> 1363 seq: 0b248a4d ack: 308003de
win: 4096 hl: 5 xsum: 0x3d8b urg: 0 flags: <ACK><PUSH>
 data (9/9): Password:

TCP: 1363 -> telnet(23) seq: 308003de ack: 0b248a56
win: 2039 hl: 5 xsum: 0xbf7d urg: 0 flags: <ACK>
```

Figure 12.12 Telnet session, password prompt.

user make changes in the current entry before sending it. It would not be a good idea to make this a requirement since some older or more limited devices, such as printer-based terminals, could not use this method. Above all, TCP/IP's goal is to offer flexible, universal connectivity.

The last segment in this figure is an acknowledgment of the previous three segments' data.

Password. In the first segment of Fig. 12.12, the Client sends screen control characters to position the next user prompt. The second segment requests that the client's user provide a password. Does this mean that we, simply by using a protocol analyzer, can also see the users' passwords as they pass on the network segment we are watching? At the risk of repeating ourselves, Absolutely! Maybe you thought native TCP/IP was a secure way of transporting data. Let us put that notion to rest. It is not secure at all, by itself. In the next chapter we will look at ways to add some security to an otherwise wide open protocol.

Segment 2 acknowledges receipt of those 9 bytes and reduces the window size from the first segment's size of 2048 to 2039.

Session password. In response to the server's request that the client provide a password for the previously entered user ID, the user types in "abc123." It is clear in this session (see Fig. 12.13), since the server agreed to echo the client's data, that when the user enters the password, the server does not echo it back to the client.

Telnet does not normally display the user's password on the client's screen. This, at least, provides protection from "shoulder surfers" who like to watch as someone logs in to the system in hopes of getting a user ID and matching password for future use. The best guideline is to create passwords that users do not need to write down *anywhere.*

The last segment is a plain acknowledgment of the previous five segments'

```
TCP: 1363 -> telnet(23) seq: 308003de ack: 0b248a56
win: 2048 hl: 5 xsum: 0xad1c urg: 0 flags: <ACK><PUSH>
 data (1/1): a

TCP: 1363 -> telnet(23) seq: 308003df ack: 0b248a56
win: 2048 hl: 5 xsum: 0xac1b urg: 0 flags: <ACK><PUSH>
 data (1/1): b

TCP: 1363 -> telnet(23) seq: 308003e0 ack: 0b248a56
win: 2048 hl: 5 xsum: 0xab1a urg: 0 flags: <ACK><PUSH>
 data (2/2): c1

TCP: 1363 -> telnet(23) seq: 308003e2 ack: 0b248a56
win: 2048 hl: 5 xsum: 0xa94b urg: 0 flags: <ACK><PUSH>
 data (1/1): 2

TCP: 1363 -> telnet(23) seq: 308003e3 ack: 0b248a56
win: 2048 hl: 5 xsum: 0xa84a urg: 0 flags: <ACK><PUSH>
 data (1/1): 3

TCP: telnet(23) -> 1363 seq: 0b248a56 ack: 308003e4
win: 4096 hl: 5 xsum: 0xe872 urg: 0 flags: <ACK>
```

Figure 12.13 Telnet session, session password.

6 bytes of data. Each vendor's software may acknowledge individual segments or multiple packets of data (as the server does in Fig. 12.13).

Screen control and banner. The first segment in Fig. 12.14 contains screen control characters to position the cursor or data on the user's screen. The decimal 13 (hex 0d) indicates a carriage return (CR) that moves the cursor to the left side of the screen (first position). The decimal 01 designates the start of a heading (SOH). The hex 0A (not translated to decimal by this particular analyzer) represents a line feed (LF) that moves to the next line. Most systems combine the LF with the CR (so we see 130A in Fig. 12.14).

The second segment is the beginning of the banner that the server presents to users as they log on. It is typically generic for all users, though some vendors allow for user- or group-specific banners. Note the embedded CR and LF: 13 0A. Segment 3 is a continuation of the banner and segment 4 starts a new screen line.

Again, the last segment acknowledges receipt of the 317 bytes of data and indicates that the client has processed all but the last two, since the window size is now 2 bytes less than the original size of 2048.

Session prompt. After all the packets and segments we have reviewed, it all comes down to the system prompt (see Fig. 12.15). The "$" and its immediate

```
TCP: 1363 -> telnet(23) seq: 308003e4 ack: 0b248a56
win: 2048 hl: 5 xsum: 0xa670 urg: 0 flags: <ACK><PUSH>
 data (4/4): 1301130A

TCP: telnet(23) -> 1363 seq: 0b248a58 ack: 308003e8
win: 4096 hl: 5 xsum: 0x8328 urg: 0 flags: <ACK><PUSH>
 data (46/269): ESIX System V/386 Release 4.0 Version A130ACopyr

TCP: telnet(23) -> 1363 seq: 0b248b65 ack: 308003e8
win: 4096 hl: 5 xsum: 0x06ed urg: 0 flags: <ACK><PUSH>
 data (44/44): Last login: Fri May 29 17:49:37 on console130A

TCP: telnet(23) -> 1363 seq: 0b248b91 ack: 308003e8
win: 4096 hl: 5 xsum: 0x9764 urg: 0 flags: <ACK><PUSH>
 data (2/2): 130A

TCP: 1363 -> telnet(23) seq: 308003e8 ack: 0b248b93
win: 2046 hl: 5 xsum: 0x737c urg: 0 flags: <ACK>
```

Figure 12.14 Telnet session, screen control characters and banner.

trailing space, are typical for a UNIX server. They indicate that the server is ready for the remote user's instructions.

From this point, the user could have a Telnet session with another host, FTP to another server, connect to a mail server, or perform any application that the prompting server has the capability (and allows this user) to perform. This can be very handy in a large network.

Of course, there is another edge to this flexible sword. When we leave this server for another location, the next receiving server does not know where we originated. It looks as if we are only coming in from the last server. As a possible security issue, this is also the reason that the Internet overseers tend to be so cooperative in helping track back remote users by any footprints they leave. We will look into this in the next chapter.

```
TCP: telnet(23) -> 1363 seq: 0b248b93 ack: 308003e8
win: 4096 hl: 5 xsum: 0x7f4d urg: 0 flags: <ACK><PUSH>
 data (2/2): $

TCP: 1363 -> telnet(23) seq: 308003e8 ack: 0b248b95
win: 2046 hl: 5 xsum: 0x717c urg: 0 flags: <ACK>
```

Figure 12.15 Telnet session, session prompt.

```
TCP: 1363 -> telnet(23) seq: 308003e8 ack: 0b248b95
win: 2048 hl: 5 xsum: 0x6617 urg: 0 flags: <ACK><PUSH>
 data (1/1): e

TCP: 1363 -> telnet(23) seq: 308003e9 ack: 0b248b95
win: 2048 hl: 5 xsum: 0x6504 urg: 0 flags: <ACK><PUSH>
 data (2/2): xi

TCP: 1363 -> telnet(23) seq: 308003eb ack: 0b248b95
win: 2048 hl: 5 xsum: 0x6308 urg: 0 flags: <ACK><PUSH>
 data (3/3): t1301

TCP: telnet(23) -> 1363 seq: 0b248b95 ack: 308003ee
win: 4096 hl: 5 xsum: 0xa371 urg: 0 flags: <ACK>
```

Figure 12.16 Telnet session, user ends the session.

User ends the session. The command to stop the Telnet session is EXIT (see Fig. 12.16). Other commands, such as BYE or QUIT, can work on some systems, but EXIT is practically universal. If the two systems did not fully establish the Telnet session, EXIT probably will not work to end the nonexistent session. We are not talking with the server's Telnet application but the client's Telnet software. In that case, many vendors support QUIT as a more accepted command.

Notice that the third segment carries the CR (13) and SOH (01) characters as well. It is not unusual for that to be CR LF instead. The last segment in Fig. 12.16 also signifies that the server has received the 6 bytes, processed them, and returned its window size to its initial value.

TCP ends the session. After the instructions from the user to end the session, the server ends Telnet and, in the first segment in Fig. 12.17, begins

```
TCP: telnet(23) -> 1363 seq: 0b248b95 ack: 308003ee
win: 4096 hl: 5 xsum: 0xa071 urg: 0 flags: <ACK><FIN>

TCP: 1363 -> telnet(23) seq: 308003ee ack: 0b248b96
win: 0 hl: 5 xsum: 0xa081 urg: 0 flags: <ACK>

TCP: 1363 -> telnet(23) seq: 308003ee ack: 0b248b96
win: 0 hl: 5 xsum: 0x9f81 urg: 0 flags: <ACK><FIN>

TCP: telnet(23) -> 1363 seq: 0b248b96 ack: 308003ef
win: 4096 hl: 5 xsum: 0x9f71 urg: 0 flags: <ACK>
```

Figure 12.17 Telnet session, TCP ends the session.

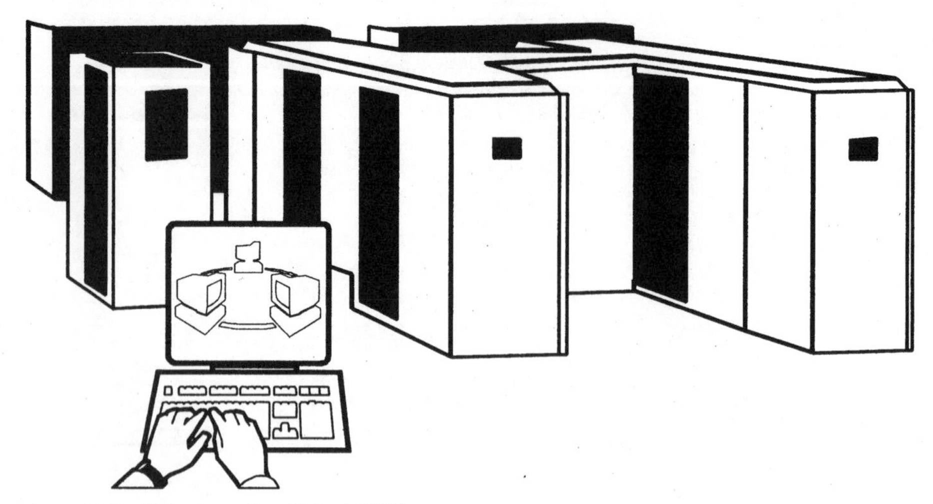

Figure 12.18 Telnet session, Telnet 3270.

the process of ending the TCP session. As we saw in our discussion of TCP, this is a four-step process. In the first segment, the server sets the final data sent (FIN) bit to indicate that it has sent (or is now sending) the last data that it has for this TCP session. In the second segment the client TCP acknowledges the FIN bit by increasing the server's sequence number by an increment of 1.

Notice that the client returns a window size of 0. Once the client receives and processes the final data sent bit, it frees the reserved buffer space for other uses.

In the next segment, the client sends its FIN bit. The last segment has the server's acknowledgment and adds to the client's sequence number in increments to reflect that it recognizes the end of the TCP session. As we said before, some vendors combine the functions of segments 2 and 3 into one segment. In this case segment 2 was not necessary.

Telnet 3270

Though many of the IBM hosts on the Internet operate with the regular version of Telnet, at times we need to use Telnet to communicate with a host that only uses IBM's 3270 family of terminal products (see Fig. 12.18). RFCs 1647, 1646, 1576, 1205, and 1041 lay out the methods and rules that Telnet uses to support these (and the 5250) products.

Negotiation code 29 identifies 3270 negotiations. Telnet requires that we make the next adjustments to the negotiation rules. To support 3270, both client and server must be able to

- Exchange binary data (negotiation code 0)
- Use well-defined delimiters in outbound and inbound data streams (negotiation code 25)
- Agree with each other on what type of terminal the session will use (negotiation code 24)

Chapter

13

Network Security

Security Threats

Just as a reality check, if you have a computer, a network, a modem, or any other network resource such as people, you are at risk. You do not have to have information of interest to others. Security risks can start with mild curiosity from hackers about your systems and how they work. By accepted definition, hackers are curious.

The problems come at the other end of the scale. Crackers (hackers' evil twins), for their own personal or political reasons, take control of your organization's systems from outside. You can quickly tell which you have by the category of threat you face.

While security threats fall into many categories, in this book we will place them in the following four:

- *Availability attacks.* Threaten network resource access. This includes disability or destruction of hardware resources whether from natural or human intervention.
- *Confidentiality threats.* Are from unauthorized access of stored or transmitted data. This area includes wiretapping or Trojan horse programs to intercept information.
- *Integrity risks.* Go beyond the confidentiality threat by tampering with a network resource. Included here are modifying password files and altering programs.
- *Authenticity attacks.* Are another step beyond the confidentiality threat. Attackers add false information to network resources.

Availability attacks

While all of these threats are valid areas of concern, there is only so much we can do to prepare for a disaster that destroys circuits, systems, or worse: staff. In that case (and for the purposes of this book) we will leave that to redundancy and disaster recovery planning. Let's focus on those threats that come from areas that we can control.

An organization must keep the physical access to its computers as open as possible to those with authorized access. To protect those resources, many organizations institute policies that limit physical access to the facilities that house these systems. While this can keep intruders from direct access to the internal systems, there is still the threat of connecting into these systems from outside the facilities by modem or some other telecommunications method. Here we need to be careful about being so secure that authorized users lose availability while they are away from the facility.

Confidentiality threats

The first problem that comes to most minds when we raise the issue of security is hacking. While susceptibility to hacking is the weakest point in the security of most networks, it is also one of the most documented and so has many preventative procedures.

Other confidentiality problem areas include

- *Dumpster diving.* Hackers continue to find it helpful to dig through organizational trash to find user IDs and passwords.
- *Trojan horse program.* These programs carry a hidden program that collects user IDs and passwords on the local system.
- *Human nature.* This is the toughest to solve. People like passwords that are easy to remember and tend to write them down in predictable places.
- *Protocol analyzer.* If the wrong people get access to the raw data flowing on a network, they also get access (as we have seen) to the logins and passwords.

Integrity risks

Once a hacker has compromised the system, she or he may just want to see what we have in it. Usually the hacker has no intention of changing anything. However, if that hacker is a disgruntled worker, a person working to "improve the plight of the masses," someone who is looking to improve his or her own situation, or just someone with a twisted mind, you can be sure that she or he wants to make some changes (and maybe a lot more). Those changes can be to modify a grade achieved in a course, alter a program to change parameters to provide chargeable features without cost, or increase the monetary value in an account. As you can see, these are basically "something for nothing" motives.

Authenticity attacks

If the attacker has compromised the system, that person may be a cracker (someone intent on causing problems for the sake of causing problems or to brag about being in the system). The alternatives are no better: a representative of an opposing organization, a disgruntled former employee, someone looking for capabilities to sell, or worse.

These people may want to add false information or one of the following:

Trap door. An undocumented entrance into a system or network

Logic bomb. A malicious instruction set that triggers on some future event

Worm. A program that copies itself onto other systems across a network

Virus. A set of instructions that inserts copies of itself into other programs

Bacterium. A program that replicates itself and consumes resource on a host

Other types of computer crime

A number of other potential computer crimes to watch out for include

Scavenging. An external client accesses a large network to steal data from competitors on the same network.

Leakage. Interested people can pick up data that radiates from the cables in a network with just a limited amount of "James Bond" equipment.

Piggybacking. You meet a person with both arms full of tapes, printouts, and supplies waiting to enter the secure door to the computer department. As a kind person, you help them by using your key card or combination without first verifying theirs.

Impersonation. Security administrators have caught hackers impersonating magazine writers, claiming to be gathering research for an article on the target network. Hackers can use one terminal to impersonate another secure or privileged terminal by being on the same line.

Salami slicing. It is often easier to take small bits of information at a time and not get caught than to try to grab a lot of data and get nabbed in the act.

Simulation and modeling. By writing an electronic simulation of a program and substituting it for the real thing, a programmer can gather any data on the network. That can include passwords, credit card numbers, and other sensitive information.

Security Procedures

Security procedures are any combination of hardware and software features and operating or management methods that neutralize the security threats.

Procedures	Availability	Confidentiality	Integrity	Authenticity
Access control	Y	Y		
Authentication	Y			Y
Data integrity			Y	
Digital signature			Y	Y
Encryption		Y	Y	Y
Routing control		Y		
Traffic padding		Y		

Figure 13.1 Threat categories and procedures

As you can see in Fig. 13.1, each procedure targets one or more categories of threat to provide protection.

We talked in this chapter about the threats to integrity and authenticity as being the result of breached confidentiality or availability. In other words, someone gained physical (availability) or virtual (confidentiality) access before compromising integrity or authenticity.

In the next few sections we will look at each procedure, what it is, and offer some generic examples of security use. Following that, we will look at steps that we can take to implement these procedures.

Access control

The most common area of security compromise is access control. That can come from physical access to a facility or virtual access to the network and its resources. Blocking physical access to the facility sounds very easy. However, in a world where we learn to be polite, it can be a challenge. Piggybacking and impersonation can get a hacker into your facility and from there it only takes a little effort to jot down one or more user IDs and passwords.

We have found passwords taped to the bottom of keyboards, telephones, and mice as well as on the slide-out writing shelf of a desk. Some are bold enough to write it on a sticky note and attach it to their monitor. Others are less obvious when they write it on the metal strip that supports suspended or soundproofed ceiling tiles. Once a hacker has access to a system, through copying the ID and password or by being left alone near a terminal, it only takes a few minutes to set up a new server account for later remote access.

Passwords, questions, key cards, voice recognition, fingerprints, smart cards, handwriting and retinal scans protect physical access. Different passwords (such as those from a smart card), account numbers, questions, and user IDs are some of the various virtual access protections. The best protection is worthless without well trained, security-conscious users' participation. The chain is only as strong as the weakest link.

Passwords

The best access security uses Smart Cards to provide the user password so there is no need to write it down. But since many implementations of access control security still hinge so strongly on passwords, let's look at some of the weak areas.

The first, and weakest, area is that passwords (and Smart Cards) rely on the discretion of the people using them. In the case of Smart Cards, some people will loan them to a co-worker, leave them lying around, or just lose them.

With the use of passwords, the challenge goes further. We need a password that is long enough and complex enough to make it work while keeping it easy to remember so that users do not feel the need to write it down. Of course, we should not use names, birthdays, or common words like "xyzzy," "Matilda," or "open sesame."

As a rule of thumb, the password should be eight or more characters long, use keyboard characters, and include at least one nonalphanumeric and at least one shifted key. Tests have shown that whimsical words or phrases make it easier for the user to remember the password. You can see some in Fig. 13.2. We may see them as custom license plates (as long as the plate belongs to someone else). A popular password building alternative is to use four-letter words and combine them with some numbers, for example: $47HensEars or LostHair39.

Another good way of building passwords is to use the first one or two letters of each word in phrases or a sentence. For example: "Off the keyboard, Over the bridge, Through the router, Nothing but net!" gives us a password of OtkOtbTtrNbn!. Substituting the number zero for the letter Os while keeping the shift changes it to)tk)tbTtrNbn!

Authentication

The checksum and CRC are forms of authentication of the data that is flowing over a network. They ensure that nothing in the network corrupts the data that a system sent. They do not authenticate the sending user or system. At the transport layer (and below) it is possible to reproduce the authentication.

To move beyond that level of authentication the National Computer Security Center (NCSC) established multiple security levels. From the top down they are

A1 Mandatory control through a proven mathematical model

B3 Mandatory control using a viable mathematical model

KFrog&MsPigE	gr8Ba!sOfyr	Hey!10sNe1	1beLL2Nsr

Figure 13.2 Sample passwords.

B2	Mandatory control by way of a guaranteed path between user and secure system
B1	Mandatory control through Department of Defense clearances
C2	Discretionary control, by way of a user ID, with password and audit capability
C1	Discretionary control with a user or group ID
D	Not secure

Data integrity

Data integrity uses many of the same methods as authentication. Authentication ensures that the data is from a known and trusted source to a known destination. Data integrity adds the assurance that no one (or nothing) changed the data en route. It is possible to provide integrity protection without authentication.

For example, extortionists must prove that they have something worth the requested price (some information or picture as proof). There is no source authentication because the ransom note has no return address. To protect the amount and location of the drop from other thieves, everyone involved must keep the message secret. Integrity would include protecting the proof, amount, and location of the payment to be sure that only the extortionists profit.

It does not take a rocket scientist to see that when there is no data corruption, there is little reason to have data integrity without authentication. Anyone could pick the data off the line and substitute their own data.

As we see in the security procedures table (Fig. 13.1), two of the main ways to enforce data integrity are digital signature and encryption.

Digital signature. With the growing level of commerce being transacted over the Internet, the Digital Signature Standard (DSS) is a technique that will continue to expand. The software bases a digital signature on the bits of the message combined with a secret key to create a message authentication code (MAC).

While authentication uses a secret key for verification, the digital signature uses a public key encryption. This way, digital signatures can authenticate as well as provide data integrity. The authentication should prove very helpful in financial transactions across the Internet. By adding third-party verification of digital signatures through a process known as notarization, the third party certifies the signature and so the identity of the sender. That way, we can bill financial transactions more accurately (to the correct user or organization) with an extremely limited possibility of falsification.

A variation of this procedure is currently in use in many of the Smart Card implementations that share the load by doing some of the mathematical work before the actual login. The message could carry an encoded, though visible, digital signature.

Encryption. Of all the security procedures, the best-known and most widely used is encryption. If we could implement one security procedure, most of us would choose this one. Encryption translates a string of data into another string by using a mathematical formula, or a translation table. The result is data that is meaningless to anyone who does not have the decoding piece.

Some of the common methods include

The Data Encryption Standard (DES), which is available in four modes:

- Electronic Code Book (ECB) performs a simple, block-at-a-time encryption and so cannot recognize added or deleted blocks
- Cipher Block Chaining (CBC) encrypts each block along with the previous block's encrypted output to prevent added or deleted blocks
- Cipher Feedback (CFB) recycles the output into process along with the new input, which requires the decrypt process to start at a predetermined position ahead of the desired data
- Output Feedback (OFB) is rarely used as it feeds the output into the process and creates very long key bit strings

A patented public key system such as

- Diffie-Hellman key exchange protocol
- EIGamal public key encryption with digital signatures
- The RSA (Rivest, Shamir, Adleman) public key encryption algorithm with digital signatures
- The Digital Signature Algorithm (DSA), which only supports digital signatures though it has encryption capability

A private key system such as

- Kerberos from the MIT, IBM, and DEC Project Athena operation
- The International Data Encryption Algorithm (IDEA), which was developed in Switzerland

A combination private key and public key program such as Pretty Good Privacy (PGP), developed by Phil Zimmermann, uses IDEA as its private key encryption algorithm and RSA for its public key encryption (in Version 2). It was released on the Internet and improved by many programmers before Zimmermann turned distribution over to MIT. It is available via anonymous FTP from net-dist.mit.edu.

Routing control

There are two primary methods to attack a network resource through routing. Both of these methods involve the attacking machine impersonating a real system in the network. This usually happens when the system's real user is away from the office and so does not notice the change in connectivity.

The attacker can implement loose source routing to tell the routers what path to use to reach the resource (server). RFC 1122 says the resource must use the inverse route to return to the source, no matter how strange it may be. The hacker can indicate the route back to a falsely addressed system outside the organization's network. Most router vendors allow configurations that will reject packets that contain this option.

The common alternative is to send the router instructions to load a bogus host-specific routing table entry. This is especially easy to do if the attacking system is closer to the resource than the real source. To restrict this function, administrators can choose to use RIP Version 2 or OSPF, adding authentication of the routing information source.

Traffic padding

If a security attacker can get access to the traffic flow on a network, it is possible to see which systems and users are sending and receiving the encrypted data. While the sender encrypts the data portion, the network must pass the headers "in the clear" to get the traffic across bridges and through routers to its destination.

The only protection that has worked here is to encrypt everything above the network access layer. That requires the great expense of having the same encryption function in every system on the network. This costs in two ways: money for the product and time for the encrypting and decrypting of each datagram.

By monitoring the volume of traffic, an attacker can also determine how often the "partners" communicate. Knowing that means she can know when something important like an international incident requires a sudden change in the amount of traffic being sent. To protect a network from this eavesdropping, many organizations have turned to traffic padding. This requires sending a continuous stream of random data (or cyphertext) so there is only a minor increase in volume in the most extreme situations. Traffic padding and encryption also prevents inside staff from passing data across a network that they know is being monitored by an outsider. The insider could otherwise use a code to identify important information that an outsider may want to capture.

Security servers

When most people talk about security and servers, the conversation frequently comes down to three areas: firewalls, proxy servers, and SOCKS servers.

Firewalls stand in the middle between an organization's network and the rest of the world to control the traffic that can go in either direction. They keep internal users from freely exchanging information with outsiders (users not on the organization's network). They also keep outsiders from compromising or attacking the company network.

In Fig. 13.3 we see an outside network on the left connecting to Router A, which acts as the outer barrier. In that capacity, it filters out all traffic except

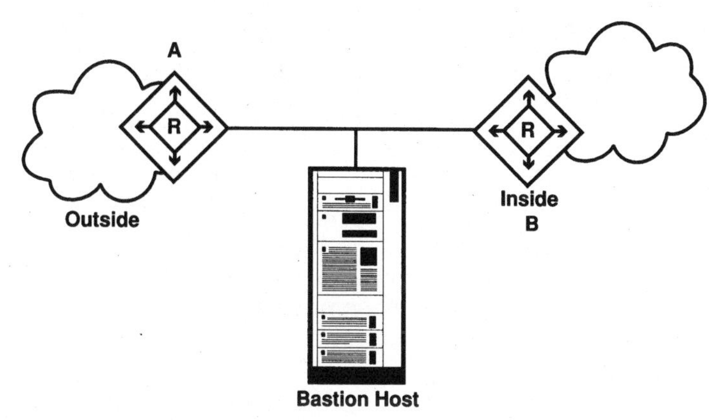

Figure 13.3 Firewall implementation.

datagrams with a target of the Bastion Host. Router B is the inner barrier which isolates organizational network users from outsiders by filtering out datagrams that do not originate in the Bastion Host.

The Bastion Host is a proxy server application gateway. It acts as the client for outside information. For example, someone inside the organization's network (on the right in Fig. 13.3) must first contact the Bastion Host to request a file that is available via FTP. The Bastion Host will call outside and retrieve the file. Once the Bastion Host has the file, the internal user can retrieve it.

Note that the safety of the firewall operation depends on the safety Bastion Host. If an outsider can access the Bastion Host, she can exploit the Bastion Host's operating system or applications to access the inside network.

Sockets Secure (SOCKS) servers intercept and redirect all IP datagrams at the firewall. To run a SOCKS server, the user stations must use the client software that works with the SOCKS server software.

Computer Security Organizations

While many entrepreneurs have begun offering security support, the following groups have been doing it better for a longer time.

FIRST

The Forum of Incident Response and Security Teams (FIRST) is a global coalition of government and private sector organizations. Its Uniform Resource Locator (URL) is http://www.first.org. Its mission is to

- Provide members with technical information, tools, methods, assistance, and guidance

- Coordinate proactive liaison activities and analytical support
- Encourage quality product and service development
- Improve national and international information security for government, private industry, academia, and the private individual

CERT Coordination Center

The CERT (Computer Emergency Response Team) Coordination Center at Carnegie Mellon University is a team of anonymous experts who help network managers deal with security attacks (see Fig. 13.4). Its URL is http://www.cert.org.

Much of its fame comes from the CERT advisories that pass information about potential security threats, along with tools and procedures that can eliminate or limit their effect. The CERT archive site is available through its web site (as well as by anonymous FTP at ftp.cert.org). It contains the files seen in Fig. 13.5.

COAST

Computer Operations, Audit and Security Technology (COAST) is a multiple project, multiple investigator laboratory in computer security research at Purdue University. It functions with close ties to researchers and engineers in major companies and government agencies. Its URL is http://www.cs.purdue.edu/coast.

> If you believe that your system has been compromised, contact the CERT Coordination Center or your representative in FIRST (Forum of Incident Response and Security Teams).
>
> Internet e-mail: cert@cert.org
> Telephone: 412-268-7090 (24-hour hotline)
>
> CERT personnel answer the phone 7:30 a.m. to 6:00 p.m. EST (GMT-5)/EDT (GMT-4); on call for emergencies during other hours.
>
> CERT Coordination Center
> Software Engineering Institute
> Carnegie Mellon University
> Pittsburgh, PA 15213-3890
>
> Past advisories, information about FIRST representatives, and other information related to computer security are available for anonymous FTP from cert.org (192.88.209.5).

Figure 13.4 Computer emergency response team.

tech_tips/packet_filtering	Guidance on which ports should be blocked
pub/cert_advisories	Guidance on dealing with security threats
pub/tools	Tools contributed by the Internet community
pub/tech_tips	Technical guidance on security procedures
pub/tools/crack and cracklib	Password-cracking program by Alec Muffett to find out how secure your system really is
pub/tools/cops	Auditing package to spot configuration errors

Figure 13.5 CERT files.

The COAST effort builds on a record of innovative success. Associates designed and developed many of the computer security tools and techniques. Adding long-term collaboration with commercial and government organizations brings them to their high expertise. Purdue also maintains an excellent World Wide Web site with information about these organizations and much more. It is at URL http://www.cs.purdue.edu/homes/spaf/hotlists/csec-top.html.

Network Scanning

Help (and potential heartburn) is also available with network scanning software. These packages help you assess your security by scanning your network for known types of security flaws. The heartburn comes when a hacker or cracker uses one of them to scan your network. You can bet you will not get the results, but you will learn them before long.

The best known of the programs is SATAN (Security Administrator for Analyzing Networks). Though the authors (Dan Farmer and Wietse Venema) had strong credentials in network security, the software faces controversy for its external nature. A hacker can run SATAN against any system, not just one she can access.

It is interesting that SATAN uses a web browser to report the results of the scan and that it is so large that most attack response teams can quickly spot a SATAN scan. A copy of the source is available from ftp://ftp.win.tue.nl/pub/security/satan.tar.Z.

Chapter

14

File Transfer Protocol

Multiple Sessions

The File Transfer Protocol's purpose is to handle file transfers, in a variety of formats, from one host to another. The FTP can maintain multiple sessions with multiple hosts and move multiple files during each of these sessions.

Speaking of multiples, FTP uses two TCP sessions for each file transfer. We could call the first session the FTP control session. It sends commands that start, stop, and control the session. Figure 14.1 shows the client random (R) port calling the TCP well-known port 21 to establish that session.

The second session (between another random client port and well-known TCP port 20) acts as the transfer session and moves the actual file between hosts. This session sets up with a different client random port (R_1 in Fig. 14.1) for each file that the user wants transferred. That second session is only for transferring the one file and ends after the transfer is complete.

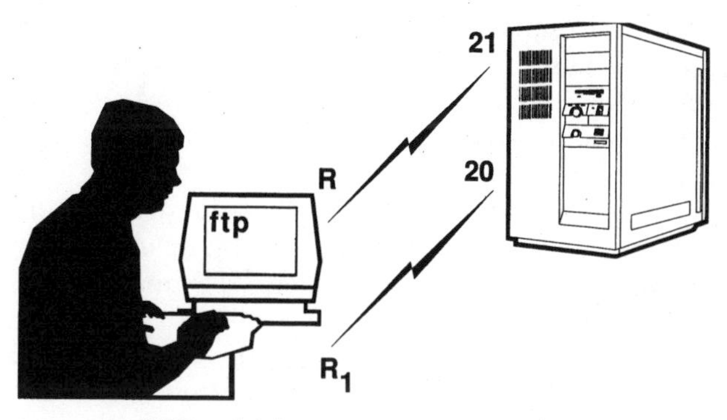

Figure 14.1 FTP multiple sessions.

UNIX/DOS	FTP	English translation
(login)	USER	User ID
(password)	PASS	Password
cd	CWD	Change working directory
ls *or* dir	NLST	Name list (from directory)
binary *or* ascii	TYPE	Set the type of transfer
get *or* mget	PORT	File destination port
get *or* mget	RETR	Retrieve file to local host
pwd *or* cd	PWD	Print working directory
mkdir	MKD	Make directory
put *or* mput	STOR	Store file on remote host
rmdir	RMD	Remove directory
quit	QUIT	Quit session

Figure 14.2 Sample UNIX, DOS, and FTP commands.

FTP Commands

When an FTP user sends commands to the application, it uses the Telnet NVT standard commands to carry the information. RFC 959 lists these commands (see the appendix at the end of this chapter). There is also a special set of additional commands for transferring, locating, and organizing the files and directories.

To streamline our understanding of FTP we examine some of the more frequently used commands from that RFC while looking at an FTP session captured from a live network (see Fig. 14.2). These FTP commands are the result of user commands. The commands fall into three categories: access control, transfer parameter, and service commands. At the end of this chapter, we identify some of the commands in each of these categories.

FTP Response Codes

When FTP sends out a request either to start a new session or to use one of the commands, the response that comes back typically carries two parts. The first, sometimes carried in a separate packet, is a three-digit decimal number that FTP uses to provide more information than a simple acknowledgment could offer. It specifies the status of the request.

The second is a brief text message to the user. Figure 14.3 shows the standard numbers and suggested messages. Figure 14.4 shows the significance of the first and second digit in each code. Each FTP author may use his own messages but *not* his own numbers.

Each category has its own definition of what these numbers mean. Let's see how vendors use them in the sample FTP session that follows.

Code	Suggested message	Code	Suggested message
110	Restart marker	332	Need account for login
120	Service ready in *n* minutes	350	Requested action pending
125	Data connection open	421	Service not available
150	Opening data connection	425	Can't open data connection
200	Command okay	426	Transfer aborted
202	Command not implemented	450	Requested file action not taken
211	System status	451	Local processing error
212	Directory status	452	Requested action not taken
213	File status	500	Command unrecognized
214	Help message	501	Syntax error
215	NAME system type	502	Command not implemented
220	Service ready for new user	503	Bad sequence of commands
221	Closing control connection	504	Command parameter error
225	Data connection open	530	Not logged in
226	Closing data connection	532	Need account to store files
227	Entering passive mode	550	Requested action not taken
230	User logged in	551	Page type unknown
250	Requested action okay	552	Requested file action aborted
257	PATHNAME created	553	Requested action not taken
331	User okay, need password		

Figure 14.3 FTP response codes and meanings.

FTP sample session

TCP handshake. In the next few sections, we will examine a decoded sample of an FTP session beginning with the TCP handshake in Fig. 14.5. We will step through this session to understand how FTP accomplishes its mission.

By now you can recognize the three-step, TCP startup sequence. You should be able to pick out the following key pieces of information:

Client port: 30849	Server port: 21
Client sequence number: 2fea6881	Server sequence number: 14eb7802
Client window size: 1336	Server window size: 4096
Client maximum segment: 668	Server maximum segment: 1024

FTP hundreds digit significance		FTP tens digit significance	
		n0n	Syntax error
1nn	Preliminary positive response	n1n	Information
2nn	Complete positive response	n2n	Command connections
3nn	Positive step response	n3n	Login and accounting
4nn	Transient negative response	n4n	Reserved
5nn	Terminal command failure	n5n	File system

Figure 14.4 Response code categories.

```
TCP: 30849 -> ftp(21) seq: 2fea6880 ack: ——
win: 1336 hl: 6 xsum: 0xcd95 urg: 0 flags: <SYN> mss: 668

TCP: ftp(21) -> 30849 seq: 14eb7801 ack: 2fea6881
win: 4096 hl: 6 xsum: 0xa3fc urg: 0 flags: <ACK><SYN> mss: 1024

TCP: 30849 -> ftp(21) seq: 2fea6881 ack: 14eb7802
win: 1336 hl: 5 xsum: 0x751d urg: 0 flags: <ACK>
```

Figure 14.5 FTP sample startup.

Service ready. At the beginning of an FTP session, after TCP has completed its handshake, the response number 220 signifies that the FTP server is ready for a new user session. As we noted in our comments on the response numbers, the vendor may (or may not) use the same words to describe the response, though the reason for using the number must remain the same.

The 220 response, in the first segment of Fig. 14.6, identifies the start of the FTP portion of the session (though the words may be different from "service ready for new user.") The software author (in Fig. 14.6) chose to use her own terminology.

You may also find the data line interesting. The (46/59) signifies that the protocol analyzer only displayed 46 of the 59 bytes of data that it received. Many protocol analyzers do this to save memory, since they can store more small packets in their finite space. Typically vendors leave off the last of the application data by keeping the first 100 or so bytes of each packet.

Login. As you should remember, Telnet defaulted to sending characters as it found them during a periodic keyboard sweep on the client system. The FTP client program works differently. It sends characters after receiving a carriage return. This gives the user the opportunity to make changes before sending an entry. The FTP also requires fewer packets to convey the same bytes of information as would a Telnet session.

In the first segment of Fig. 14.7, the USER command is carrying the FTP users login ID "fred" followed by a carriage return (Control M or ^M) and a

```
TCP: ftp(21) -> 30849 seq: 14eb7802 ack: 2fea6881
win: 4096 hl: 5 xsum: 0xc1cf urg: 0 flags: <ACK><PUSH>
 data (46/59): 220 unix FTP server (UNIX(r) System V Release

TCP: 30849 -> ftp(21) seq: 2fea6881 ack: 14eb783d
win: 1277 hl: 5 xsum: 0x751d urg: 0 flags: <ACK>
```

Figure 14.6 FTP service ready.

```
TCP: 30849 -> ftp(21) seq: 2fea6881 ack: 14eb783d
win: 1277 hl: 5 xsum: 0xe381 urg: 0 flags: <ACK><PUSH>
 data (11/11): USER fred

TCP: ftp(21) -> 30849 seq: 14eb783d ack: 2fea688c
win: 4096 hl: 5 xsum: 0x6712 urg: 0 flags: <ACK>
```

Figure 14.7 FTP login.

line feed (Control J or ^J). Note: FTP may seem less secure than Telnet. With FTP, we only have to capture one TCP segment to get the whole user ID.

Notice that this protocol analyzer is different in decoding the data than the one we used in the Telnet session. While we believe it is important to have a protocol analyzer to troubleshoot a network, we also believe a learning experience should include a number of different examples.

The second segment does not acknowledge the user ID. Instead, it is a plain acknowledgment of the 11 bytes of data that the server received. The window size on the client still shows the initial 59 bytes in the buffer.

Password. In the first segment in Fig. 14.8, the server application notifies the client application that the user name was okay, but it also needs a password. The vendor uses the 331 response code though the software author chose to use different words. Notice that the server application includes the same carriage return and line feed at the end of the response string (the password).

Segment two carries the acknowledgment of those 33 password-request bytes and shows that they are probably now in the client's buffer since the window size is exactly 33 bytes below its original size.

```
TCP: ftp(21) -> 30849 seq: 14eb783d ack: 2fea688c
win: 4096 hl: 5 xsum: 0xfb51 urg: 0 flags: <ACK><PUSH>
 data (33/33): 331 Password required for fred

TCP: 30849 -> ftp(21) seq: 2fea688c ack: 14eb785e
win: 1303 hl: 5 xsum: 0x2f1d urg: 0 flags: <ACK>

TCP: 30849 -> ftp(21) seq: 2fea688c ack: 14eb785e
win: 1303 hl: 5 xsum: 0x8188 urg: 0 flags: <ACK><PUSH>
 data (13/13): PASS abc123

TCP: ftp(21) -> 30849 seq: 14eb785e ack: 2fea6899
win: 4096 hl: 5 xsum: 0x3912 urg: 0 flags: <ACK>
```

Figure 14.8 FTP password.

The third segment is as secure as the all the others in this session have been, but if you are responsible for the security of your network, it will still make you cringe. Yes, that is the real password following the PASS command. This is just another example of the lack of security in TCP/IP itself.

The last segment shows the server's acknowledgment of the 13 bytes of the PASS command string. Notice that the server continues to process the bytes of data it receives as they arrive instead of storing them in its buffer. The window size remains at the 4096 original value.

Session started. As the first segment in Fig. 14.9 indicates, the user has logged in and the session is ready to begin the process of transferring files. Since the client FTP software and the server FTP software operate independently yet cooperatively, there is no need for the server to send a prompt to the client (like the $ that the Telnet server sent in Chap. 12).

After the client uses the second segment to acknowledge the 26 bytes that the server sent, it can send the third segment. The Print Working Directory (PWD) command is one of the few that travels by itself. The user, or the software configuration, asked for the identities of the path and directory that it is currently using. The PWD command carries that request to the server.

The server acknowledges the 5 bytes of the client's PWD command in the fourth segment in Fig. 14.9. It responds with code 257 and identifies the path and directory as /home/fred.

The last segment carries the client's acknowledgment of the 40 bytes of directory information just received from the server.

Setting the port. When the user instructs the client FTP application to transfer a file, the client software must tell the server software the client's IP

TCP: ftp(21) -> 30849 seq: 14eb785e ack: 2fea6899
win: 4096 hl: 5 xsum: 0x1529 urg: 0 flags: <ACK><PUSH>
data (26/26): 230 User fred logged in.

TCP: 30849 -> ftp(21) seq: 2fea6899 ack: 14eb7878
win: 1310 hl: 5 xsum: 0x011d urg: 0 flags: <ACK>

TCP: 30849 -> ftp(21) seq: 2fea6899 ack: 14eb7878
win: 1310 hl: 5 xsum: 0x8f7e urg: 0 flags: <ACK><PUSH>
data (5/5): PWD

TCP: ftp(21) -> 30849 seq: 14eb7878 ack: 2fea689e
win: 4096 hl: 5 xsum: 0xb437 urg: 0 flags: <ACK><PUSH>
data (40/40): 257 "/home/fred" is current directory.

TCP: 30849 -> ftp(21) seq: 2fea689e ack: 14eb78a0
win: 1270 hl: 5 xsum: 0xfc1c urg: 0 flags: <ACK>

Figure 14.9 FTP session started.

```
TCP: 30849 -> ftp(21) seq: 2fea689e ack: 14eb78a0
win: 1336 hl: 5 xsum: 0xbd65 urg: 0 flags: <ACK><PUSH>
 data (28/28): PORT 192,153,183,3,126,113

TCP: ftp(21) -> 30849 seq: 14eb78a0 ack: 2fea68ba
win: 4096 hl: 5 xsum: 0xc854 urg: 0 flags: <ACK><PUSH>
 data (30/30): 200 PORT command successful.

TCP: 30849 -> ftp(21) seq: 2fea68ba ack: 14eb78be
win: 1306 hl: 5 xsum: 0x9e1c urg: 0 flags: <ACK>
```

Figure 14.10 FTP sets the return port.

address and the port number that will send or receive the file on the client's end of the session.

As you can see in the first segment in Fig. 14.10, FTP sends the information in ASCII characters, with each byte separated by commas. The IP address, already identified in a byte-by-byte format, comes across in a readable string. We (and the software) read the port number from its 2-byte field as one decimal number.

To stay in the byte-by-byte mode, FTP sends the port number in separate bytes: 126, 113. To translate this from two ASCII values into one decimal value, we may (as in this case) have to first translate it to hex: 126 becomes 7e hex and 113 becomes 71 hex. If we translate 7e 71 hex into one decimal number it becomes

$$\begin{aligned} 7 \times 4096 &= 28672 \\ \text{e}\,(14) \times 256 &= 3584 \\ 7 \times 16 &= 112 \\ 1 \times 1 &= 1 \\ \text{Equals} &\quad 32369 \end{aligned}$$

The second segment contains the response code of 200 that says the server received (and processed) the PORT command successfully. The last segment is the client's acknowledgment of those 30 bytes received from the server.

Name list request. If the client user is unsure of the contents of the remote directory (or the client software configuration requests it automatically), then FTP requests a list of the files in the directory. On a UNIX FTP server, the user may enter "ls" at the user prompt. On a DOS FTP server, the user command is "DIR."

In either case, FTP sends an NLST command to the server, as we see in the first segment in Fig. 14.11. The user's (or software's automatic) command

```
TCP: 30849 -> ftp(21) seq: 2fea68ba ack: 14eb78be
win: 1306 hl: 5 xsum: 0xe56d urg: 0 flags: <ACK><PUSH>
 data (6/6): NLST

TCP: ftp(21) -> 30849 seq: 14eb78be ack: 2fea68c0
win: 4096 hl: 5 xsum: 0xb211 urg: 0 flags: <ACK>
```

Figure 14.11 FTP's directory request.

triggered the file transfer. The server must know where to send the directory, so the client sent the PORT command we examined in the last figure.

The second segment acknowledges the 6 bytes of client data received at the server.

Data session startup. As with the FTP control session, the FTP-data session has to go through the same three-step TCP startup sequence. In this case, the client sent its port number to the server. The server initiated the session and (due to the data-related nature of the session) set a much larger window size than in the FTP control session.

You can now recognize the three-step TCP startup sequence information in Fig. 14.12. The key pieces of information for the FTP-data session are

Client port: 32369	Server port: 20
Client sequence number: 2fea7271	Server sequence number: 14f05a02
Client window size: 4788	Server window size: 24576
Client maximum segment: 1460	Server maximum segment: 1024

Session control. In the sequence of segments in Fig. 14.13, we switch back from the FTP-data session to the FTP control session. The server informs the client that it is opening the data connection with the first segment's response code of 150 and the text identifying it as an ASCII data connection for /bin/ls (ASCII data connection for a binary UNIX directory listing).

```
TCP: ftp-data(20) -> 32369 seq: 14f05a01 ack: ——
win: 24576 hl: 6 xsum: 0x2b5d urg: 0 flags: <SYN> mss: 1024

TCP: 32369 -> ftp-data(20) seq: 2fea7270 ack: 14f05a02
win: 4788 hl: 6 xsum: 0x5806 urg: 0 flags: <ACK><SYN> mss: 1460

TCP: ftp-data(20) -> 32369 seq: 14f05a02 ack: 2fea7271
win: 24576 hl: 5 xsum: 0xc8d0 urg: 0 flags: <ACK>
```

Figure 14.12 FTP data session startup.

```
TCP: ftp(21) -> 30849 seq: 14eb78be ack: 2fea68c0
win: 4096 hl: 5 xsum: 0x9347 urg: 0 flags: <ACK><PUSH>
 data (46/72): 150 ASCII data connection for /bin/ls (192.153

TCP: 30849 -> ftp(21) seq: 2fea68c0 ack: 14eb7906
win: 1264 hl: 5 xsum: 0x7a1c urg: 0 flags: <ACK>

TCP: ftp(21) -> 30849 seq: 14eb7906 ack: 2fea68c0
win: 4096 hl: 5 xsum: 0x9174 urg: 0 flags: <ACK><PUSH>
 data (30/30): 226 ASCII Transfer complete.

TCP: 30849 -> ftp(21) seq: 2fea68c0 ack: 14eb7924
win: 1234 hl: 5 xsum: 0x7a1c urg: 0 flags: <ACK>
```

Figure 14.13 FTP session control.

After the client's acknowledgment of the 72 bytes of data in the second segment, the server sends the third segment stating that the ASCII transfer is complete. With the 226 code, it is closing the data connection. Wait a minute! Where is the file (directory) that we were expecting? The server really transferred it...to RAM for sending over the session connection. Don't worry, FTP will not dismantle the session until the client has successfully received the file.

File transfer and session end. Finally, in the first segment of Fig. 14.14 we see the 68-byte content of the server's current directory with the individual files separated by carriage returns (^M) and line feeds (^J). The second segment is the client's acknowledgment of those 68 bytes.

```
TCP: ftp-data(20) -> 32369 seq: 14f05a02 ack: 2fea7271
win: 24576 hl: 5 xsum: 0x5b96 urg: 0 flags: <ACK><PUSH>
 data (46/68): City_of_Mt.Airy^M^JDirector.txt^M^Jtester^M^J fred.ex

TCP: 32369 -> ftp-data(20) seq: 2fea7271 ack: 14f05a46
win: 4720 hl: 5 xsum: 0x151e urg: 0 flags: <ACK>

TCP: ftp-data(20) -> 32369 seq: 14f05a46 ack: 2fea7271
win: 24576 hl: 5 xsum: 0x83d0 urg: 0 flags: <ACK><FIN>

TCP: 32369 -> ftp-data(20) seq: 2fea7271 ack: 14f05a47
win: 0 hl: 5 xsum: 0x8430 urg: 0 flags: <ACK>

TCP: 32369 -> ftp-data(20) seq: 2fea7271 ack: 14f05a47
win: 0 hl: 5 xsum: 0x8330 urg: 0 flags: <ACK><FIN>

TCP: ftp-data(20) -> 32369 seq: 14f05a47 ack: 2fea7272
win: 24575 hl: 5 xsum: 0x83d0 urg: 0 flags: <ACK>
```

Figure 14.14 File transfer and session end.

```
TCP: 30849 -> ftp(21) seq: 2fea68c0 ack: 14eb7924
win: 1336 hl: 5 xsum: 0x1161 urg: 0 flags: <ACK><PUSH>
 data (28/28): PORT 192,153,183,3,207,177

TCP: ftp(21) -> 30849 seq: 14eb7924 ack: 2fea68dc
win: 4096 hl: 5 xsum: 0x2254 urg: 0 flags: <ACK><PUSH>
 data (30/30): 200 PORT command successful.

TCP: 30849 -> ftp(21) seq: 2fea68dc ack: 14eb7942
win: 1306 hl: 5 xsum: 0xf81b urg: 0 flags: <ACK>
```

Figure 14.15 FTP identifies another port.

The last four segments are the four-step shutdown of the FTP-data and TCP session. Since the server successfully transferred the file that the user requested, the session has completed its mission. That leads FTP to close the data session. In the previous figure, you saw the FTP control session reporting that it was closing the data connection, and it has now done so.

Although the data session is complete, as we will see, the FTP control session is very much alive and well.

Preparing for another file. After finding another file of interest, the user makes a request to retrieve that file. As a result, the client FTP application sends a new PORT command (in the first segment of Fig. 14.15) to tell the Server where to send the file that it will request.

The second segment does double duty by acknowledging the PORT command's 28 bytes and carrying the response code 200 to tell the client that it has received and processed the command.

The third (and last) segment is the client's acknowledgment of the 30 bytes in the PORT command response from the server.

File request and session startup. In Fig. 14.16, the user actually typed in the command "get tester." As a result, the client FTP software sent the PORT command, got a successful response, and sent the RETR command to retrieve the file named tester.

After the second segment's acknowledgment of the 14 bytes in the RETR command string, the server starts another FTP-data session by completing the three-step TCP handshake. The key information is as follows:

Client port: 53169	Server port: 20
Client sequence number: 2feacbb1	Server sequence number: 151b5202
Client window size: 4788	Server window size: 24576
Client maximum segment: 1460	Server maximum segment: 1024

```
TCP: 30849 -> ftp(21) seq: 2fea68dc ack: 14eb7942
win: 1336 hl: 5 xsum: 0x6d07 urg: 0 flags: <ACK><PUSH>
 data (14/14): RETR tester

TCP: ftp(21) -> 30849 seq: 14eb7942 ack: 2fea68ea
win: 4096 hl: 5 xsum: 0x0411 urg: 0 flags: <ACK>

TCP: ftp-data(20) -> 53169 seq: 151b5201 ack: ——
win: 24576 hl: 6 xsum: 0xc013 urg: 0 flags: <SYN> mss: 1024

TCP: 53169 -> ftp-data(20) seq: 2feacbb0 ack: 151b5202
win: 4788 hl: 6 xsum: 0xac63 urg: 0 flags: <ACK><SYN> mss: 1460

TCP: ftp-data(20) -> 53169 seq: 151b5202 ack: 2feacbb1
win: 24576 hl: 5 xsum: 0x1d2e urg: 0 flags: <ACK>
```

Figure 14.16 File request and session startup.

Session status. In the sequence of segments we see in Fig. 14.17, we switch from the FTP-data session back to the FTP control session. The server tells the client that it is opening the data connection with the first segment's response code of 150, and the text identifies it as an ASCII data connection for tester (ASCII data connection for the file named tester).

After the client's second segment acknowledgment of those 75 bytes of data, the server sent the third segment saying the ASCII transfer was complete and so (226) it is closing the data connection. As we saw in the previous NLST-based session, FTP really transferred the file to RAM for sending over the session connection.

Multiple packets. As you can see in this session's startup in Fig. 14.18, FTP set the client's window size to 4788. In the TCP segments, the server sent

```
TCP: ftp(21) -> 30849 seq: 14eb7942 ack: 2fea68ea
win: 4096 hl: 5 xsum: 0xeb16 urg: 0 flags: <ACK><PUSH>
 data (46/75): 150 ASCII data connection for tester (192.153

TCP: 30849 -> ftp(21) seq: 2fea68ea ack: 14eb798d
win: 1261 hl: 5 xsum: 0xcc1b urg: 0 flags: <ACK>

TCP: ftp(21) -> 30849 seq: 14eb798d ack: 2fea68ea
win: 4096 hl: 5 xsum: 0xe073 urg: 0 flags: <ACK><PUSH>
 data (30/30): 226 ASCII Transfer complete.

TCP: 30849 -> ftp(21) seq: 2fea68ea ack: 14eb79ab
win: 1306 hl: 5 xsum: 0x811b urg: 0 flags: <ACK>
```

Figure 14.17 Session status.

```
TCP: ftp-data(20) -> 53169 seq: 151b5202 ack: 2feacbb1
win: 24576 hl: 5 xsum: 0x09ef urg: 0 flags: <ACK>
 data (46/1024):@ \377WPCD^M^J^@^@^A^M^J^@^A^@^@^@^@\373\377^E^@2^@^D
^A^@^@\377\377^H^@^@^@B^@^@^@^F^@^P^@^@^@J^@

TCP: ftp-data(20) -> 53169 seq: 151b5602 ack: 2feacbb1
win: 24576 hl: 5 xsum: 0x5539 urg: 0 flags: <ACK>
 data (46/1024): |^@h^@@^@l^@\204^@<\#94>@<\#94>@|^@<\#94>@\274^@\20
4^@p^@|^@t^@X
^@h^@L^@\210^@h^@\234^@p^@l^@`^@

TCP: ftp-data(20) -> 53169 seq: 151b5a02 ack: 2feacbb1
win: 24576 hl: 5 xsum: 0xef70 urg: 0 flags: <ACK>
 data (46/1024): ^@\202^@\377\377\377\377S^A\377\377\243^A\377
\377^Y^B\377\377\377
\377\377\377\377\377\377\377\377\377\377\377^52?gg\226\205%22Id5C

TCP: ftp-data(20) -> 53169 seq: 151b5e02 ack: 2feacbb1
win: 24576 hl: 5 xsum: 0x3ff1 urg: 0 flags: <ACK><PUSH>
 data (46/1024): sn't. Any other condition is by definition an
```

Figure 14.18 File transfer—multiple packets.

four segments at the maximum size that the client had advertised (1024 bytes). Notice that FTP did not need to send an acknowledgment since the server had not exceeded the client's window size.

Adjusting the window size. In the first segment in Fig. 14.19, the client acknowledges receipt of the four segments of 1024 bytes (a total of 4096 bytes). At the same time, it adjusts its window size from 4788 down to 692 to show that there is that much space left in the client's session buffer.

The second segment sends the same acknowledgment sequence number as the first, but it sends it to report it has completed processing the 4096 bytes that were in the buffer. Notice that it now reports the window size as 4788 bytes, the same as the original value that it sent in the TCP session startup. This tells the server that there is plenty of room for more data.

Ending the data session. In these segments in Fig. 14.20, the server takes advantage of the new window size to send the rest of the file. Notice that not

```
TCP: 53169 -> ftp-data(20) seq: 2feacbb1 ack: 151b6202
win: 692 hl: 5 xsum: 0x697b urg: 0 flags: <ACK>

TCP: 53169 -> ftp-data(20) seq: 2feacbb1 ack: 151b6202
win: 4788 hl: 5 xsum: 0x696b urg: 0 flags: <ACK>
```

Figure 14.19 Adjusting the window size.

```
TCP: ftp-data(20) -> 53169 seq: 151b6202 ack: 2feacbb1
win: 24576 hl: 5 xsum: 0x1222 urg: 0 flags: <ACK>
 data (46/1024): (0*8 + 0*4 + 0*2 + 0*1)\301^Bh^Ph^P^X^A\3011 0 0
0\301^B^X^U^X^Uh^A\3018\301

TCP: ftp-data(20) -> 53169 seq: 151b6602 ack: 2feacbb1
win: 24576 hl: 5 xsum: 0x08d6 urg: 0 flags: <ACK>
 data (46/1024): F^Mwhen we ran out of numbers to keep consisten

TCP: ftp-data(20) -> 53169 seq: 151b6a02 ack: 2feacbb1
win: 24576 hl: 5 xsum: 0x9b2d urg: 0 flags: <ACK>
 data (46/1024): JThis technique of sixteen possible four\251bit

TCP: ftp-data(20) -> 53169 seq: 151b6e02 ack: 2feacbb1
win: 24576 hl: 5 xsum: 0x3584 urg: 0 flags: <ACK><PUSH><FIN>
 data (46/790): 0 0 0 1
0\301^B\270^K\270^K\310^@\301\301^B^P^N^P^N\360^@\3010
2\301^Bh^Ph^P^X^A\301\301^B\300^R\300

TCP: 53169 -> ftp-data(20) seq: 2feacbb1 ack: 151b7118
win: 926 hl: 5 xsum: 0x696b urg: 0 flags: <ACK>
```

Figure 14.20 Ending the data session with data present.

one of the segments the server sends exceeds the client's advertised maximum of 1024 per the TCP specifications.

In the fourth segment, we see the server send the last 790 bytes of the file and the final (FIN) data sent bit. When the server has transferred the file, the FTP-data session has finished its work. When the FTP-data session finishes, the TCP session is shut down.

In the last segment, the client sends its acknowledgment to signify the receipt of the 3862 bytes. According to the advertised window size value of 926, the data is in the client's buffer waiting for the CPU to process it. It does not acknowledge the final data sent bit in this TCP segment.

FTP-data session end. The first segment is the client's acknowledgment of the server's final data sent bit. The rest of the segments, shown in Fig. 14.21, complete the four-step TCP session shutdown process for the FTP-data session.

```
TCP: 53169 -> ftp-data(20) seq: 2feacbb1 ack: 151b7119
win: 3862 hl: 5 xsum: 0xf05f urg: 0 flags: <ACK>

TCP: 53169 -> ftp-data(20) seq: 2feacbb1 ack: 151b7119
win: 3862 hl: 5 xsum: 0xef5f urg: 0 flags: <ACK><FIN>

TCP: ftp-data(20) -> 53169 seq: 151b7119 ack: 2feacbb2
win: 24575 hl: 5 xsum: 0x060f urg: 0 flags: <ACK>
```

Figure 14.21 Completing the four-step shutdown.

```
TCP: 30849 -> ftp(21) seq: 2fea68ea ack: 14eb79ab
win: 1336 hl: 5 xsum: 0xa173 urg: 0 flags: <ACK><PUSH>
 data (6/6): QUIT

TCP: ftp(21) -> 30849 seq: 14eb79ab ack: 2fea68f0
win: 4096 hl: 5 xsum: 0xa721 urg: 0 flags: <ACK><PUSH>
 data (14/14): 221 Goodbye.

TCP: 30849 -> ftp(21) seq: 2fea68f0 ack: 14eb79b9
win: 1322 hl: 5 xsum: 0x5d1b urg: 0 flags: <ACK>

TCP: ftp(21) -> 30849 seq: 14eb79b9 ack: 2fea68f0
win: 4096 hl: 5 xsum: 0x8610 urg: 0 flags: <ACK><FIN>

TCP: 30849 -> ftp(21) seq: 2fea68f0 ack: 14eb79ba
win: 0 hl: 5 xsum: 0x8620 urg: 0 flags: <ACK>

TCP: 30849 -> ftp(21) seq: 2fea68f0 ack: 14eb79ba
win: 0 hl: 5 xsum: 0x8520 urg: 0 flags: <ACK><FIN>

TCP: ftp(21) -> 30849 seq: 14eb79ba ack: 2fea68f1
win: 4096 hl: 5 xsum: 0x8510 urg: 0 flags: <ACK
```

Figure 14.22 Ending the FTP control session.

Ending the control session. When the user is through transferring files, that user issues the quit command at either the command line prompt or by clicking on a Windows screen button. This generates the FTP QUIT command we see in the first segment of Fig. 14.22.

The server responds with the 221 code that tells the client that the FTP control session is closing. This vendor chose to use the word *goodbye* as the text instead. The client then acknowledges the 14 bytes in the third segment.

The last four segments are the TCP shutdown process for the FTP control session. Since they follow the same procedural steps that we examined before, we will not review them here. The appendix that follows is an excerpt of RFC 959. It lists the FTP commands, defines the syntax (as necessary), and explains each command. We changed only the text format to align with the rest of this book. We made no other changes to the RFC text.

Appendix FTP Commands from RFC 959 (File Transfer Protocol, October 1985)

4.1 FTP Commands

4.1.1 Access Control Commands The following commands specify access control identifiers (command codes are shown in parentheses).

User Name (USER) The argument field is a Telnet string identifying the user. The user identification is that which is required by the server for access to its file system. This command will normally be the first command transmitted by the user after the control connections are made (some servers may require this). Additional identification information in the form of a password and/or an account command may also be required by some servers. Servers may allow a new USER command to be entered at any point in order to change the access control and/or accounting information. This has the effect of flushing any user, password, and account information already supplied and beginning the login sequence again. All transfer parameters are unchanged and any file transfer in progress is completed under the old access control parameters.

Password (PASS) The argument field is a Telnet string specifying the user's password. This command must be immediately preceded by the user name command, and, for some sites, completes the user's identification for access control. Since password information is quite sensitive, it is desirable in general to "mask" it or suppress typeout. It appears that the server has no foolproof way to achieve this. It is therefore the responsibility of the user-FTP process to hide the sensitive password information.

Account (ACCT) The argument field is a Telnet string identifying the user's account. The command is not necessarily related to the USER command, as some sites may require an account for login and others only for specific access, such as storing files. In the latter case the command may arrive at any time.

There are reply codes to differentiate these cases for the automation: when account information is required for login, the response to a successful PASSword command is reply code 332. On the other hand, if account information is NOT required for login, the reply to a successful PASSword command is 230; and if the account information is needed for a command issued later in the dialogue, the server should return a 332 or 532 reply depending on whether it stores (pending receipt of the ACCounT command) or discards the command, respectively.

Change Working Directory (CWD) This command allows the user to work with a different directory or dataset for file storage or retrieval without altering his login or accounting information. Transfer parameters are similarly unchanged. The argument is a pathname specifying a directory or other system dependent file group designator.

Change to a Parent Directory (CDUP) This command is a special case of CWD, and is included to simplify the implementation of programs for transferring directory trees between operating systems having different syntaxes for naming the parent directory. The reply codes shall be identical to the reply codes of CWD. See Appendix II for further details.

Structure Mount (SMNT) This command allows the user to mount a different file system data structure without altering his login or accounting information. Transfer parameters are similarly unchanged. The argument is a pathname specifying a directory or other system dependent file group designator.

Reinitialize (REIN) This command terminates a USER, flushing all I/O and account information, except to allow any transfer in progress to be completed. All parameters are reset to the default settings and the control connection is left open. This is identical to the state in which a user finds himself immediately after the control connection is opened. A USER command may be expected to follow.

Logout (QUIT) This command terminates a USER and if file transfer is not in progress, the server closes the control connection. If file transfer is in progress, the connection will remain open for result response and the server will then close it. If the user-process is transferring files for several USERs but does not wish to close and then reopen connections for each, then the REIN command should be used instead of QUIT.

An unexpected close on the control connection will cause the server to take the effective action of an abort (ABOR) and a logout (QUIT).

4.1.2 Transfer Parameter Commands All data transfer parameters have default values, and the commands specifying data transfer parameters are required only if the default parameter values are to be changed. The default value is the last specified value, or if no value has been specified, the standard default value is as stated here. This implies that the server must "remember" the applicable default values. The commands may be in any order except that they must precede the FTP service request. The following commands specify data transfer parameters:

Data Port (PORT) The argument is a HOST-PORT specification for the data port to be used in data connection. There are defaults for both the user and server data ports, and under normal circumstances this command and its reply are not needed. If this command is used, the argument is the concatenation of a 32-bit internet host address and a 16-bit TCP port address. This address information is broken into 8-bit fields and the value of each field is transmitted as a decimal number (in character string representation). The fields are separated by commas. A port command would be:

PORT h1,h2,h3,h4,p1,p2

where h1 is the high order 8 bits of the Internet host address.

Passive (PASV) This command requests the server-DTP to "listen" on a data port (which is not its default data port) and to wait for a connection rather than initiate one upon receipt of a transfer command. The response to this command includes the host and port address this server is listening on.

Representation Type (TYPE) The argument specifies the representation type as described in the Section on Data Representation and Storage. Several types take a second parameter. The first parameter is denoted by a single Telnet character, as is the second Format parameter for ASCII and EBCDIC; the second parameter for local byte is a decimal integer to indicate Byte size. The parameters are separated by a <SP> (Space, ASCII code 32).

The following codes are assigned for type:

A	ASCII
N	Nonprint
T	Telnet format effectors
E	EBCDIC
C	Carriage control (ASA)
I	Image
L <byte size>	Local-byte Byte size

The default representation type is ASCII Nonprint. If the Format parameter is changed, and later just the first argument is changed, Format then returns to the Nonprint default.

File Structure (STRU) The argument is a single Telnet character code specifying file structure described in the section on Data Representation and Storage.

The following codes are assigned for structure:

F	File (no record structure)
R	Record structure
P	Page structure

The default structure is File.

Transfer Mode (MODE) The argument is a single Telnet character code specifying the data transfer modes described in the section on Transmission Modes.

The following codes are assigned for transfer modes:

S	Stream
B	Block
C	Compressed

The default transfer mode is Stream.

4.1.3 FTP Service Commands The FTP service commands define the file transfer or the file system function requested by the user. The argument of an FTP service command will normally be a pathname. The syntax of pathnames must conform to server site conventions (with standard defaults applicable), and the language conventions of the control connection. The suggested default handling is to use the last specified device, directory or file name, or the standard default defined for local users. The commands may be in any order except that a "rename from" command must be followed by a "rename to" command and the restart command must be followed by the interrupted service command (e.g., STOR or RETR). The data, when transferred in response to FTP service commands, shall always be sent over the data connection, except for certain informative replies. The following commands specify FTP service requests:

Retrieve (RETR) This command causes the server-DTP to transfer a copy of the file, specified in the pathname, to the server- or user-DTP at the other end of the data connection. The status and contents of the file at the server site shall be unaffected.

Store (STOR) This command causes the server-DTP to accept the data transferred via the data connection and to store the data as a file at the server site. If the file specified in the pathname exists at the server site, then its contents shall be replaced by the data being transferred. A new file is created at the server site if the file specified in the pathname does not already exist.

Store Unique (STOU) This command behaves like STOR except that the resultant file is to be created in the current directory under a name unique to that directory. The 250 Transfer Started response must include the name generated.

Append (with Create) (APPE) This command causes the server-DTP to accept the data transferred via the data connection and to store the data in a file at the server site. If the file specified in the pathname exists at the server site, then the data shall be appended to that file; otherwise the file specified in the pathname shall be created at the server site.

Allocate (ALLO) This command may be required by some servers to reserve sufficient storage to accommodate the new file to be transferred. The argument shall be a decimal integer representing the number of bytes (using the logical byte size) of storage to be reserved for the file. For files sent with record or page structure a maximum record or page size (in logical bytes) might also be necessary; this is indicated by a decimal integer in a second argument field of the command. This second argument is optional, but when present should be separated from the first by the three Telnet characters <SP> R <SP>. This command shall be followed by a STORe or APPEnd command. The ALLO command should be treated as a NOOP (no operation) by those servers which do not require that the maximum size of the file be declared beforehand, and those servers interested in only the maximum record or page size should accept a dummy value in the first argument and ignore it.

Restart (REST) The argument field represents the server marker at which file transfer is to be restarted. This command does not cause file transfer but skips over the file to the specified data checkpoint. This command shall be immediately followed by the appropriate FTP service command which shall cause file transfer to resume.

Rename From (RNFR) This command specifies the old pathname of the file which is to be renamed. This command must be immediately followed by a "rename to" command specifying the new file pathname.

Rename To (RNTO) This command specifies the new pathname of the file specified in the immediately preceding "rename from" command. Together the two commands cause a file to be renamed.

Abort (ABOR) This command tells the server to abort the previous FTP service command and any associated transfer of data. The abort command may require "special action," as discussed in the Section on FTP Commands, to force recognition by the server. No action is to be taken if the previous command has been completed (including data transfer). The control connection is not to be closed by the server, but the data connection must be closed.

There are two cases for the server upon receipt of this command: (1) the FTP service command was already completed, or (2) the FTP service command is still in progress.

> In the first case, the server closes the data connection (if it is open) and responds with a 226 reply, indicating that the abort command was successfully processed.
>
> In the second case, the server aborts the FTP service in progress and closes the data connection, returning a 426 reply to indicate that the service request terminated abnormally. The server then sends a 226 reply, indicating that the abort command was successfully processed.

Delete (DELE) This command causes the file specified in the pathname to be deleted at the server site. If an extra level of protection is desired (such as the query, "Do you really wish to delete?"), it should be provided by the user-FTP process.

Remove Directory (RMD) This command causes the directory specified in the pathname to be removed as a directory (if the pathname is absolute) or as a subdirectory of the current working directory (if the pathname is relative). See Appendix II.

Make Directory (MKD) This command causes the directory specified in the pathname to be created as a directory (if the pathname is absolute) or as a subdirectory of the current working directory (if the pathname is relative). See Appendix II.

Print Working Directory (PWD) This command causes the name of the current working directory to be returned in the reply. See Appendix II.

List (LIST) This command causes a list to be sent from the server to the passive DTP. If the pathname specifies a directory or other group of files, the server should transfer a list of files in the specified directory. If the pathname specifies a file then the server should send current information on the file. A null argument implies the user's current working or default directory. The data transfer is over the data connection in type ASCII or type EBCDIC. (The user must ensure that the TYPE is appropriately ASCII or EBCDIC). Since the information on a file may vary widely from system to system, this information may be hard to use automatically in a program, but may be quite useful to a human user.

Name List (NLST) This command causes a directory listing to be sent from server to user site. The pathname should specify a directory or other system-specific file group descriptor; a null argument implies the current directory. The server will return a stream of names of files and no other information. The data will be transferred in ASCII or EBCDIC type over the data connection as valid pathname strings separated by <CRLF> or <NL>. (Again the user must ensure that the TYPE is correct.) This command is intended to return information that can be used by a program to further process the files automatically. For example, in the implementation of a "multiple get" function.

Site Parameters (SITE) This command is used by the server to provide services specific to his system that are essential to file transfer but not sufficiently universal to be included as commands in the protocol. The nature of these services and the specification of their syntax can be stated in a reply to the HELP SITE command.

System (SYST) This command is used to find out the type of operating system at the server. The reply shall have as its first word one of the system names listed in the current version of the Assigned Numbers document [4].

Status (STAT) This command shall cause a status response to be sent over the control connection in the form of a reply. The command may be sent during a file transfer (along with the Telnet IP and Synch signals—see the Section on FTP Commands) in which case the server will respond with the status of the operation in progress, or it may be sent between file transfers. In the latter case, the command may have an argument field. If the argument is a pathname, the command is analogous to the "list" command except that data shall be transferred over the control connection. If a partial pathname is given, the server may respond with a list of file names or attributes associated with that specification. If no argument is given, the server should return general status information about the server FTP process. This should include current values of all transfer parameters and the status of connections.

Help (HELP) This command shall cause the server to send helpful information regarding its implementation status over the control connection to the user. The command may take an argument (e.g., any command name) and return more specific information as a response. The reply is type 211 or 214. It is suggested that HELP be allowed before entering a USER command. The server may use this reply to specify site-dependent parameters, e.g., in response to HELP SITE.

Noop (NOOP) This command does not affect any parameters or previously entered commands. It specifies no action other than that the server send an OK reply.

Chapter

15

Simple Mail Transfer Protocol

Overview

E-mail, the second most commonly used service on the Internet following the World Wide Web, gives people a fast way to drop off messages (large or small) to anyone in the Internet. The advantage is that the person receiving the message does not have to be on-line to receive it. Since the message goes to the mail server, the receiving user can pick it up the next time she logs on. If the recipient has logged into the mail server, the mail delivery time is very short (usually minutes). This is the reason that you may hear experienced Internet users refer to the alternative, the U.S. Postal Service, as *snail mail.*

E-mail Names

One of the keys to using electronic mail, as sender or receiver, is understanding e-mail addresses. On some e-mail systems, the addressing scheme is so complex that it is almost impossible to understand. On the Internet, the e-mail name format is much easier to understand. It is so easy that it is rapidly becoming a feature on a business card (as in Fig. 15.1). Business and professional people, particularly those in high-tech areas, find that others judge how current they keep their skills by the presence of an e-mail address on their business card and letterhead.

We build the e-mail address by adding the recipient's user name to his or her e-mail server's domain name. Convention and SMTP specifications call for separating these two pieces with an @. Some vendors use different separators (such as :, !, or %) for other purposes like mail forwarding or local handling. An e-mail address is pronounced "user at domain dot name" (for user@domain.name). For example, we could tell you about one mail address,

International Nauga Ranches

Clint Anderson

clint@unix.inr.com

Figure 15.1 E-mail name on a business card.

President at White House dot gov (president@whitehouse.gov). If you send mail to that address you will get a response from the highest levels of government.

Mail Server

In electronic mail it is rarely practical to create a live session with the target of the mail since that user may have logged in to the network or the mail server. Simple Mail Transfer Protocol gets around this difficulty by routing the e-mail message through the chain of IP gateways to the destination network.

There, SMTP (with the help of the DNS) delivers the message to the target user's e-mail server. In many networks the mail server holds the mail in the server for three to five days. If the user does not log in by then, the server returns the mail to the sender. In many corporations, where mail is a major part of business communications, the mail holding time is much longer.

A big advantage to this way of handling messages is that the user could be physically on the other side of the Internet and remotely login to the "home" network to pick up the mail. That is very helpful to road warriors who have to travel a lot but need to stay in touch as well. Equally helpful is that a user may save messages for as long as the user wishes. (The system administrator may have a different idea in some situations.)

SMTP Commands

Early in the development of the Internet, the e-mail system used was *Mail Transfer Protocol* (MTP); SMTP is a simpler version. Simple Mail Transfer

Command	Purpose	Req/rec
HELO domain	Starting point; client sets the domain sending mail.	Required
MAIL FROM: user@host	Start sending SMTP mail to the SMTP server.	Required
RCPT TO: user@host	Sets the mail addressee; one for each addressee.	Required
DATA	Begin the body of the mail message.	Required
SEND TO: user@host	SEND the mail to the user's terminal, if logged in.	Recommended
SOML TO: user@host	Send Or MaiL if not logged in, send to server.	Recommended
SAML TO: user@host	Send And MaiL; combined SEND and SOML.	Recommended
RSET	Abort this mail and dump the data received so far.	Required
VRFY user@host	VeRiFY that the specified user is a valid mailbox.	Recommended
EXPN mail_list	EXPaNd the appropriate mail list.	Recommended
HELP command	Provide help to the user of the mail server.	Recommended
NOOP	NO OPeration; keeps the connection open.	Required
QUIT	Client ends the session with the server.	Required
TURN	Client offers to let the server send queued mail.	Recommended

Figure 15.2 SMTP commands.

Protocol sends the e-mail messages over IP and TCP to provide reliable mail delivery. As we will see later in this chapter when we review an SMTP message, it uses TCP's well-known port 25.

To accomplish its primary function of moving readable text between Internet mailboxes, SMTP uses a client/server relationship that is very similar to the one we examined when we discussed FTP. Like FTP, SMTP also uses a command request and a numbered response system. You can see the commands, what they do, and their recommended or required status in Fig. 15.2. These will make more sense as we go through a decoded sample session.

SMTP Response Codes

When SMTP sends out a command request, the response that comes back typically carries two parts (see Fig. 15.3). The first, at times carried in a separate packet, is a three-digit decimal number. The SMTP uses it to provide more details than an acknowledgment could offer alone: It expresses the status of the request.

If the codes look familiar, remember that transferring an e-mail message is a lot like transferring a file with FTP. As with FTP, we can find the hidden meanings in the response codes. These categories and meanings are almost identical to FTP categories and meanings, as Fig. 15.4 shows.

Code	Purpose	Code	Purpose
211	System status or help reply	500	Command unknown
214	Help message	501	Bad parameters
220	SMTP service ready	502	Command not implemented
221	SMTP closing session	503	Bad command sequence
250	Requested action OK	504	Parameter not implemented
251	User not local; will forward	550	Unknown mailbox
354	Start your mail input	551	Not a local user
421	Service no longer available	552	Exceeded local storage
450	Mailbox busy	553	Invalid mailbox syntax
451	Local error; mail aborted	554	Transaction failed
452	Insufficient space		

Figure 15.3 SMTP response codes.

The second part is a brief text message to the user. Figure 15.3 shows the standard numbers and purposes. Each SMTP author may use his or her own messages, but *not* his or her own numbers. Let's see how they work in the sample SMTP session that follows.

To begin our review of a sample SMTP session, we start with the three-step TCP handshake in Fig. 15.5. As we did for Telnet and FTP, we will step through this session to understand how SMTP performs its task.

We can recognize the following key pieces of session information:

Client port: 20281
Client sequence number: 04db017a
Client window size: 2048
Client maximum segment: 1460

Server port: 25
Server sequence number: bda8bc02
Server window size: 4096
Server maximum segment: 1024

FTP hundreds digit significance		FTP tens digit significance	
		n0n	Syntax error
1nn	Preliminary positive response	n1n	Information
2nn	Positive completion response	n2n	Command connections
3nn	Positiveintermediate response	n3n	Reserved
4nn	Transient negative response	n4n	Reserved
5nn	Permanent negative completion	n5n	Mail system

Figure 15.4 Response code categories.

```
TCP: 20281 -> smtp(25) seq: 04db0179 ack: ——
win: 2048 hl: 6 xsum: 0x454b urg: 0 flags: <SYN> mss: 1460

TCP: smtp(25) -> 20281 seq: bda8bc01 ack: 04db017a
win: 4096 hl: 6 xsum: 0x3ecb urg: 0 flags: <ACK><SYN> mss: 1024

TCP: 20281 -> smtp(25) seq: 04db017a ack: bda8bc02
win: 2048 hl: 5 xsum: 0x47e9 urg: 0 flags: <ACK>
```

Figure 15.5 The TCP session handshake.

SMTP service start

The SMTP session in Fig. 15.6 starts with a response code of 220 in the first segment. This is the server's way of telling the client that the service is ready for a new user. Instead of using those words, this vendor chose to identify the SMTP server as a UNIX system.

In the second segment, the HELO command is the client's announcement that new mail is coming from the domain known as hq11.inr.com. At the same time, this segment carries the acknowledgment of the 15 bytes of data that the client received from the server in the first segment.

The server responds, in the last segment in Fig. 15.6, with the code 250 that tells the client it received the HELO command and the requested action is okay.

Mail origin

In the first segment of Fig. 15.7, the client tells the server that it has incoming mail from a user named Amy. Her e-mail mailbox is on the mail server in

```
TCP: smtp(25) -> 20281 seq: bda8bc02 ack: 04db017a
win: 4096 hl: 5 xsum: 0x42d8 urg: 0 flags: <ACK><PUSH>
data (15/15): 220 unix SMTP

TCP: 20281 -> smtp(25) seq: 04db017a ack: bda8bc11
win: 2033 hl: 5 xsum: 0x3cea urg: 0 flags: <ACK><PUSH>
data (19/19): HELO hq11.inr.com

TCP: smtp(25) -> 20281 seq: bda8bc11 ack: 04db0190
win: 4096 hl: 5 xsum: 0xca92 urg: 0 flags: <ACK><PUSH>
data (10/10): 250 unix
```

Figure 15.6 SMTP service starts.

```
TCP: 20281 -> smtp(25) seq: 04db0190 ack: bda8bc1b
win: 2023 hl: 5 xsum: 0x7821 urg: 0 flags: <ACK><PUSH>
data (30/30): MAIL FROM:<amy@hq11.inr.com>

TCP: smtp(25) -> 20281 seq: bda8bc1b ack: 04db01b1
win: 4096 hl: 5 xsum: 0x3d22 urg: 0 flags: <ACK><PUSH>
data (8/8): 250 OK^M^J
```

Figure 15.7 Mail origin.

the domain hq11.inr.com. This segment also acknowledges the 10 bytes the client just received from the server.

The second segment carries the numeric response that tells the client that the server received the mail from command and processed it. It also adjusts the acknowledgment sequence number to account for the 33 bytes it received in that command.

Mail recipient

The client sends the next command in the first segment of Fig. 15.8, to tell the server that the user addressed the mail to clinta@unix.inr.com. If the mail has multiple recipients, the software will send this command multiple times rather than listing those multiple recipients in one command.

The layout of this particular RCPT TO: command is a bit unusual. Most e-mail packages that use SMTP are more direct in identifying the receiving user's e-mail address. The second segment also acknowledges the 37 bytes of the command and tells the client that it has received and processed the command.

Carrying the mail

The DATA command that appears in the first segment of Fig. 15.9 tells the server that the body of the e-mail message is the next data sent to the server. You can see that the third segment carries that message.

```
TCP: 20281 -> smtp(25) seq: 04db01b1 ack: bda8bc23
win: 2015 hl: 5 xsum: 0x921b urg: 0 flags: <ACK><PUSH>
data (37/37): RCPT TO:<@unix.inr.com:clinta@unix>

TCP: smtp(25) -> 20281 seq: bda8bc23 ack: 04db01d6
win: 4096 hl: 5 xsum: 0x1022 urg: 0 flags: <ACK><PUSH>
data (8/8): 250 OK
```

Figure 15.8 Mail recipient.

```
TCP: 20281 -> smtp(25) seq: 04db01d6 ack: bda8bc2b
win: 2007 hl: 5 xsum: 0x5143 urg: 0 flags: <ACK><PUSH>
data (6/6): DATA

TCP: smtp(25) -> 20281 seq: bda8bc2b ack: 04db01dc
win: 4096 hl: 5 xsum: 0xb2b4 urg: 0 flags: <ACK><PUSH>
data (46/46): 354 Start mail input; end with <CRLF>.<CRLF>

TCP: 20281 -> smtp(25) seq: 04db01dc ack: bda8bc59
win: 1961 hl: 5 xsum: 0xf795 urg: 0 flags: <ACK><PUSH>
data (46/92): Date: Sat 30 May 1992 16:01:10^M^JFrom: amy@head

TCP: smtp(25) -> 20281 seq: bda8bc59 ack: 04db0238
win: 4096 hl: 5 xsum: 0x32e0 urg: 0 flags: <ACK>
```

Figure 15.9 Carrying the mail message.

The second segment carries the instruction that the server is ready for the client's message and that the client should start the mail input. It also adds instructions that it expects the client to end the e-mail message with a sequence of carriage return, line feed, period, carriage return, line feed. That is a period on a line by itself.

As we have seen throughout this session, each of the segments also carries acknowledgment of the previous segment's data. The last segment performs that function only.

SMTP message end

In the previous figure, the server told the client to signify the end of the e-mail message by sending a sequence of a carriage return (^M), a line feed (^J), a period, a carriage return (^M), and a line feed (^J). As instructed, the client has done so in the first segment of Fig. 15.10.

The 250 response from the server (in the second segment) tells the client that the server got the end of the message and processed it. It also acknowledges the 5 bytes of data in that message.

```
TCP: 20281 -> smtp(25) seq: 04db0238 ack: bda8bc59
win: 1961 hl: 5 xsum: 0x65a3 urg: 0 flags: <ACK><PUSH>
data (5/5): ^M^J.^M^J

TCP: smtp(25) -> 20281 seq: bda8bc59 ack: 04db023d
win: 4096 hl: 5 xsum: 0x7321 urg: 0 flags: <ACK><PUSH>
data (8/8): 250 OK
```

Figure 15.10 SMTP message end.

```
TCP: 20281 -> smtp(25) seq: 04db023d ack: bda8bc61
win: 1953 hl: 5 xsum: 0xc340 urg: 0 flags: <ACK><PUSH>
data (6/6): QUIT

TCP: smtp(25) -> 20281 seq: bda8bc61 ack: 04db0243
win: 4096 hl: 5 xsum: 0x5c4c urg: 0 flags: <ACK><PUSH>
data (22/22): 221 unix Terminating
```

Figure 15.11 Ending the SMTP session.

Ending the SMTP session

As with FTP, the command to end an SMTP session is QUIT (see Fig. 15.11). That is both the command the user sends from a command line prompt and the SMTP command that the client software sends to the server to notify it that there is no more need for this particular SMTP session.

The server responds with 221 to let the client know that the server has received (and processed) the QUIT command. In the same segment, the server acknowledges the 6 bytes of data from the client.

Ending the TCP session

Now that the SMTP session has finished its business, the next step is to close the TCP session. The four segments in Fig. 15.12 are the TCP shutdown process.

Multipurpose Internet Mail Extensions

As this chapter has explained, SMTP was designed to carry text messages. RFC 1341 added Multipurpose Internet Mail Extensions (MIME) to the

```
TCP: 20281 -> smtp(25) seq: 04db0243 ack: bda8bc77
win: 1931 hl: 5 xsum: 0x7de8 urg: 0 flags: <ACK><FIN>

TCP: smtp(25) -> 20281 seq: bda8bc77 ack: 04db0244
win: 4096 hl: 5 xsum: 0x08e0 urg: 0 flags: <ACK>

TCP: smtp(25) -> 20281 seq: bda8bc77 ack: 04db0244
win: 4096 hl: 5 xsum: 0x07e0 urg: 0 flags: <ACK><FIN>

TCP: 20281 -> smtp(25) seq: 04db0244 ack: bda8bc78
win: 1931 hl: 5 xsum: 0x7ce8 urg: 0 flags: <ACK>
```

Figure 15.12 Ending the TCP session.

SMTP text handling and the race was on. At last count 16 RFCs identify MIME information with RFCs 1521 and 1522 leading the pack, thereby rendering RFC 1341 obsolete.

The specifications call for a MIME header at the top of the SMTP message that identifies the type of MIME message so that MIME-compliant routers and systems can recognize them. If there are multiple parts, the header defines a boundary string to mark the start of each part. Each art also has its own header to describe the details of each part.

MIME content-type headers

Multipurpose Internet Mail Extensions can carry many different types of content, and each of them has one or more subtypes. As Fig. 15.13 shows, that provides a wide array of MIME contents. Since more are being added with new RFCs, the Assigned Numbers RFC (1700 as of this writing) will have the current MIME content types.

The MIME sample in Fig. 15.14 gives us a basic understanding of where the MIME pieces (we explored in this chapter) may fit in an electronic mail message. We numbered the lines to make it easier to follow the explanation of each piece. These numbers are not part of the message and do not appear in the message or a protocol analyzer's message decode. The lines are as follows:

- Line 1 represents the mail headers that would get us to this point.
- Line 2 identifies the version of MIME we are using.
- Line 3 announces that it is multipart and identifies the boundary statement that will be used.
- Line 4 is intentionally blank to mark the end of the introductory header.
- Line 5 is the boundary statement with its introductory hyphens.
- Line 6 starts this part with the indication that plain text follows in ASCII.
- Line 7 is blank for the same reason that line 4 was left blank: to identify the end of this part's header.
- Line 8 is the introduced plain text.
- Line 9 sets another part's boundary. Line 10 provides the plain text, its size, and the name of the file that is attached.
- Line 11 sets the end of this part's header so the contents can begin.
- Line 12 is the plain text contents of the file teststrp.txt.
- Line 13 is the final boundary as identified by the ending hyphens.

Type	Subtype	Description
Application	activemessage	Binary or application-formatted data
	andrew-insert	
	atomicmail	
	applefile	
	dec-dx	Digital Equipment Corp. document exchange
	dca-rft	IBM's Document Content Architecture-Revisable Format
	mac-binhex40	Converted Macintosh file
	macwriteii	
	msword	
	news-message-id	
	news-transmission	
	octet-stream	
	oda	Office document architecture
	pdf	Adobe Acrobat PostScript
	postscript	
	remote-printing	
	rtf	Richtext format
	slate	
	wita	Wang information transfer application
	wordperfect5.1	
	zip	Compressed file
Audio	basic	
Image	gif	Graphics interchange format
	ief	Image exchange format
	jpeg	Joint photographic experts group
	tiff	Tag image file format
Message	external-body	Document pointer only
	news	Usenet news format
	partial	Part of a larger message
	rfc822	Classic electronic mail
Multipart	alternative	User choice
	appledouble	
	digest	Multiple messages combined
	header-set	
	mixed	
	parallel	Combinations that work together
Text	plain	
	richtext	Word processor format
	tab-separated-values	
Video	mpeg	Motion Picture Experts Group
	quicktime	

Figure 15.13 MIME content types and subtypes.

```
1
2    Mime-Version: 1.0
3    Content-Type: MULTIPART/MIXED; BOUNDARY = "mac.neurolink.com:9612241310"
4
5    —mac.neurolink.com:9612241310
6    Content-Type: TEXT/plain; charset = US-ASCII
7
8    Here is the promised attachment.
9    —mac.neurolink.com:9612241310
10   Content-Type: TEXT/plain; size = 37; name = "teststrp.txt"; charset = US-ASCII
11
12   This is a test, this is only a test.
13   —mac.neurolink.com:9612241310—
```

Figure 15.14 MIME headers.

Encoding

A file may be encoded for different reasons, including

- To assure its privacy without securing it with encryption
- To encapsulate it in an archive with other files before sending it
- To compress it so the transmission will take less time
- To send a binary, sound, executable program, or graphic file using an ASCII method like mail
- A combination of these reasons

Wisely, encoding is separate from SMTP (though it often happens with MIME), as it can also be helpful in FTP or a news feed. We saw one of those encoding methods identified in Fig. 15.13 as an application subtype: mac-binhex40. By converting the MIME attachment from binary to hex, Version 4.0 can send the file using an ASCII transfer method. The receiving station will have to convert it back to binary to use it.

A popular PC-encoding software, uuencode, also lets you convert a presentation file or a word-processing document to an ASCII file for sending it across a network. In its encoded form the file is ASCII and attaches quite well to an e-mail message. While you could read it this way it would make no sense. When it arrives at its destination, another UUCP program, uudecode, translates it back to it original binary form.

Chapter

16

Centrally Managed TCP/IP Addressing

Overview

Most TCP/IP systems can remember their own hardware and protocol addresses. Diskless workstations (like an X-window station) are different. These systems know their own hardware address because the NIC vendor coded it into a chip on the card. However, they must rely on an external resource for their protocol (IP) address.

This is where Reverse Address Resolution Protocol (RARP), Boot Protocol (BootP), and Dynamic Host Configuration Protocol (DHCP) come into play. Each of them, separately, provides a server that stores the protocol address with a cross reference to the hardware address. That server responds to a formatted (Who am I?) request with the protocol (in this case IP) address of the diskless workstation.

The differences between them become clearer when we consider that, of the three, RARP was the first protocol to offer this service. In that case, as is usual for technology, BootP improved the process and is largely replacing RARP. The future is where we see DHCP.

RARP Overview

Reverse Address Resolution Protocol is just what it sounds like, a reverse version of ARP. While ARP's job is getting the remote (target) hardware address to match a known protocol address, RARP's job is to find its own protocol address by supplying its hardware address as a key.

While the idea is simple, designing a RARP server is system dependent and can get complex quickly. The more diskless the device (such as a remote monitoring probe) on the specific network (or subnet), the more complicated the situation. Lest you think that this is a great, inexpensive networking solution, there are a few drawbacks. If the RARP server goes down while many systems are searching for their IP addresses, the results can be disastrous.

The situation is equally challenging when the power returns and all the nodes want their IP addresses at the same time.

The hardware-to-protocol address mapping is usually in the etc/ethers file on a UNIX system. Unlike ARP, RARP is not part of the TCP/IP kernel. The network administrator must load it on the server as a user application. In addition, the server must be able to recognize the RARP identification (8035) in the Ethernet protocol type field.

Like the ARP, RARP goes out as a hardware broadcast. As such, it can only travel on the local side of a router. Backup RARP servers must be on the same subnet or cable segment to support the diskless workstations if the primary RARP server fails. Some routers can be configured to forward RARPs to other subnets. At this point, with reconfiguring time or extra costs, inexpensive is not a word that comes to mind.

RARP request

The RARP request, which is shown in Fig. 16.1, uses the same format as the ARP request. Like the ARP request, a broadcast packet carries it. The RARP's Ether type is 8035.

Differences occur also in the field values. The most obvious is that the request operation field value is 3 instead of ARP's 1. The RARP reply follows that with an operation value of 4.

RARP also differs in that the source and target hardware address fields both contain the sending station's physical (Network Interface Card) address. The sender does this because he or she does not know the RARP server's physical address. This also ensures that the reply will come back to the sender.

```
ff  ff  ff  ff  ff  ff  00  00  -  c0  93  19  00  80  35  00  01
08  00  06  04  00  03  00  00  -  c0  93  19  00  00  00  00  00
00  00  c0  93  19  00  ff  ff  -  ff  ff  00  00  c5  00  10  07
00  a5  00  7b  00  00  8f  a3  -  00  f2  e9  26
```

2 bytes	2 bytes	1 byte	1 byte	2 bytes	Set	Set	Set	Set
Hardware type	Protocol type	Hardware length	Protocol length	Operation code	Source hardware address	Source protocol address	Target hardware address	Target protocol address
00 01	08 00	06	04	00 03	00 00 c0 93 19 00	00 000 00 00	00 00 c0 93 19 00	ff ff ff ff

Figure 16.1 A sample RARP request.

```
00  00  c0  93  19  00  00  00  -  c0  7d  4d  2c  80  35  00  01
08  00  06  04  00  04  00  00  -  c0  7d  4d  2c  c0  99  ba  47
00  00  c0  93  19  00  c0  99  -  ba  71  00  00  c5  00  10  07
00  a5  00  7b  00  00  8f  a3  -  00  f2  e9  26
```

2 bytes	2 bytes	1 byte	1 byte	2 bytes	Set	Set	Set	Set
Hardware type	Protocol type	Hardware length	Protocol length	Operation code	Source hardware address	Source protocol address	Target hardware address	Target protocol address
00 01	08 00	06	04	00 04	00 00 c0 7d 4d 2c	c0 99 ba 47	00 00 c0 93 19 00	c0 99 ba 71

Figure 16.2 A sample RARP reply.

Since the protocol does not know its own IP address, it sets the source protocol address to 0.0.0.0. After all, it is sending a RARP request to find its own identity. The booting system must make sure that the request reaches the RARP server. To do that, it sets the target protocol address to the IP broadcast value of 255.255.255.255 (ff ff ff ff hex).

RARP reply

The RARP reply (see Fig. 16.2) uses the same format as the ARP reply. Like the ARP reply, it goes back to the request sender by using a unicast packet header. It uses the RARP Ether type of 8035.

In returning the requested information, the RARP reply places the requested information in the target protocol address field. The source hardware and protocol addresses contain the RARP server's address.

The diskless workstation can then use its IP address and hardware address to send a TFTP read request for the bootstrap file that the network administrator usually loads on the RARP server as well. The bootstrap file helps the diskless system begin its start-up process.

Since the requesting system has no disk for the file, it must rely on the server's disk to store the file. Most servers store multiple files and so must have a quick and easy way to recall the right file for the requesting system. Many implementations use the hardware address (in its hex form) as part of the name of that file (for example, 0000c0a05124.boot).

No RARP server

If there is no RARP server on the portion of the network containing the diskless workstation, the sender will continue to transmit RARP requests until a preset timeout occurs. Typically, the request sender uses a binary backoff formula to calculate the delay it will use between the retransmissions of the RARP request (with a limit on the size of the interval).

The binary-backoff formula that calculates the interval approximately doubles the interval value before resending the RARP request. This happens until the sender reaches some preset (vendor-defined) timer value. Then the software may retry or notify the user that there is no response.

BootP Overview

The Bootstrap protocol (or Boot Protocol) is a draft standard protocol. It is an alternative to the network-specific standard, RARP. Its name comes from the software design that causes it to fit on a diskless workstation's bootstrap ROM chip. BootP goes beyond RARP by retrieving the path and file name of the bootstrap file as well as the booting system's default router and IP address, and the BootP server's IP address.

One of the reasons that many systems are using BootP instead of RARP is BootP's capability of reaching its server beyond a router. The BootP server can be several hops (routers) away from the BootP client system. BootP's advantages over RARP stem, in part, from the way it travels on the network. RARP uses the Network Access (or Data Link) Layer (Ethernet, 802.3, 802.5, etc.). BootP, as a TCP/IP application, sends its messages in UDP headers enclosed in IP datagrams. The BootP client uses UDP port 68 while the BootP server uses UDP port 67.

BootP request

The BootP message format in Fig. 16.3 applies to both the request and the reply. In the request, the client fills with 0s the fields he or she does not have the information to complete the software to position the rest of the client data fields correctly. The IP header carries a source address of 0.0.0.0 and a target address of 255.255.255.255.

As with ARP, BootP sets the 1-byte Op(eration) code to 1 for the request. BootP also uses the same hardware type and hardware length field values. The fourth byte of the BootP header is the option to identify the maximum number of routers that this request can pass through on its way to the server.

The Transaction ID lets the client verify that the reply it gets is in response to its request. The seconds field tells the server how much time has elapsed since the client began its bootstrap process. Early BootP software fills the 2-byte unused field with 0s. More recent implementations call the same field flags and use the first bit to tell the server to reply to the broadcast address.

This way, clients who cannot receive unicast IP datagrams until they have an IP address are still able to receive the BootP reply. Those BootP clients set the first bit in the flags field to 1, requesting the broadcast reply. Otherwise, that bit (and the other 15 bits in the field) will contain a 0.

Op code (1)	Hardware type (1)	Hardware length (1)	Hops (1)
Transaction ID (4)			
Seconds (2)		Unused or flags (2)	
Client IP address (4)			
Your IP address (4)			
Server IP address (4)			
Gateway IP address (4)			
Client hardware address (16)			
Server host name (64)			
Boot path and file name (128)			
Vendor-specific area (64)			

Figure 16.3 BootP request format. Byte length in parentheses.

If the client has an IP address loaded in its ROM, it provides this in the client IP address field. The last field, usually provided by the client, is the hardware address. BootP uses it as the key to identifying the data that the server will return (just as the protocol address did with ARP). If the client knows the gateway IP address or the server host name and provides it in the request, that is the only server that may respond. Otherwise, any server that receives the broadcast request can reply.

BootP response

As you can see in Fig. 16.4, the BootP reply is in the same message format as the request, and yet the fields that the server fills are, in most cases, different. That starts with the Op(eration) code of 2 to identify this as a BootP reply. The next seven fields return the same values that the client placed there.

Since the BootP server's task is to provide information, it does just that throughout much of the remainder of the message. The server identifies the client's correct IP address, its (the server's) own IP address, and the IP address of the default router or gateway that the client should use for its traffic. If the server record has no specific gateway, the field carries 0s.

The server may send the path and file name of the Bootstrap file, as well as any vendor-specific boot or initial configuration information, back to the client. Note that these optional fields have RFC-specified lengths (rather

Op code (1)	Hardware type (1)	Hardware length (1)	Hops (1)
Transaction ID (4)			
Seconds (2)		Unused or flags (2)	
Client IP address (4)			
Your IP address (4)			
Server IP address (4)			
Gateway IP address (4)			
Client hardware address (16)			
Server host name (64)			
Boot path and file name (128)			
Vendor-specific area (64)			

Figure 16.4 BootP reply format. Byte length in parentheses.

than variable lengths) to make loading this application into ROM an easier (more standard) process.

DHCP Overview

As the mobility of this society grows (and the use of notebook computers with it), we find that one issue that has repeatedly appeared is the need for dynamic assignment of IP addresses. Though TCP/IP's design was around the static assignment of IP addresses, it can handle the newly required flexibility with a recent addition to the family of protocols: Dynamic Host Configuration Protocol.

This protocol is a bit different from RARP and BootP in that its focus is not on diskless workstations but on any TCP/IP systems. It centralizes network administration by automatically assigning valid IP addresses from a pool. When the client is through with the address, it releases that address to return to the pool.

In this way, the client rents or leases the address from the server. The server can impose a time limit and renew the lease as needed. Multiple DHCP servers can each offer an IP address. The client then selects a valid address and confirms it to the server. The rest of the offered addresses return to their prospective servers' pools.

As we will see, the reason we include DHCP is that its roots are in BootP. It simply extends BootP capabilities to perform DHCP functions. In fact, it

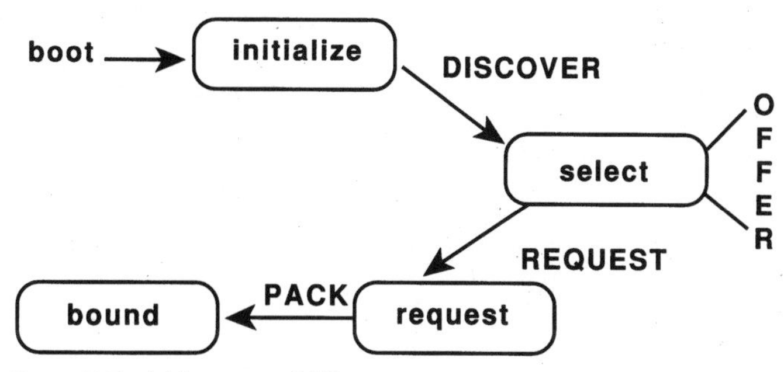

Figure 16.5 Address acquisition process.

uses the BootP client and server port numbers in the UDP header. It is backwardly compatible with BootP so that a DHCP server can answer either a BootP or a DHCP request.

DHCP address acquisition

When a DHCP client first boots, it enters the INITIALIZE state (Fig. 16.5). Once the DHCP software initializes, it sends a DISCOVER message and moves to the SELECT state. All the DHCP servers within the hop count range on the network receive the DISCOVER message.

While in the SELECT state, the client receives offers from all the DHCP servers that the network administrator has configured to respond to the discovering client. In this way, the client can receive multiple offers. Each of the offers contains an IP address and configuring information. With multiple offers, the client must choose one. The software may base that choice on which was received first, configuration parameters, or some other preset value.

Once the client has selected an offer, it sends a REQUEST message to the selected server and enters the request state to negotiate with that server. The server acknowledges the request and starts the lease by sending the client a PACK message. The client moves to the BOUND state on receiving the pack message and thereafter uses the IP address and configuration information.

Client address release

The bound state (see Fig. 16.6) is the operational state for a DHCP client. If it is a disk-equipped system, it may record its IP address onto that disk and request the same IP address the next time it needs one.

When the client determines that it no longer needs the IP address, it sends a RELEASE message to the server. The release signifies that the client will no longer attempt to use the leased address. The client exits the BOUND state and does not enter another state until it needs to initialize for another lease.

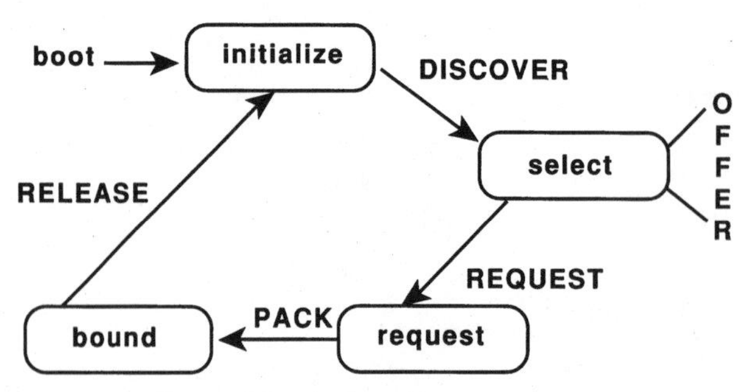

Figure 16.6 Client address release process.

If the systems sent that release in error or the user wants to do something else that requires an IP address, the client must reenter the initialize state and begin the acquisition process again. This can mean leasing a different IP address.

Lease renewal/rebinding

When the client enters the bound state in Fig. 16.7, the protocol sets three timers to control the lease. If the server does not specify values, the client uses defaults. The DHCP sets the first timer for the length of the lease. It specifies the minimum of one hour and the maximum of infinite time (not recommended).

The second timer is one-half of the lease period. When this timer expires, the client must attempt to renew the lease by sending a REQUEST message to the server that provided the lease and go to the RENEW state to wait for the response.

The server can tell the client, by sending a PACK message, that it has renewed the lease and provides new timers for the bound state. If the server sends a NACK, the client must stop using the address and return to the initialize state.

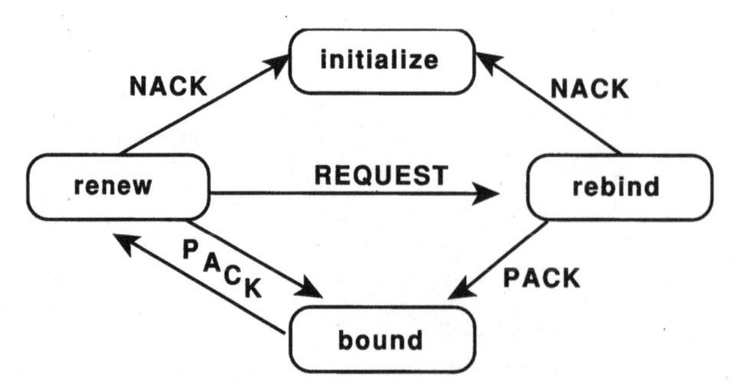

Figure 16.7 Lease renewal and rebinding process.

Op code (1)	Hardware type (1)	Hardware length (1)	Hops (1)
Transaction ID (4)			
Seconds (2)		Unused or flags (2)	
Client IP address (4)			
Your IP address (4)			
Server IP address (4)			
Router IP address (4)			
Client hardware address (16)			
Server host name (64)			
Boot path and file name (128)			
DHCP options (variable)			

Figure 16.8 DHCP message format. Byte length in parentheses.

If no response comes from the server, the third timer (87.5 percent of the lease period) takes over. When it expires, the client moves to the REBIND state and asks any server to renew the lease by sending a REQUEST message. The response to this may continue the lease (PACK) or deny renewal (NACK). No response is equal to a nack.

DHCP message format

As Fig. 16.8 shows, DHCP makes slight modifications to the BootP message format. What BootP defined as a 2-byte reserved field becomes a flags field. Currently, DHCP sets only the flag in the first (leftmost) bit position. All other bits in this field must be 0.

If the first bit is 1, the client is requesting the response by way of a hardware broadcast. DHCP usually does this until the client has an IP address and can revert to the default of a hardware unicast. With one exception, the rest of the fields are the same in DHCP as they are in BootP. For example, requests still carry an Op code of 1 and replies an Op code of 2. The exception is that the BootP vendor-specific area is now the DHCP options field.

DHCP options

Instead of using the flags field for message type identification, DHCP does this in a variable-length options field. While it may change the name of the

Message	Type
Discover	1
Offer	2
Request	3
Decline	4
Pack	5
Nack	6
Release	7

Figure 16.9 DHCP options and message types.

field, DHCP will honor any BootP vendor-specific information included in this field.

Like the vendor-specific area that it replaces, DHCP sets the first 3 bytes in this field to identify the information being carried. DHCP uses the 1-byte code value of 53 to indicate that it is going to set the message type. The 1-byte length field specifies that it will take 1 byte to identify the type of message. As you can see from Fig. 16.9, a number, in the range of 1 through 7, defines the message.

If they are not in use, the server host name and boot path and file name fields waste a lot of bytes. The DHCP has a solution for that. By defining an option overload, the message sender tells the receiver to expect options in that space. Option overload and other BootP and DHCP parameters are listed in Fig. 16.10, an excerpt from RFC 1700.

Virtual IP Networking Issues

With DHCP offering the flexibility, virtual IP networking may not be far behind. Notebook users may no longer need to worry about getting a static IP address from the network administrator for the short (one hour to a few days) time that they would need network access.

It makes the network administrator's job easier. The work of administering the network addresses can drop from knowing where each one is on each separate system to maintaining the DHCP servers that will keep track. As with anything this new, however, one should not run out and jump on the virtual network bandwagon just yet. For example, the DNS uses static entries and DHCP does not interact with DNS (yet). Once it does, the mobile hosts will probably have a fixed name that they pass to the domain name server when the DHCP server assigns an IP address. Some vendors offer third-party packages to accomplish this.

Security personnel will also want a way to track who had use of a particular IP address at a certain time of a certain day. We may need some client logging method to be able to protect the network that way. A good authentication procedure would prevent just anyone with a computer from getting a local IP address and access to the network.

To reinforce the learning experience, Figs. 16.11 (page 291) and 16.12 (page 292) contain protocol analyzer decoded BootP traffic for your review.

Figure 16.10 BootP and DHCP parameters from RFC 1700.

BOOTP AND DHCP PARAMETERS

The Bootstrap Protocol (BOOTP) [RFC951] describes an IP/UDP bootstrap protocol (BOOTP) which allows a diskless client machine to discover its own IP address, the address of a server host, and the name of a file to be loaded into memory and executed. The Dynamic Host Configuration Protocol (DHCP) [RFC1531] provides a framework for automatic configuration of IP hosts. The "DHCP Options and BOOTP Vendor Information Extensions" [RFC1533] describes the additions to the Bootstrap Protocol (BOOTP) which can also be used as options with the Dynamic Host Configuration Protocol (DHCP).

BOOTP Vendor Extensions and DHCP Options are listed below:

Tag	Name	Data length	Meaning
0	Pad	0	None
1	Subnet Mask	4	Subnet Mask Value
2	Time Offset	4	Time Offset in Seconds from UTC
3	Gateways	N	N/4 Gateway addresses
4	Time Server	N	N/4 Timeserver addresses
5	Name Server	N	N/4 IEN-116 Server addresses
6	Domain Server	N	N/4 DNS Server addresses
7	Log Server	N	N/4 Logging Server addresses
8	Quotes Server	N	N/4 Quotes Server addresses
9	LPR Server	N	N/4 Printer Server addresses
10	Impress Server	N	N/4 Impress Server addresses
11	RLP Server	N	N/4 RLP Server addresses
12	Hostname	N	Hostname string
13	Boot File Size	2	Size of boot file in 512-byte chunks
14	Merit Dump File	N	Client to dump and name the file
15	Domain Name	N	The DNS domain name of the client
16	Swap Server	N	Swap Server addeess
17	Root Path	N	Path name for root disk
18	Extension File	N	Path name for more BOOTP info
19	Forward On/Off	1	Enable/Disable IP Forwarding
20	SrcRte On/Off	1	Enable/Disable Source Routing
21	Policy Filter	N	Routing Policy Filters
22	Max DG Assembly	2	Max Datagram Reassembly Size
23	Default IP TTL	1	Default IP Time to Live
24	MTU Timeout	4	Path MTU Aging Timeout
25	MTU Plateau	N	Path MTU Plateau Table
26	MTU Interface	2	Interface MTU Size
27	MTU Subnet	1	All Subnets are Local
28	Broadcast Address	4	Broadcast Address
29	Mask Discovery	1	Perform Mask Discovery
30	Mask Supplier	1	Provide Mask to Others
31	Router Discovery	1	Perform Router Discovery
32	Router Request	4	Router Solicitation Address
33	Static Route	N	Static Routing Table
34	Trailers	1	Trailer Encapsulation
35	ARP Timeout	4	ARP Cache Timeout
36	Ethernet	1	Ethernet Encapsulation
37	Default TCP TTL	1	Default TCP Time to Live
38	Keepalive Time	4	TCP Keepalive Interval
39	Keepalive Data	1	TCP Keepalive Garbage

Tag	Name	Data length	Meaning
40	NIS Domain	N	NIS Domain Name
41	NIS Servers	N	NIS Server Addresses
42	NTP Servers	N	NTP Server Addresses
43	Vendor Specific	N	Vendor Specific Information
44	NETBIOS Name Srv	N	NETBIOS Name Servers
45	NETBIOS Dist Srv	N	NETBIOS Datagram Distribution
46	NETBIOS Note Type	1	NETBIOS Note Type
47	NETBIOS Scope	N	NETBIOS Scope
48	X Window Font	N	X Window Font Server
49	X Window Manmager	N	X Window Display Manager
50	Address Request	4	Requested IP Address
51	Address Time	4	IP Address Lease Time
52	Overload	1	Overload "sname" or "file"
53	DHCP Msg Type	1	DHCP Message Type
54	DHCP Server Id	4	DHCP Server Identification
55	Parameter List	N	Parameter Request List
56	DHCP Message	N	DHCP Error Message
57	DHCP Max Msg Size	2	DHCP Maximum Message Size
58	Renewal Time	4	DHCP Renewal (T1) Time
59	Rebinding Time	4	DHCP Rebinding (T2) Time
60	Class Id	N	Class Identifier
61	Client Id	N	Client Identifier
62	Netware/IP Domain	N	Netware/IP Domain Name
63	Netware/IP Option	N	Netware/IP sub Options
64	NIS-Domain-Name	N	NIS + v3 Client Domain Name
65	NIS-Server-Addr	N	NIS + v3 Server Addresses
66	Server-Name	N	TFTP Server Name
67	Bootfile-Name	N	Boot File Name
68	Home-Agent-Addrs	N	Home Agent Addresses
69	SMTP-Server	N	Simple Mail Server Addresses
70	POP3-Server	N	Post Office Server Addresses
71	NNTP-Server	N	Network News Server Addresses
72	WWW-Server	N	WWW Server Addresses
73	Finger-Server	N	Finger Server Addresses
74	IRC-Server	N	Chat Server Addresses
75	StreetTalk-Server	N	StreetTalk Server Addresses
76	STDA-Server	N	ST Directory Assistance Addresses
77	User-Class	N	User Class Information
78–127	Unassigned		
128–254	Reserved		
255	End	0	None

REFERENCES

[RFC951] Croft, B., and J. Gilmore, "BOOTSTRAP Protocol (BOOTP)", RFC-951, Stanford and SUN Microsytems, September 1985.

[RFC1531] Droms, R., "Dynamic Host Configuration Protocol", Bucknell University, October 1993.

[RFC1533] Alexander, S., and R. Droms, "DHCP Options and BOOTP Vendor Extensions", Lachman Technology, Inc., Bucknell University, October 1993.

URL = ftp://ftp.isi.edu/in-notes/iana/assignments/bootp-and-dhcp-parameters

Figure 16.10 *(Conclusion)*

Ethernet Header	
Destination	= Broadcast
Source	NAT_a80a77
Type	= IP
IP Header	
Version	$45
Type of Service	$00
Total Length	328
Ident	$0110
Flags	$0000(May Fragment, Last Fragment)
Time to Live	255
Protocol	17 (UDP)
Checksum	$baa5
Src. Addr.	0.0.0.0
Dest. Addr.	255.255.255.255
UDP Header	
Src. Port	68 (BOOTPc)
Dest. Port	67 (BOOTPs)
Length	308
Checksum	$%baf
BOOTP Header	
Op Code	1 (BOOTREQUEST)
HW Type	1 (Ethernet)
HW Length	6
Hops	0
Trans ID	$ela80a77
Seconds	16
Client	IP 0.0.0.0
Your IP	0.0.0.0
Server IP	0.0.0.0
Gateway IP	0.0.0.0
Client HW	NAT____80a77
Server Host	
Boot File	
Packet Data	
00 00 00 00 00 00 00 00	....
00 00 00 00 00 00 00 00	....
00 00 00 00 00 00 00 00	....
00 00 00 00 00 00 00 00	,,,,,,,,
00 00 00 00 00 00 00 00	....
Pad/CRC Bytes	
43 1d 47 2c	C?G,

Figure 16.11 Decoded BootP request.

Ethernet Header	
Destination	NAT____a80a77
Source	NAT____16120e
Type	= IP
IP Header	
Version	$45
Type of Service	$00
Total Length	328
Ident	$0000
Flags	$0000 (May Fragment, Last Fragment)
Time to Live	255
Protocol	17 (UDP)
Checksum	$c4e3
Src. Addr.	192.153.186.70
Dest. Addr.	192.153.186.71
UDP Header	
Src. Port	67 (BOOTPs)
Desdt. Port	68 (BOOTPc)
Length	308
Checksum	$0000
BOOTP Header	
Op Code	2 (BOOTREPLY)
HW Type	1 (Ethernet)
HW Length	6
Hops	0
TransID	$ela80a77
Seconds	16
Client	IP 0.0.0.0
Your IP	192.153.186.71
Server IP	192.153.186.70
Gateway IP	192.153.186.100
Client HW	NAT[fru3]180a77
Server Host	
Boot File	
Packet Data	
63 82 53 63 80 01 02 81	c.sc....
05 c0 99 ba 46 04 85 04	....F...
c0 99 ba 47 ff 00 00 00	...G....
00 00 00 00 00 00 00 00	....
00 00 00 00 00 00 00 00	....
Pad/CRC Bytes	
f4 28 51 01	.(Q.

Figure 16.12 Decoded BootP reply.

Chapter

17

Trivial File Transfer Protocol

Overview

Trivial File Transfer Protocol did not get its name by handling unimportant files, but rather from the job that we can expect from the application. Some file transfer situations need the speed and simplicity that UDP can provide more than they need the overhead and reliability of TCP.

To use FTP and TCP would be cumbersome. Imagine loading them into a batch file for updating multiple hosts with new software, new database entries, or new configuration data. Since TFTP uses UDP as its transport layer function, it is small enough to do the job. This is also why diskless workstations use TFTP coded into their ROM chips to load configuration files.

You may remember that UDP is connectionless and does not guarantee delivery. To compensate for this, TFTP uses a timeout retransmission on both ends of the communication.

If a participating host does not get the acknowledgment (or data) within a timer period, the application must send another copy of the data or the acknowledgment. In other words, for each block of data sent there must be an acknowledgment before the transfer can continue.

Client and Server Ports

TFTP uses a variation of the client-server relationship that FTP uses. Instead of setting up another UDP session (since UDP does not have the setup handshake that TCP uses), TFTP adds a server port to free the well-known port for another request.

In our example in Fig. 17.1, you can see the originating request to read or write a file comes from the client's random port number (R_c) and goes to the

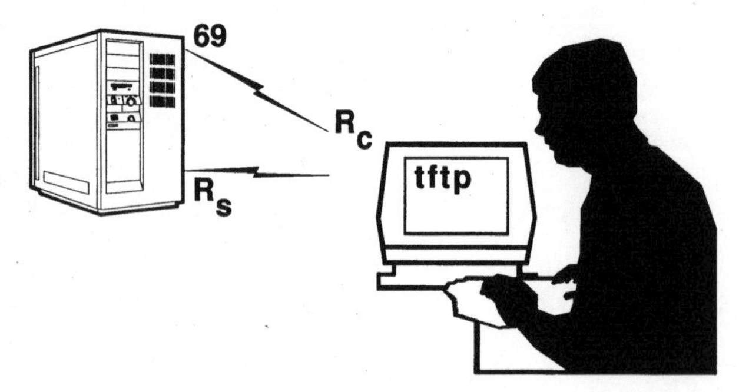

Figure 17.1 TFTP client and server ports.

server's well-known port, 69 (decimal). When the server's software responds, it does so from another client-style random port (R_s) number to carry the data back to the client's original random port. In this way, instead of the four ports that FTP uses, TFTP uses three.

Operation Codes

The operation code (opcode) field indicates the type of TFTP message the software is using in this datagram. The valid opcodes are shown in Fig. 17.2.

Inasmuch as TCP/IP's driving consideration is flexibility, UDP's is speed. By using these operation codes, we can reduce the number of characters necessary to provide information on the content of each message and stay in the "short and sweet" mode that UDP provides. As we will see in the next few sections, the opcodes have their own TFTP message format. The exception that proves the rule is that the read request and write request share the same message format.

Read Request/Write Request Layout

The opcode of 1, read request (RRQ), and 2, write request (WRQ), share the same header format. In keeping with the name Trivial, you can see the simplicity of the header in Fig. 17.3.

Opcode	Type	Purpose
1	RRQ	Read request
2	WRQ	Write request
3	Data	File data
4	Ack	Acknowledgment
5	Error	Error handling

Figure 17.2 TFTP operational codes.

2 bytes	Variable string	1 byte	Variable string	1 byte
Opcode	Filename	00	Data mode	00

Figure 17.3 Opcode 1 and 2—Read/write request layout.

After the 2-byte opcode is the string that identifies the file name the user requests. The TFTP marks the end of this variable length field with an end-of-file name (EOF) marker: a 0-filled byte.

The data mode tells the other end of the link what type of file the application is transferring. Like the file name, TFTP marks the end of this variable length field with an end-of-mode marker: a 0-filled byte. The three valid data mode choices are

netascii This identifies the file's data as standard ASCII text

binary The data in the file is a binary image file

mail The file name is the recipient's address, the data is ASCII text (WRQ only)

Opcode 3—Data

The TFTP data message (opcode 3) transfers information (see Fig. 17.4). After the 2-byte opcode, a 2-byte block number uniquely identifies the particular 512-byte block of data that is in this datagram. The 512-byte arrangement is important in that a block of less than 512 bytes (0 through 511) signifies the end of the file and so the session. If a file ends with the last block of exactly 512 bytes, TFTP sends an extra block with no data. That way the last file block is always less than 512 bytes.

Opcode 4—Ack

The TFTP Ack message (opcode 4) acknowledges receipt of a block of information (see Fig. 17.5). After the 2-byte opcode, a 2-byte block number identifies the particular 512-byte block of data that TFTP has received.

Since UDP does not have the TCP windowing capability, the application must acknowledge each block of data before the other end of the session can send the next block of data. We call this *lock-step* or *stop-and-wait* acknowledgment.

2 bytes	2 bytes	2 bytes
Opcode	Block number	Data

Figure 17.4 Opcode 3—Data message layout.

2 bytes	2 bytes
Opcode	Block number

Figure 17.5 Opcode 4—Ack message layout.

Each time TFTP sends a block, it sets a timer. If the Ack message does not return by the timeout, the data-sending side retransmits the block. With a retransmit timer involved, this can become a challenge as we will see later in this chapter.

Error Handling

There are eight possible error codes (see Fig. 17.6). The TFTP defines seven codes and leaves one undefined to cover any other possibility. Most are self-explanatory but error code 7 may need some additional clarification. The software can use it in response to a TFTP write request that is carrying mail that the sender addressed to an e-mail user the receiving system does not know.

The error message acknowledges any of the other four message types by pinpointing the trouble area. After the 2-byte opcode field's value of 5 is the 2-byte error code field. Figure 17.7 depicts an opcode 5 error message layout.

A variable length field follows the error code field. It contains the vendor's choice of text to define the error. As we saw earlier, these variable length fields in TFTP end with a 0 byte as an end-of-text marker.

TFTP Challenges

As we said previously, there are some challenges with TFTP. First is the question of reliability: There is no error detection except UDP's optional checksum. Second is the security issue. TFTP has no user ID (login) or password to identify the person requesting a file or adding a new file to the target host.

Error code	Error text
0	Undefined error
1	File not found
2	Access violation
3	Disc space exceeded
4	Illegal operation
5	Unknown transfer ID
6	File exists
7	No such user

Figure 17.6 Opcode 5—Error codes.

2 bytes	2 bytes	Variable string	1 byte
Opcode	Error code	Error text	00

Figure 17.7 Opcode 5—Error message layout.

Third, there is a potential for excessive traffic. This is the result of an undocumented feature known as the Sorcerer's Apprentice Bug (see Fig. 17.8). If the retransmit timer expires before the Ack message reaches the file block sender, the sender retransmits the last unacknowledged block. When the Ack finally arrives, the sender generates the next block. When the same block arrives a second time at the receiving host, it sends another Ack message. When the second Ack for the same block arrives, the sender transmits another copy of the next block. In other words, this doubling can happen over and over and build unnecessary traffic on the network because someone set a timer to too low of a value.

A Sample Read Session

Now let's use samples of sessions to look at how TFTP does its job on a network. Since each session is so brief, we will look at three different sessions to understand different aspects of TFTP's methods in file transfer.

In the first UDP datagram in Fig. 17.9, we see that the read request is for a netascii file named *testfile.* The client uses random port 5342 to make the request.

Figure 17.8 Sorcerer's apprentice.

```
UDP: 5342 -> tftp(69) len: 29 xsum: ——
TFTP: Read request file name: testfile mode: netascii

UDP: 1041 -> 5342 len: 140 xsum: 0xf0dc
TFTP: Data block: 1

UDP: 5342 -> 1041 len: 12 xsum: ——
TFTP: Ack block: 1
```

Figure 17.9 TFTP sample read session.

The second datagram carries the data from the server to the client. Notice that the server is using random port 1041 to respond to the client's random port. It is carrying the 132 bytes of block 1 data that follows the 8-byte UDP header. The UDP header length identifies its own length as well as the TFTP length in the 140 value it carries. Since it is less than 512 bytes, that datagram must be the end of the file, so TFTP has completed the file transfer. The Ack datagram, last in this figure, acknowledges the receipt of that block 1 data.

A Sample Write Session

In the first datagram in Fig. 17.10, the client sends a write request for a binary file named *tftptest.* The client uses random port 5343 to make that request.

The server responds in the second datagram to okay the write request with a message acknowledging block 0. There was no block 0 message. Instead, TFTP uses this message to identify that the server's random port 1042 will receive the data from the client's random port 5343. It carries only the 4 bytes of data in the Ack message that follows the UDP header.

The third datagram carries the 132 bytes of data in block 1 to the server. Notice that this vendor chose not to use the UDP checksum here or in the first datagram in Fig. 17.10.

```
UDP: 5343 -> tftp(69) len: 27 xsum: ——
TFTP: Write request file name: tftptest mode: binary

UDP: 1042 -> 5343 len: 12 xsum: 0xb7cb
TFTP: Ack block: 0

UDP: 5343 -> 1042 len: 140 xsum: ——
TFTP: Data block: 1

UDP: 1042 -> 5343 len: 12 xsum: 0xb7ca
TFTP: Ack block: 1
```

Figure 17.10 TFTP sample write session.

```
UDP: 5344 -> tftp(69) len: 25 xsum: ——
TFTP: Read request file name: \tftp mode: netascii

UDP: 1043 -> 5344 len: 41 xsum: 0x7497
TFTP: Error: Access violation Error message: unau-
thorized file
```

Figure 17.11 TFTP sample session error message.

Ending the session is the Ack datagram, last in this figure. It responds by acknowledging the receipt of that block 1 data. Since that block is less than 512 bytes, it must be the end of the file, so the TFTP has finished the write session.

A Sample Session Error Message

In Figure 17.11's first datagram, the read request is to retrieve a netascii file named *\tftp*. The client uses random port 5344 to make that request.

The server responds in the second datagram to tell the client that it found an error as a result of that request. Apparently the server had locked the file or directory or otherwise barred the client from accessing the file. Since the server cannot send a file, the TFTP session ends there (in most cases). Some vendors have built in an error handling capability that allows for a change in the file name the user is requesting. It will then start a new TFTP session to make the new request.

Chapter

18

Simple Network Management Protocol

Basics

A network manager's need for two things drove the development of SNMP: first, a method of managing internetworks that used the TCP/IP protocol suite; and second, a single, common way to manage many diverse vendor products. SNMP lets network managers view (get) and modify (set) a wide variety of network parameters.

Jeffrey Case, James Davin, Mark Fedor, and Martin Schoffstall designed Version 1 of SNMP, which appears in RFC 1157 and RFC 1098. Keith McCloghrie and Marshall Rose edited the MIB I and MIB II RFCs (1066 and 1156, and 1158 and 1213, respectively) and wrote the Structure of Management Information RFCs (1065 and 1155).

The first commercial SNMP implementations appeared in late 1988. As originally planned, SNMP would be a short-term solution to ease the transition from Simple Gateway Monitoring Protocol (SGMP) (RFC 1028) on through Common Management Information Protocol (CMIP) over TCP/IP (CMOT) (RFC 1214) to CMIP as networks migrated to OSI from TCP/IP.

A Model

In the manager-agent system we use in SNMP, the manager user interface hides all the work that goes on behind the scenes. It is usually a graphical display program that a vendor has designed to make information retrieval and display easier. It can present the retrieved data through one or more management applications. It uses the command structure of SNMP to make it easier to understand network conditions and events.

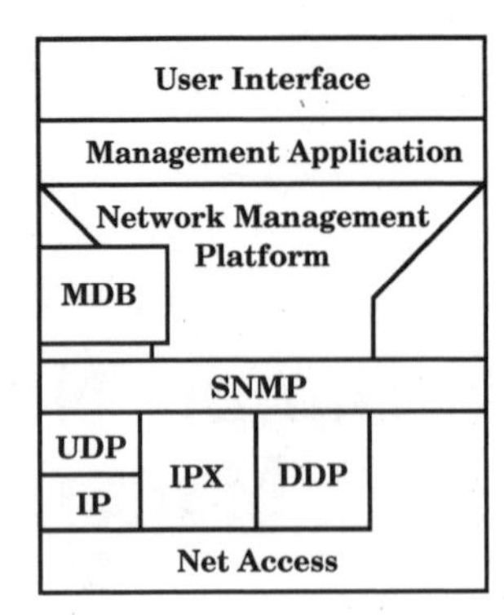

Figure 18.1 An SNMP manager model.

If you use a management application, it will organize and format the retrieved data. It will also offer a method of telling the network management station what to retrieve and add an extra layer of user control to the NMS functions. This way, the management station can act as a data platform to retrieve and display data to the user.

The NMS uses the available manageable objects database (MDB) to find the questions it can ask each agent. The MDB is a data dictionary that the SNMP system has compiled from multiple management information bases (MIBs). MIBs contain lists of manageable objects that an NMS user can retrieve or modify. For example: Is that interface up? How long has this system been running? What is in the ARP cache in that router? Who is running what application on the server?

The structure of management information (SMI) identifies the exact data object in each MIB, in terms that the managed device (agent) can understand. It also identifies the type of data the agent will store and a way to identify what the SNMP message is carrying.

As Fig. 18.1 shows, SNMP can work over protocols and communication methods other than TCP/IP. In its basic configuration it uses UDP/IP to communicate with the managed, IP-addressed host. RFCs specify the other protocols as AppleTalk's Datagram Delivery Protocol (DDP), Novell NetWare's Internet Packet eXchange (IPX), and directly over Ethernet.

The Agent

An SNMP agent is very similar to a CIA agent. It works in the area that we want to know about and provides the inside information we ask it to find. The SNMP agent is a software piece that runs on the managed device when that device is network enabled (see Fig. 18.2). It uses the same network layer protocol as the management station. As we will see later in this chapter, both systems (NMS and agent) also support common management information base (MIB) objects.

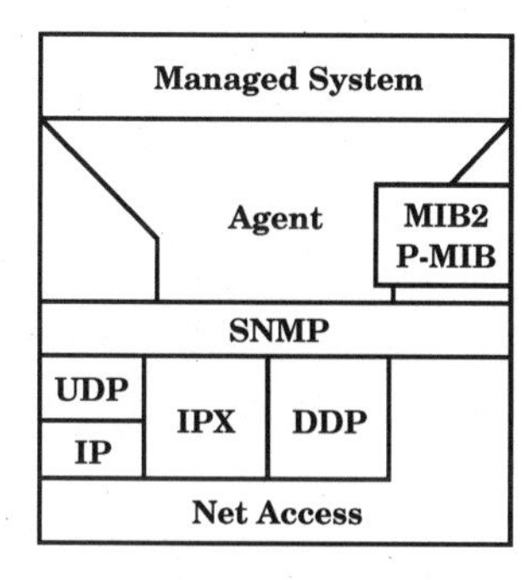

Figure 18.2 An SNMP agent model.

After the agent receives the NMS's request for data, the SNMP software in the agent uses the SMI to decode the message. The SMI must figure out which of its MIBs contains the object(s) the NMS requested, a request number, the access authority, the version of SNMP being used, the type of request being made, and so on.

For example, using the correct MIB, it must match the request to the specific MIB item (object) and retrieve that information. That means the agent must know how and where to retrieve every MIB object that it supports. That can be hundreds or thousands of objects. The agent does not have to support all the data that a MIB contains. It makes no sense for an agent in a bridge to support any of the Exterior Gateway Protocol group (that we will see is part of MIB II) since EGP does not run on a bridge.

After retrieving the data (or finding the data it requested is not available), the agent uses the SMI and ASN.1 to encode a response. The return path through UDP/IP and the network takes the answer back to the NMS, which displays it for the user.

The Structure of Management Information

RFC 1065 first defined the structure of management information in August 1988. RFC 1155 later modified it to its present form in May 1990 (see Fig. 18.3). The SMI performs three functions to make the managed entities' information logically accessible from the NMS.

The SMI first identifies the network management tree and the information "leaves" (or objects) on that tree that the manager may retrieve or modify. In doing so, it requires that each managed object will have a unique object identifier (OID), a name, an ASN.1 syntax, and a means of encoding the value for transmission. Management information bases are groups of tree-structured OIDs.

The SMI also identifies the fields needed to define correctly each object type in the appropriate MIB. These fields must include a name with the OID, syntax, access, status, and description. Other possible fields include a reference, an index for tabled information, and a default value.

```
                  ISO (1)
                     |
        Identified Organizations (3)
                     |
                  DOD (6)
                     |
                Internet (1)
                     |
Management (2) -----Experimental (3)-----Private (4)
     |                                      |
  MIB (1)                            Enterprise (1)
```

Figure 18.3 SNMPv1 MIB tree.

Third, it identifies the data types for each of the ASN.1 tag classes. We will look into this aspect as we explore ASN.1 and its relationship to SNMP.

Abstract Syntax Notation One

Abstract Syntax Notation One offers a standard, platform-independent method of representing data across any network. It also hides its functions in many different aspects of SNMP. ASN.1 sets the data types of messages and the objects in the MIBs. The ASN.1 basic encoding rules (BER) encode and decode the messages to send them on the network.

For all of its complexity, ASN.1 (shown in Fig. 18.4) handles platform differences in several areas:

- It helps resolve data type combinations like EBCDIC and ASCII.
- It resolves "big endian" versus "little endian" incompatibilities.
- It sets structure alignments on byte versus word boundaries.
- It deals with the size (number of octets and/or bytes) of the data types.

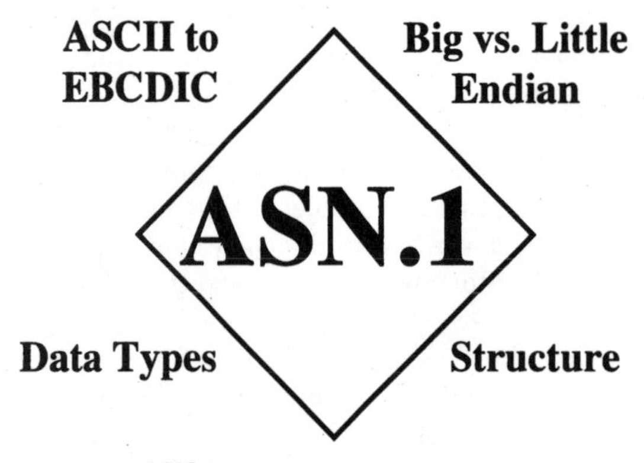

Figure 18.4 ASN.1 support.

Management Information Base

A MIB is a virtual database that identifies each manageable object and then organizes these objects into similar groups for flexibility and easier management. The NMS uses MIBs as a data dictionary to identify the objects that it will manage in the agent. The agent uses the MIB to store a value for each manageable object.

RFC 1066 first defined an open MIB in August 1988, and RFC 1156 modified it in May 1990 to contain 114 manageable objects in eight groups: system, interfaces, address translation, IP, ICMP, TCP, UDP, and EGP. We call the result MIB I.

In May 1990, RFC 1158 introduced the current open MIB (known as MIB II), which RFC 1213 modified in March 1991. MIB II added three new groups (CMOT, transmission, and SNMP) and, as a superset of MIB I, it now contains 171 manageable objects (see Fig. 18.5). Vendor-specific MIBs often arrange their objects logically under these or similar groups.

The IETF planned CMOT (Common Management Information Protocol Over TCP/IP) as the migration from SNMP (along with the change from TCP/IP to OSI) to CMIP. This group will probably be empty in most implementations of NMSs or agents. RFC 1214 defined it, and the current standards RFC (RFC 1920) classifies RFC 1214 as a historical protocol (its time has come and gone).

RFC 1213 does not define any objects in the transmission group. Instead, separate RFCs (see Fig. 18.6) identify the objects that a network manager can choose to compile under the transmission group. The IETF considers those RFCs experimental until they are proven, when they are then promoted to standard status. This way each NMS can have a transmission group that is appropriate for the managed network.

OID	MIB group	Provides
1.3.6.1.2.1.1	System	Model, type, and software info
1.3.6.1.2.1.2	Interfaces	Data on the logical I/O ports
1.3.6.1.2.1.3	Address translation	Table of IP to MAC addresses
1.3.6.1.2.1.4	IP	Data about the IP processes
1.3.6.1.2.1.5	ICMP	Info about ICMP functions
1.3.6.1.2.1.6	TCP	TCP-oriented data
1.3.6.1.2.1.7	UDP	UDP-related information
1.3.6.1.2.1.8	EGP	Information about EGP activities
1.3.6.1.2.1.9	CMOT	OSI support (historical)
1.3.6.1.2.1.10	Transmission	Physical I/O port info
1.3.6.1.2.1.11	SNMP	Data about SNMP processes

Figure 18.5 SNMP MIB II groups.

OID and name	Syntax	Access	Description
1.3.6.1.2.1.10 Transmission	Group	None	Data for managing each interface's underlying transmission media.
1.3.6.1.2.1.10.5 X.25 Packet Layer	Group	None	The ITU-T X.25 packet layer. (RFC1382)
1.3.6.1.2.1.10.7 802.3	Group	None	Ethernet-like media access. (RFC1623)
1.3.6.1.2.1.10.8 802.4	Group	None	IEEE 802.4 media access. (RFC1230, RFC1239)
1.3.6.1.2.1.10.9 802.5	Group	None	IEEE 802.5 media access. (RFC1231, RFC1239)
1.3.6.1.2.1.10.15 FDDI	Group	None	Fiber Distributed Data Interface media access. (RFC1285, RFC1512)
1.3.6.1.2.1.10.16 LAPB	Group	None	Link Access Procedure Balanced (LAPB) media access. (RFC1381)
1.3.6.1.2.1.10.18 DS-1	Group	None	Digital Signal and Cross Connect, Level One media access. (RFC1406)
1.3.6.1.2.1.10.19 E-1	Group	None	European equivalent to DS-1 media access. (RFC1406)
1.3.6.1.2.1.10.23 PPP	Group	None	Point-to-Point Protocol media access. (RFC1471)
1.3.6.1.2.1.10.30 DS-3	Group	None	Digital Signal and Cross Connect, Level Three media access. (RFC1407)
1.3.6.1.2.1.10.31 SMDS IP	Group	None	Switched Multimegabit Data Services Interface Protocol. (RFC1694)
1.3.6.1.2.1.10.32 Frame Relay	Group	None	Frame Relay DTE media access. (RFC1315)
1.3.6.1.2.1.10.33 RS-232	Group	None	RS-232-like media access. (RFC1659)
1.3.6.1.2.1.10.34 Parallel MIB	Group	None	Parallel-Printer-like media access. (RFC1660)
1.3.6.1.2.1.10.35 ARCnet	Group	None	Attached Resource Computer Network (ARCNet) media access.
1.3.6.1.2.1.10.36 ARCnet-plus	Group	None	Attached Resource Computer Network Plus media access.
1.3.6.1.2.1.10.37 ATM	Group	None	Asynchronous Transfer Mode media access. (RFC1695)
1.3.6.1.2.1.10.38 MIOX25	Group	None	Multiprotocol Interconnect on X.25 media access. (RFC1461)
1.3.6.1.2.1.10.39 SONET	Group	None	Synchronous Optical Network media access. (RFC1595)
1.3.6.1.2.1.10.44 Frame Relay Svc.	Group	None	Frame Relay Service for DCE. (RFC1604)

Figure 18.6 Transmission group.

Decimal	Name	Decimal	Name
0	Reserved	8	PSI
1	Proteon	9	cisco
2	IBM	10	NSC
3	CMU	11	HP
4	UNIX	12	Epilogue
5	ACC	13	U of Tennessee
6	TWG	14	BBN
7	CAYMAN	15	Xylogics, Inc.

Figure 18.7 A sample of private MIB numbers. Prefix: 1.3.6.1.4.1.

Private MIBs

Beyond the sample in Fig. 18.7, over 1800 organizations (or people) have chosen to create their own MIB. There are two main reasons for this decision. One reason is that vendors create MIBs to provide a better way to manage uniquely their product and service offerings. A second reason is that organizations, particularly colleges and universities, create their own MIBs for common (platform-independent) objects that they can use in managing a diverse array of systems and software.

We can get many of these MIBs via anonymous FTP from ftp.isi.edu and in the /mib directory. Some organizations, instead of placing their MIB in this directory, provide a pointer file that indicates the location of their MIB. Those pointer files typically end with the .txt suffix.

The current list of organizations that have their own MIB can be found in the appendix at the end of this chapter. RFC 1700 also has a limited list of approximately 1200 private MIBs. Since these are the enterprise numbers that identify the separate private MIBs, their object ID numbers all begin with the sequence 1.3.6.1.4.1.*n* (where the *n* is the IANA-assigned enterprise number). By comparing this OID with Fig. 18.3, we can see how to build an OID.

Note that these different organizations have different needs. We cannot expect any interoperability between different vendors' MIBs. That means vendor A's MIB will not gather the expected information from vendor B's agent. However, MIB designers do try to organize their objects logically under groups such as those in MIB II.

Network Management Applications

Network management applications (NMAs) are one of the hot topics in SNMP these days. Vendors design them as enhancements to the network manager. The first generation of NMAs included audible alarm reporting and trouble ticket generation.

The next and current generation added knowledge-based, expert systems (artificial intelligence) that can alert management personnel (by paging or other means) when one or more specific events occur that indicate potential problems. These expert systems can also support network planning and operations, provide information in groupings that managers need to save the time of multiple requests, and take steps to correct specific problems in the network.

Many NMAs have graphical user interfaces (GUIs). The GUIs present the management of a network from that vendor's perspective. While the information may be the same (and even use the same NMS platform), each NMA can combine the retrieved data and display the results completely differently.

SNMP Version 1 and 2 Protocol Data Units

The IETF released RFCs 1901 to 1908 in January 1996 as the updates to RFCs 1441 to 1452. During that time, we have been working toward the potential of SNMPv2 becoming the standard for network management on the Internet. In fact, RFC 1905 covers the formats of SNMPv2 message types the same way that RFC 1157 does for SNMPv1. We list the current and proposed message types (protocol data units) in Figure 18.8.

The first four protocol data units (PDUs) are in both SNMP versions. SNMPv2 changes the GetResponse name (only) to Response. SNMPv2 replaces the SNMPv1 trap response with a new format and calls it the SNMPv2 Trap so that PDU number 4 becomes obsolete.

The Desktop Management Task Force

Desktop Management Task Force (DMTF) is a consortium of vendors who have been working since 1992 to define a method for managing desktop workstations. The specification requires a network operating system, a network management protocol, and platform independence for flexibility and interoperability. It also specifies these through SNMP access.

Message	PDU number	SNMP version	Purpose
Get request	0	1 and 2	Request primitive MIB values
Get next request	1	1 and 2	Request table MIB values
Get response	2	1 and 2	Agent response to all requests
Set request	3	1 and 2	Change agent information
Trap response	4	1	Agent event-triggered alarm
Get bulk request	5	2	Combined request
Inform request	6	2	NMS to NMS data transfers
SNMPv2 trap	7	2	Version 2 trap

Figure 18.8 SNMP PDUs.

The information about each desktop system includes managed objects in the operating system software, application software versions and license data, and workstation hardware information including add-in boards (network interface cards, video cards, multimedia cards, etc.), drives (hard disks, diskettes, CD ROMs, etc.), and CPU and other chips (math coprocessor, ROM, and RAM).

The DMTF plans a desktop management interface (DMI), which will include a management interface, a service layer, and a component interface. The DMI will offer a standard interface to protocols like SNMP (with others to follow) and the service layer. The service layer will handle requests and responses for information between the management interface and the component interface. The component interface offers easy implementation for common access to various vendors' components.

The DMTF announced DMI 2.0 in March 1996. For more information point your web browser to http://www.dmtf.org.

Appendix SMI Network Management Private Enterprise Numbers

Prefix: iso.org.dod.internet.private.enterprise (1.3.6.1.4.1)

This file is

ftp://ftp.isi.edu/in-notes/iana/assignments/enterprise-numbers

Decimal	Name		References
0	Reserved	Joyce K. Reynolds	<jkrey@isi.edu>
1	Proteon	Avri Doria	<avri@proteon.com>
2	IBM	Bob Moore	remoore@ralvm6.vnet.ibm.com
3	CMU	Steve Waldbusser	<sw01+@andrew.cmu.edu>
4	Unix	Keith Sklower	<sklower@okeeffe.berkeley.edu>
5	ACC	Art Berggreen	<art@SALT.ACC.COM>
6	TWG	John Lunny	<jlunny@eco.twg.com> (703) 847-4500
7	CAYMAN	Beth Miaoulis	beth@cayman.com
8	PSI	Marty Schoffstahl	schoff@NISC.NYSER.NET
9	cisco	Greg Satz	satz@CISCO.COM
10	NSC	Geof Stone	geof@NETWORK.COM
11	Hewlett Packard	R. Dwight Schettler	rds%hpcndm@HPLABS.HP.COM
12	Epilogue	Karl Auerbach	karl@cavebear.com
13	U of Tennessee	Jeffrey Case	case@CS.UTK.EDU
14	BBN	David Waitzman	djw@bbn.com
15	Xylogics, Inc.	Jim Barnes	barnes@xylogics.com
16	Timeplex	Laura Bridge	laura@uunet.UU.NET
17	Canstar	Sanand Patel	sanand@HUB.TORONTO.EDU
18	Wellfleet	Caralyn Brown	cbrown@wellfleet.com
19	TRW	Mathew Lew	mlew@venice.dh.trw.com
20	MIT	Jon Rochlis	jon@ATHENA.MIT.EDU
21	EON	Michael Waters	—none—
22	Fibronics	Jakob Apelblat	jakob@fibronics.co.il
23	Novell	Steve Bostock	steveb@novell.com

Decimal	Name		References
24	Spider Systems	Peter Reid	peter@spider.co.uk
25	NSFNET	Hans-Werner Braun	HWB@MCR.UMICH.EDU
26	Hughes LAN Systems	Keith McCloghrie	KZM@HLS.COM
27	Intergraph	Guy Streeter	guy@guy.bll.ingr.com
28	Interlan	Bruce Taber	taber@europa.InterLan.COM
29	Vitalink Communications		
30	Ulana	Bill Anderson	wda@MITRE-BEDFORD.ORG
31	NSWC	Stephen Northcutt	SNORTHC@RELAY-NSWC.NAVY.MIL
32	Santa Cruz Operation	Keith Reynolds	keithr@SCO.COM
33	Xyplex	Bob Stewart	STEWART@XYPLEX.COM
34	Cray	Hunaid Engineer	hunaid@OPUS.CRAY.COM
35	Bell Northern Research	Glenn Waters	gwaters@BNR.CA
36	DEC	Ron Bhanukitsiri	rbhank@DECVAX.DEC.COM
37	Touch	Brad Benson	—none—
38	Network Research Corp.	Bill Versteeg	bvs@NCR.COM
39	Baylor College of Medicine	Stan Barber	SOB@BCM.TMC.EDU
40	NMFECC-LLNL	Steven Hunter	hunter@CCC.MFECC.LLNL.GOV
41	SRI	David Wolfe	ctabka@TSCA.ISTC.SRI.COM
42	Sun Microsystems	Dennis Yaro	yaro@SUN.COM
43	3Com	Jeremy Siegel	jzs@NSD.3Com.COM
44	CMC	Dave Preston	—none—
45	SynOptics	David Perkins	dperkins@synoptics.com
46	Cheyenne Software	Reijane Huai	sibal@CSD2.NYU.EDU
47	Prime Computer	Mike Spina	WIZARD%enr.prime.com@RELAY.CS.NET
48	MCNC/North Carolina Data Network	Ken Whitfield	ken@MCNC.ORG
49	Chipcom	John Cook	cook@chipcom.com
50	Optical Data Systems	Josh Fielk	—none—
51	gated	Jeffrey C. Honig	jch@gated.cornell.edu
52	Cabletron Systems	Roger Dev	—none—
53	Apollo Computers	Jeffrey Buffun	jbuffum@APOLLO.COM
54	DeskTalk Systems, Inc.	David Kaufman	—none—
55	SSDS	Ron Strich	—none—
56	Castle Rock Computing	John Sancho	—none—
57	MIPS Computer Systems	Charles Marker II	marker@MIPS.COM
58	TGV, Inc.	Ken Adelman	Adelman@TGV.COM
59	Silicon Graphics, Inc.	Ronald Jacoby	rj@SGI.COM
60	University of British Columbia	Don McWilliam	mcwillm@CC.UBC.CA
61	Merit	Bill Norton	wbn@MERIT.EDU
62	NetEdge	Dave Minnich	dave_minnich@netedge.com
63	Apple Computer Inc	Jim Hayes	Hayes@APPLE.COM
64	Gandalf	Henry Kaijak	—none—
65	Dartmouth	Philip Koch	Philip.Koch@DARTMOUTH.EDU
66	David Systems	Kathryn de Graaf	degraaf@davidsys.com
67	Reuter	Bob Zaniolo	—none—
68	Cornell	Laurie Collinsworth	ljc1@cornell.edu
69	LMS	L. Michael Sabo	Sabo@DOCKMASTER.NCSC.MIL
70	Locus Computing Corp.	Arthur Salazar	lcc.arthur@SEAS.UCLA.EDU
71	NASA	Steve Schoch	SCHOCH@AMES.ARC.NASA.GOV
72	Retix	Alex Martin	—none—
73	Boeing	Jerry Geisler	—none—
74	AT&T	Rich Bantel	Richard.Bantel@att.com
75	Ungermann-Bass	Didier Moretti	—none—

Decimal	Name		References
76	Digital Analysis Corporation	Skip Koppenhaver	stubby!skip@uunet.UU.NET
77	LAN Manager	Doug Karl	KARL-D@OSU-20.IRCC.OHIO-STATE.EDU
78	Netlabs	Jonathan Biggar	jon@netlabs.com
79	ICL	Jon Infante	—-none—-
80	Auspex Systems	Rod Livingood	rodo@auspex.com
81	Lannet Company	Efrat Ramati	—-none—-
82	Network Computing Devices	Dave Mackie	lupine!djm@UUNET.UU.NET
83	Raycom Systems	Bruce Willins	—-none—-
84	Pirelli Focom Ltd.	Sam Lau	—-none—-
85	Datability Software Systems	Larry Fischer	lfischer@dss.com
86	Network Application Technology	Jim Kinder	jkinder@nat.com
87	LINK (Lokales Informatik-Netz Karlsruhe)	Guenther Schreiner	snmp-admin@ira.uka.de
88	NYU	Bill Russell	russell@cmcl2.NYU.EDU
89	RND	Rina Nethaniel	—-none—-
90	InterCon Systems Corporation	Amanda Walker	AMANDA@INTERCON.COM
91	Coral Network Corporation	Jason Perreault	jason@coral.com
92	Webster Computer Corporation	Robert R. Elz	kre@munnari.oz.au
93	Frontier Technologies Corporation	Prakash Ambegaonkar	—-none—-
94	Nokia Data Communications	Douglas Egan	—-none—-
95	Allen-Bradely Company	Bill King	abvax!calvin.icd.ab.com!wrk@uunet.UU.NET
96	CERN	Jens T. Rasmussen	jenst%cernvax.cern.ch@CUNYVM.CUNY.EDU
97	Sigma Network Systems, Inc.	Ken Virgile	signet!ken@xylogics.COM
98	Emerging Technologies, Inc.	Dennis E. Baasch	etinc!dennis@uu.psi.com
99	SNMP Research	Jeffrey Case	case@SNMP.COM
100	Ohio State University	Shamim Ahmed	ahmed@nisca.ircc.ohio-state.edu
101	Ultra Network Technologies	Julie Dmytryk	Julie_Dmytryk.MKT@usun.ultra.com
102	Microcom	Annmarie Freitas	—-none—-
103	Lockheed Martin	David Rageth	dave@lmco.com
104	Micro Technology	Mike Erlinger	mike@lexcel.com
105	Process Software Corporation	Bernie Volz	VOLZ@PROCESS.COM
106	Data General Corporation	Joanna Karwowska	karwowska@dg-rtp.dg.com
107	Bull Company	Anthony Berent	berent@rdgeng.enet.dec.com
108	Emulex Corporation	Jeff Freeman	—-none—-
109	Warwick University Computing Services	Israel Drori	raanan@techunix.technion.ac.il
110	Network General Corporation	James Davidson	ngc!james@uunet.UU.NET
111	Oracle	John Hanley	jhanley@oracle.com

Decimal	Name		References
112	Control Data Corporation	Nelluri L. Reddy	reddy@uc.msc.umn.edu
113	Hughes Aircraft Company	Keith McCloghrie	KZM@HLS.COM
114	Synernetics, Inc.	Jas Parmar	jas@synnet.com
115	Mitre	Bede McCall	bede@mitre.org
116	Hitachi, Ltd.	Hirotaka Usuda	—none—
117	Telebit	Mark S. Lewis	mlewis@telebit.com
118	Salomon Technology Services	Paul Maurer II	—none—
119	NEC Corporation	Yoshiyuki Akiyama	kddlab!ccs.mt.nec.co. jp!y-akiyam@uunet.uu.net
120	Fibermux	Michael Sung	msung@ccrelay.fibermux.com
121	FTP Software Inc.	Stev Knowles	stev@vax.ftp.com
122	Sony	Takashi Hagiwara	Hagiwara@Sm.Sony.Co.Jp
123	Newbridge Networks Corporation	James Watt	james@newbridge.com
124	Racal-Datacom	Frank DaCosta	frank_dacosta@usa.racal.com
125	CR SYSTEMS	Soren H. Sorensen	—none—
126	DSET Corporation	Dan Shia	dset!shia@uunet.UU.NET
127	Computone	Nick Hennenfent	nick@computone.com
128	Tektronix, Inc.	Dennis Thomas	dennist@tektronix.TEK.COM
129	Interactive Systems Corporation	Steve Alexander	stevea@i88.isc.com
130	Banyan Systems Inc.	Deepak Taneja	eepak = Taneja%Eng%Banyan@Thing. banyan.com
131	Sintrom Datanet Limited		
132	Bell Canada	Mark Fabbi	markf@gpu.utcs.utoronto.ca
133	Crosscomm Corporation	Reuben Sivan	crossc!rsivan@uunet.UU.NET
134	Rice University	Catherine Foulston	cathyf@rice.edu
135	OnStream Networks	Annie Dang	annie@onstream.com
136	Concurrent Computer Corporation	John R. LoVerso	loverso@westford.ccur.com
137	Basser	Paul O'Donnell	paulod@cs.su.oz.au
138	Luxcom		
139	Artel	Jon Ziegler	Ziegler@Artel.com
140	Independence Technologies, Inc. (ITI)	Gerard Berthet	gerard@indetech.com
141	Frontier Software Development	Narendra Popat	—none—
142	Digital Computer Limited	Osamu Fujiki	—none—
143	Eyring, Inc.	Ron Holt	ron@Eyring.COM
144	Case Communications	Peter Kumik	—none—
145	Penril DataComm, Inc.	Keith Hogan	keith%penril@uunet.uu.net
146	American Airlines	Bill Keatley	—none—
147	Sequent Computer Systems	Louis Fernandez	lfernandez@sequent.com
148	Bellcore	Kaj Tesink	kaj@nvuxr.cc.bellcore.com
149	Konkord Communications	Ken Jones	konkord!ksj@uunet.uu.net
150	University of Washington	Christopher Wheeler	cwheeler@cac.washignton.edu
151	Develcon	Sheri Mayhew	zaphod!sherim@herald.usask.ca
152	Solarix Systems	Paul Afshar	paul@solar1.portal.com
153	Unifi Communications Corp.	Yigal Hochberg	yigal@unifi.com
154	Roadnet	Dale Shelton	—none—
155	Network Systems Corp.	Nadya K. El-Afandi	nadya@khara.network.com

Decimal	Name		References
156	ENE (European Network Engineering)	Peter Cox	—none—
157	Dansk Data Elektronik A/S	Per Bech Hansen	pbh@dde.dk
158	Morning Star Technologies	Karl Fox	karl@MorningStar.Com
159	Dupont EOP	Oscar Rodriguez	—none—
160	Legato Systems, Inc.	Jon Kepecs	kepecs@Legato.COM
161	Motorola	Vince Enriquez	vince@sundevil.sps.mot.com
162	European Space Agency (ESA)	Eduardo	EDUATO%ESOC.BITNET@CUNYVM.CUNY.EDU
163	BIM	Bernard Lemercier	bl@sunbim.be
164	Rad Data Communications Ltd.	Brighitte Shalom	brighitte@radmail.rad.co.il
165	Intellicom	Paul Singh	—none—
166	Shiva Corporation	Andrew Rodwin	arodwin@shiva.com
167	Fujikura America	Debbie Reed	—none—
168	Xlnt Designs, Inc. (XDI)	Mike Anello	mike@xlnt.com
169	Tandem Computers	Rex Davis	—none—
170	BICC	David A. Brown	fzbicdb@uk.ac.ucl
171	D-Link Systems, Inc.	Henry P. Nagai	—none—
172	AMP, Inc.	Rick Downs	—none—
173	Netlink	Mauro Zallocco	—none—
174	C. Itoh Electronics	Larry Davis	—none—
175	Sumitomo Electric Industries (SEI)	Kent Tsuno	tsuno@sumitomo.com
176	DHL Systems, Inc.	David B. Gurevich	dgurevic@rhubarb.ssf-sys.dhl.com
177	Network Equipment Technologies	Raj Bhatia	rbhatia@net.com
178	APTEC Computer Systems	Larry Burton	ssds!larryb@uunet.UU.NET
179	Schneider & Koch & Co, Datensysteme GmbH	Thomas Ruf	tom@rsp.de
180	Hill Air Force Base	Russell G. Wilson	rwilson@oodis01.af.mil
181	ADC Kentrox	Mike Witt	mwitt@kentrox.com
182	Japan Radio Co.	Nagayuki Kojima	nkojima@lab.nihonmusen.co.jp
183	Versitron	Matt Harris	—none—
184	Telecommunication Systems	Hugh Lockhart	—none—
185	Interphase	Gil Widdowson	—none—
186	Toshiba Corporation	Mike Asagami	toshiba@mothra.nts.uci.edu
187	Clearpoint Research Corp.		
188	Ascom	Andrew Smith	andrew@hasler.ascom.ch
189	Fujitsu America	Chung Lam	—none—
190	NetCom Solutions, Inc.	Dale Cabell	—none—
191	NCR	Cheryl Krupczak	clefor@secola.columbia.ncr.com
192	Dr. Materna GmbH	Torsten Beyer	tb@Materna.de
193	Ericsson Business Communications	Thomas Hansen	lmethan@lme.ericsson.se
194	Metaphor Computer Systems	Paul Rodwick	—none—
195	Patriot Partners	Paul Rodwick	—none—
196	The Software Group Limited (TSG)	Ragnar Paulson	tsgfred!ragnar@uunet.UU.NET
197	Kalpana, Inc.	Anil Bhavnani	—none—

Decimal	Name		References
198	University of Waterloo	R. J. White	snmp-tech@watmath.waterloo.edu
199	CCL/ITRI	Ming-Perng Chen	N100CMP0%TWNITRI1.BITNET @CUNYVM.CUNY.EDU
200	Coeur Postel	Professor Kynikos	Special Consultant
201	Mitsubish Cable Industries, Ltd.	Masahiko Hori	—none—
202	SMC	Lance Sprung	—none—
203	Crescendo Communication, Inc.	Prem Jain	prem@cres.com
204	Goodall Software Engineering	Doug Goodall	goodall@crl.com
205	Intecom	Patrick Deloulay	pdelou@intecom.com
206	Victoria University of Wellington	Jonathan Stone	jonathan@isor.vuw.ac.nz
207	Allied Telesis, Inc.	Scott Holley	SCOTT_CLINTON_HOLLEY@cup.portal.com
208	Cray Communications A/S	Hartvig Ekner	hj@craycom.dk
209	Protools	Glen Arp	—none—
210	Nippon Telegraph and Telephone Corp.	Toshiharu Sugawara	sugawara%wink.ntt.jp@RELAY.CS.NET
211	Fujitsu Limited	Ippei Hayashi	hayashi@sysrap.cs.fujitsu.co.jp
212	Network Peripherals Inc.	Creighton Chong	cchong@fastnet.com
213	Netronix, Inc.	Jacques Roth	—none—
214	University of Wisconsin—Madison	Dave Windorski	DAVID.WINDORSKI@MAIL.ADMIN.WISC.EDU
215	NetWorth, Inc.	Craig Scott	—none—
216	Tandberg Data A/S	Harald Hoeg	haho%huldra.uucp@nac.no
217	Technically Elite Concepts, Inc.	Russell S. Dietz	Russell_Dietz@Mcimail.com
218	Labtam Australia Pty. Ltd.	Michael Podhorodecki	michael@labtam.oz.au
219	Republic Telcom Systems, Inc.	Steve Harris	rtsc!harris@boulder.Colorado.edu
220	ADI Systems, Inc.	Paul Liu	—none—
221	Microwave Bypass Systems, Inc.	Tad Artis	—none—
222	Pyramid Technology Corp.	Richard Rein	rein@pyramid.com
223	Unisys_Corp	Lawrence Brow	—none—
224	LANOPTICS LTD., Israel	Israel Drori	raanan@techunix.technion.ac.il
225	NKK Corporation	J. Yoshida	—none—
226	MTrade UK Ltd.	Peter Delchiappo	—none—
227	Acals	Patrick Cheng	pcheng@dill.ind.trw.com
228	ASTEC, Inc.	Hiroshi Fujii	fujii@astec.co.jp
229	Delmarva Power	John K. Scoggin, Jr.	scoggin@delmarva.com
230	Telematics International, Inc.	Kevin Smith	—none—
231	Siemens Nixdorf Informations Syteme AG	Gunther Kroenert	—none—
232	Compaq		
233	NetManage, Inc.	William Dunn	netmanage@cup.portal.com
234	NCSU Computing Center	David Joyner	david@unity.ncsu.edu
235	Empirical Tools and Technologies	Karl Auerbach	karl@empirical.com
236	Samsung Group	Hong K. Paik	paik@samsung.com
237	Takaoka Electric Mfg. Co., Ltd.	Hidekazu Hagiwara	hagiwara@takaoka.takaoka-electric.co.jp

Decimal	Name		References
238	Netrix Systems Corporation	Eldon S. Mast	esm@netrix.com
239	WINDATA	Bob Rosenbaum	—none—
240	RC International A/S	Carl H. Dreyer	chd@rci.dk
241	Netexp Research	Henk Boetzkes	—none—
242	Internode Systems Pty Ltd	Simon Hackett	simon@ucs.adelaide.edu.au
243	netCS Informationstechnik GmbH	Oliver Korfmacher	okorf@bunt.netcs.com
244	Lantronix	Greg Wheeler	gregw@lantronix.com
245	Avatar Consultants	Kory Hamzeh	ames!avatar.com!kory@harvard.harvard.edu
246	Furukawa Electoric Co. Ltd.	Shoji Fukutomi	kddlab!polo.furukawa.co.jp!fuku @uunet.UU.NET
247	Nortel Dasa Network Systems GmbH & Co. KG	R.Osten	(07531)86-2586
248	Richard Hirschmann GmbH & Co.	Heinz Nisi	mia@intsun.rus.uni-stuttgart.de
249	G2R Inc.	Khalid Hireche	—none—
250	University of Michigan	Tim Howes	Tim.Howes@terminator.cc.umich.edu
251	Netcomm, Ltd.	W.R. Maynard-Smith	—none—
252	Sable Technology Corporation	Rodney Thayer	rodney@sabletech.com
253	Xerox	Fonda Lix Pallone	Fonda_Lix_Pallone.PARC@Xerox.Com
254	Conware Computer Consulting GmbH	Michael Sapich	sapich@conware.de
255	Compatible Systems Corp.	John Gawf	gawf@compatible.com
256	Scitec Communications Systems Ltd.	Stephen Lewis	—none—
257	Transarc Corporation	Pat Barron	Pat_Barron@TRANSARC.COM
258	Matsushita Electric Industrial Co., Ltd.	Nob Mizuno	mizuno@isl.mei.co.jp
259	ACCTON Technology	Don Rooney	—none—
260	Star-Tek, Inc.	Carl Madison	carl@startek.com
261	Codenoll Tech. Corp.	Dan Willie	—none—
262	Formation, Inc.	Carl Marcinik	—none—
263	Seiko Instruments, Inc. (SII)	Yasuyoshi Watanabe	—none—
264	RCE (Reseaux de Communication d'Entreprise S.A.)	Etienne Baudras-Chardigny	—none—
265	Xenocom, Inc.	Sean Welch	welch@raven.ulowell.edu
266	KABELRHEYDT	Hubert Theissen	—none—
267	Systech Computer Corporation	Brian Petry	systech!bpetry@uunet.UU.NET
268	Visual	Brian O'Shea	bos@visual.com
269	SDD (Scandinavian Airlines Data Denmark A/S)	Per Futtrup	—none—
270	Zenith Electronics Corporation	David Lin	—none—
271	TELECOM FINLAND	Petri Jokela	—none—
272	BinTec Computersystems	Marc Sheldon	ms@BinTec.DE
273	EUnet Germany	Marc Sheldon	ms@Germany.EU.net
274	PictureTel Corporation	Oliver Jones	oj@pictel.com
275	Michigan State University	Lih-Er Wey	WEYLE@msu.edu

Decimal	Name		References
276	GTE Government Systems—Network Management Organization	Grant Gifford	gifford_grant@nmo.gtegsc.com
277	Cascade Communications Corp.	Chikong Shue	alpo!chi@uunet.uu.net
278	Hitachi Cable, Ltd.	Takahiro Asai	—none—
279	Olivetti	Marco Framba	framba@orc.olivetti.com
280	Vitacom Corporation	Parag Rastogi	parag@cup.portal.com
281	INMOS	Graham Hudspith	gwh@inmos.co.uk
282	AIC Systems Laboratories Ltd.	Glenn Mansfield	glenn@aic.co.jp
283	Cameo Communications, Inc.	Alan Brind	—none—
284	Diab Data AB	Mats Lindstrom	mli@diab.se
285	Olicom A/S	Lars Povlsen	krus@olicom.dk
286	Digital-Kienzle Computersystems	Hans Jurgen Dorr	—none—
287	CSELT (Centro Studi E Laboratori Telecomunicazioni)	Paolo Coppo	coppo@cz8700.cselt.stet.it
288	Electronic Data Systems	Mark Holobach	holobach@tis.eds.com
289	McData Corporation	Glenn Levitt	gpl0363@mcmail.mcdata.com
290	Harris Corporation	Tom Georges	tgeorges@harris.com
291	Technology Dynamics, Inc.	Chip Standifer	TDYNAMICS@MCIMAIL.COM
292	DATAHOUSE Information Systems Ltd.	Kim Le	—none—
293	Securicor 3net (NDL) Ltd	Tony van der Peet	vanderPeetT@securicor.co.nz
294	Texas Instruments	Blair Sanders	Blair_Sanders@mcimail.com
295	PlainTree Systems Inc.	Paul Chefurka	chefurka@plntree.UUCP
296	Hedemann Software Development	Stefan Hedemann	100015.2504@compuserve.com
297	Fuji Xerox Co., Ltd.	Hiroshi Kume	Kume%KSPB%Fuji_Xerox@tcpgw.netg.ksp.fujixerox.co.jp
298	Asante Technology	Hsiang Ming Ma	—none—
299	Stanford University	RL "Bob" Morgan	morgan@jessica.stanford.edu
300	Digital Link	Jimmy Tu	jimmy@dl.com
301	Raylan Corporation	Mark S. Lewis	mlewis@telebit.com
302	Datacraft	Alan Lloyd	alan@datacraft.oz
303	Hughes	Keith McCloghrie	KZM@HLS.COM
304	Farallon Computing, Inc.	Sam Roberts	sroberts@farallon.com
305	GE Information Services	Steve Bush	sfb@ncoast.org
306	Gambit Computer Communications	Zohar Seigal	—none—
307	Livingston Enterprises, Inc.	Steve Willens	steve@livingston.com
308	Star Technologies	Jim Miner	miner@star.com
309	Micronics Computers Inc.	Darren Croke	dc@micronics.com
310	Basis, Inc.	Heidi Stettner	heidi@mtxinu.COM
311	Microsoft	John M. Ballard	jballard@microsoft.com
312	US West Advance Technologies	Donna Hopkins	dmhopki@uswat.uswest.com
313	University College London	Shaw C. Chuang	S.Chuang@cs.ucl.ac.uk
314	Eastman Kodak Company	W. James Colosky	wjc@tornado.kodak.com
315	Network Resources Corporation	Kathy Weninger	—none—
316	Atlas Telecom	Bruce Kropp	ktxc8!bruce@uunet.UU.NET

Decimal	Name		References
317	Bridgeway	Umberto Vizcaino	—none—
318	American Power Conversion Corp.	Peter C. Yoest	apc!yoest@uunet.uu.net
319	DOE Atmospheric Radiation Measurement Project	Paul Krystosek	krystosk@eid.anl.gov
320	VerSteeg CodeWorks	Bill Versteeg	bvs@NCR.COM
321	Verilink Corp	Bill Versteeg	bvs@NCR.COM
322	Sybus Corportation	Mark T. Dauscher	mdauscher@sybus.com
323	Tekelec	Bob Grady	—none—
324	NASA Ames Research Cente	Nick Cuccia	cuccia@nas.nasa.gov
325	Simon Fraser University	Robert Urquhart	quipu@sfu.ca
326	Fore Systems, Inc.	Eric Cooper	ecc@fore.com
327	Centrum Communications, Inc.	Vince Liu	—none—
328	NeXT Computer, Inc.	Lennart Lovstrand	Lennart_Lovstrand@NeXT.COM
329	Netcore, Inc.	Skip Morton	—none—
330	Northwest Digital Systems	Brian Dockter	—none—
331	Andrew Corporation	Ted Tran	—none—
332	DigiBoard	Dror Kessler	dror@digibd.com
333	Computer Network Technology Corp.	Bob Meierhofer	—none—
334	Lotus Development Corp.	Bill Flanagan	bflanagan@lotus.com
335	MICOM Communication Corporation	Donna Beatty	SYSAD@prime.micom.com
336	ASCII Corporation	Toshiharu Ohno	tony-o@ascii.co.jp
337	PUREDATA Research	Tony Baxter	tony@puredata.com
338	NTT DATA	Yasuhiro Kohata	kohata@rd.nttdata.jp
339	Empros Systems International	David Taylor	dtaylor@ems.cdc.ca
340	Kendall Square Research (KSR)	Dave Hudson	tdh@uunet.UU.NET
341	ORNL	Gary Haney	hny@ornl.gov
342	Network Innovations	Pete Grillo	pl0143@mail.psi.net
343	Intel Corporation	Brady Orand	borand@pcocd2.intel.com
344	Proxar	Ching-Fa Hwang	cfh@proxar.com
345	Epson Research Center	Richard Schneider	rschneid@epson.com
346	Fibernet	George Sandoval	—none—
347	Box Hill Systems Corporation	Tim Jones	tim@boxhill.com
348	American Express Travel Related Services	Jeff Carton	jcarton@amex-trs.com
349	Compu-Shack	Tomas Vocetka	OPLER%CSEARN.bitnet@CUNYVM.CUNY.EDU
350	Parallan Computer, Inc.	Charles Dulin	—none—
351	Stratacom	Clyde Iwamoto	cki@strata.com
352	Open Networks Engineering, Inc.	Russ Blaesing	rrb@one.com
353	ATM Forum	Keith McCloghrie	KZM@HLS.COM
354	SSD Management, Inc.	Bill Rose	—none—
355	Automated Network Management, Inc.	Carl Vanderbeek	—none—
356	Magnalink Communications Corporation	David E. Kaufman	—none—

Decimal	Name		References
357	Kasten Chase Applied Research	Garry McCracken	pdxgmcc@rvax.kasten.on.ca
358	Skyline Technology, Inc.	Don Weir	—none—
359	Nu-Mega Technologies, Inc.	Dirk Smith	—none—
360	Morgan Stanley & Co. Inc.	Victor Kazdoba	vsk@katana.is.morgan.com
361	Integrated Business Network	Michael Bell	—none—
362	L & N Technologies, Ltd.	Steve Loring	—none—
363	Cincinnati Bell Information Systems, Inc.	Deron Meranda	dmeranda@cbis.COM
364	OSCOM International	Farhad Fozdar	f_fozdar@fennel.cc.uwa.edu.au
365	MICROGNOSIS	Paul Andon	pandon@micrognosis.co.uk
366	Datapoint Corporation	Lee Ziegenhals	lcz@sat.datapoint.com
367	RICOH Co. Ltd.	Toshio Watanabe	watanabe@godzilla.rsc.spdd.ricoh.co.jp
368	Axis Communications AB	Martin Gren	martin@axis.se
369	Pacer Software	Wayne Tackabury	wft@pacersoft.com
370	Axon Networks Inc.	Robin Iddon	robini@axon.com
371	Brixton Systems, Inc.	Peter S. Easton	easton@brixton.com
372	GSI	Etienne Demailly	etienne.demailly@gsi.fr
373	Tatung Co., Ltd.	Chih-Yi Chen	TCCISM1%TWNTTIT.BITNET@pucc. Princeton.EDU
374	DIS Research LTD.	Ray Compton	rayc@command.com
375	Quotron Systems, Inc.	Richard P. Stubbs	richard@atd.quotron.com
376	Dassault Electronique	Olivier J. Caleff	caleff@dassault-elec.fr
377	Corollary, Inc.	James L. Gula	gula@corollary.com
378	SEEL, Ltd.	Ken Ritchie	—none—
379	Lexcel	Mike Erlinger	mike@lexcel.com
380	Sophisticated Technologies, Inc.	Bill Parducci	70262.1267@compuserve.com
381	OST	A. Pele	—none—
382	Megadata Pty Ltd.	Andrew McRae	andrew@megadata.mega.oz.au
383	LLNL Livermore Computer Center	Dan Nessett	nessett@ocfmail.ocf.llnl.gov
384	Dynatech Communications	Graham Welling	s8000!gcw@uunet.uu.net
385	Symplex Communications Corp.	Cyrus Azar	—none—
386	Tribe Computer Works	Ken Fujimoto	fuji@tribe.com
387	Taligent, Inc.	Lorenzo Aguilar	lorenzo@taligent.com
388	Symbol Technologies, Inc.	John Kramer	+1-408-369-2679 jkramer@psd. symbol.com
389	Lancert	Mark Hankin	—none—
390	Alantec	Paul V. Fries	pvf@alantec.com
391	Ridgeback Solutions	Errol Ginsberg	bacchus!zulu!errol@uu2.psi.com
392	Metrix, Inc.	D. Venkatrangan	venkat@metrix.com
393	Symantec Corporation	Jim Hill	jhill@symantec.com
394	NRL Communication Systems Branch	R. K. Nair	nair@itd.nrl.navy.mil
395	I.D.E. Corporation	Rob Spade	—none—
396	Matsushita Electric Works, Ltd.	Claude Huss	claude@trc.mew.mei.co.jp
397	MegaPAC	Ian George	—none—
398	Pinacl Communication Systems Ltd	Dave Atkinson	dave_a@pinacl.co.uk
399	Hitachi Computer Products (America), Inc.	Masha Golosovker	masha@hicomb.hi.com

Decimal	Name		References
400	METEO FRANCE	Remy Giraud	Remy.Giraud@meteo.fr
401	PRC Inc.	Jim Noble	noble_jim@prc.com
402	Wal-Mart Stores, Inc.	Mike Fitzgerel	mlfitzg@wal-mart.com
403	Nissin Electric Company, Ltd.	Aki Komatsuzaki	(408) 737-0274
404	Distributed Support Information Standard	Mike Migliano	<mike@uwm.edu>
405	SMDS Interest Group (SIG)	Elysia C. Tan	<ecmt1@sword.bellcore.com>
406	SolCom Systems Ltd.	Hugh Evans	0506 873855
407	Bell Atlantic	Colin deSa	socrates!bm5ld15@bagout. BELL-ATL.COM
408	Advanced Multiuser Technologies Corporation		
409	Mitsubishi Electric Corporation	Yoshitaka Ogawa	<ogawa@nkai.cow.melco.co.jp>
410	C.O.L. Systems, Inc.	Frank Castellucci	(914) 277-4312
411	University of Auckland	Nevil Brownlee	<n.brownlee@aukuni.ac.nz>
412	Desktop Management Task Force (DMTF)	Dave Perkins	<dperkins@synoptics.com>
413	Klever Computers, Inc.	Tom Su	408-735-7723 kci@netcom.com
414	Amdahl Corporation	Steve Young	sy@uts.admahl.com
415	JTEC Pty, Ltd.	Edward Groenendaal	eddyg@jtec.com.au
416	Matra Communcation	Hong-Loc Nguyen	(33.1) 34.60.85.25
417	HAL Computer Systems	Michael A. Petonic	petonic@hal.com
418	Lawrence Berkeley Laboratory	Russ Wright	wright@lbl.gov
419	Dale Computer Corporation	Dean Craven	1-800-336-7483
420	IPTC, Universitaet of Tuebingen	Andreas J. Haug	<ahaug@mailserv.zdv. uni-tuebingen.de>
421	Bytex Corporation	Mary Ann Burt	<bytex!ws054!maryann@uunet. UU.NET>
422	Cogwheel, Inc.	Brian Ellis	bri@Cogwheel.COM
423	Lanwan Technologies	Thomas Liu	(408) 986-8899
424	Thomas-Conrad Corporation	Karen Boyd	512-836-1935
425	TxPort	Bill VerSteeg	bvs@ver.com
426	Compex, Inc.	Andrew Corlett	BDA@ORION.OAC.UCI.EDU
427	Evergreen Systems, Inc.	Bill Grace	(415) 897-8888
428	HNV, Inc.	James R. Simons	jrs@denver.ssds.COM
429	U.S. Robotics, Inc.	Chris Rozman	chrisr@usr.com
430	Canada Post Corporation	Walter Brown	+1 613 722-8843
431	Open Systems Solutions, Inc.	David Ko	davidk@ossi.com
432	Toronto Stock Exchange	Paul Kwan	(416) 947-4284
433	Mamakos\TransSys Consulting	Louis A. Mamakos	louie@transsys.com
434	EICON	Vartan Narikian	vartan@eicon.qc.ca
435	Jupiter Systems	Russell Leefer	rml@jupiter.com
436	SSTI	Philip Calas	(33) 61 44 19 51
437	Grand Junction Networks	Randy Ryals	randyr@grandjunction.com
438	Anasazi, Inc.	Chad Larson	(chad@anasazi.com)
439	Edward D. Jones and Company	John Caruso	(314) 851-3422

Decimal	Name		References
440	Amnet, Inc.	Richard Mak	mak@amnet.COM
441	Chase Research	Kevin Gage	—none—
442	PEER Networks	Randy Presuhn	randy@peer.com
443	Gateway Communications, Inc.	Ed Fudurich	—none—
444	Peregrine Systems	Eric Olinger	eric@peregrine.com
445	Daewoo Telecom	SeeYoung Oh	oco@scorpio.dwt.co.kr
446	Norwegian Telecom Research	Paul Hoff	paalh@brage.nta.no
447	WilTel	Anil Prasad	anil_prasad@wiltel.com
448	Ericsson-Camtec	Satish Popat	—none—
449	Codex	Thomas McGinty	—none—
450	Basis	Heidi Stettner	heidi@mtxinu.COM
451	AGE Logic	Syd Logan	syd@age.com
452	INDE Electronics	Gordon Day	gday@inde.ubc.ca
453	ISODE Consortium	Steve Kille	S.Kille@isode.com
454	J.I. Case	Mike Oswald	mike@helios.uwsp.edu
455	Trillium	Jeff Lawrence	j_lawrence@trillium.com
456	Bacchus Inc.	Errol Ginsberg	bacchus!zulu!errol@uu2.psi.com
457	MCC	Doug Rosenthal	rosenthal@mcc.com
458	Stratus Computer	Dave Snay	dks@sw.stratus.com
459	Quotron	Richard P. Stubbs	richard@atd.quotron.com
460	Beame & Whiteside	Carl Beame	beame@ns.bws.com
461	Cellular Technical Services	Keith Gregoire	keith@celtech.com
462	Shore Microsystems, Inc.	Gordon Elam	(309) 229-3009
463	Telecommunications Techniques Corp.	Tom Nisbet	nisbet@tt.com
464	DNPAP (Technical University Delft)	Jan van Oorschot	<bJan.vOorschot@dnpap.et.tudelft.nl>
465	Plexcom, Inc.	Bruce Miller	(805) 522-3333
466	Tylink	Stavros Mohlulis	(508) 285-0033
467	Brookhaven National Laboratory	Dave Stampf	drs@bach.ccd.bnl.gov
468	Computer Communication Systems	Gerard Laborde	<Gerard.Laborde@sp1.y-net.fr>
469	Norand Corp.	Rose Gorrell	319-269-3100
470	MUX-LAP	Philippe Labrosse	514-735-2741
471	Premisys Communications, Inc	Harley Frazee	harley@premisys.com
472	Bell South Telecommunications	Johnny Walker	205-988-7105
473	J. Stainsbury PLC	Steve Parker	44-71-921-7550
474	Ki Research Inc	Toni Barckley	410-290-0355x220
475	Wandel and Goltermann Technologies	David Walters	919-941-5730x4203 <walter@wg.com>
476	Emerson Computer Power	Roger Draper	714-457-3638 rdraper@cerf.net
477	Network Software Associates	Leslie Santiago	SANTIAGL@netsoft.com
478	Procter and Gamble	Peter Marshall	513-983-1100x5988
479	Meridian Technology Corporation	Kenneth B. Denson	<kdenson@magic.meridiantc.com>
480	QMS, Inc.	Bill Lott	lott@imagen.com
481	Network Express	Tom Jarema	313-761-5051 ITOH@MSEN.COM
482	LANcity Corporation	Pam Yassini	pam@lancity.com

Decimal	Name		References
483	Dayna Communications, Inc.	Sanchaita Datta	datta@signus.utah.edu
484	kn-X Ltd.	Sam Lau	44 943 467007
485	Sync Research, Inc.	Alan Bartky	(714) 588-2070
486	PremNet	Ken Huang	HuangK@rimail.interlan.com
487	SIAC	Peter Ripp	(212) 383-9061
488	New York Stock Exchange	Peter Ripp	(212) 383-9061
489	American Stock Exchange	Peter Ripp	(212) 383-9061
490	FCR Software, Inc.	Brad Parker	brad@fcr.com
491	National Medical Care, Inc.	Robert Phelan	(617) 466-9850
492	Dialogue Communication Systemes, S.A.	Klaus Handke	+(49) 30 802 24 97
493	NorTele	Bjorn Kvile	+47 2 48 89 90
494	Madge Networks, Inc.	Duncan Greatwood	dgreatwo@madge.mhs.compuserve.com
495	Memotec Communications	Graham Higgins	ghiggins@teleglobe.com
496	CTON	Nick Hennenfent	nicholas@cton.com
497	Leap Technology, Inc.	George Economou	george@leap.com
498	General DataComm, Inc.	William Meltzer	meltzer@gdc.com
499	ACE Communications, Ltd.	Danny On	972-3-570-1423
500	Automatic Data Processing (ADP)	Alex Rosin	(201) 714-3982
501	Programa SPRITEL	Alberto Martinez	Martinez_Alberto_SPRITEL@euskom.spritel.es
502	Adacom	Aial Haorch	972-4-899-899
503	Metrodata Ltd	Nick Brown	100022.767@compuserve.com
504	Ellemtel Telecommunication Systems Laboratories	Richard G Bruvik	Richard.Bruvik@eua.ericsson.se
505	Arizona Public Service	Duane Booher	DBOOHER@APSC.COM
506	NETWIZ, Ltd.,	Emanuel Wind	eumzvir@techunix.technion.ac.il
507	Science and Engineering Research Council (SERC)	Paul Kummer	P.Kummer@daresbury.ac.uk
508	The First Boston Corporation	Kevin Chou	csfb1!dbadmin4!kchou@uunet.UU.NET
509	Hadax Electronics Inc.	Marian Kramarczyk	73477.2731@compuserve.com
510	VTKK	Markku Lamminluoto	lamminluoto@vtkes1.vtkk.fi
511	North Hills Israel Ltd.	Carmi Cohen	carmi@north.hellnet.org
512	TECSIEL	R. Burlon	sr@teculx.tecsiel.it
513	Bayerische Motoren Werke (BMW) AG	Michael Connolly	mconnolly@net.bmw.de
514	CNET Technologies	Nelson Su	408-954-8000
515	MCI	Anil Prasad	7502332@mcimail.com
516	Human Engineering AG (HEAG)	Urs Brunner	ubrunner@clients.switch.ch
517	FileNet Corporation	Joe Raby	raby@filenet.com
518	NFT-Ericsson	Kjetil Donasen	+47 2 84 24 00
519	Dun & Bradstreet	Vic Smagovic	908-464-2079
520	Intercomputer Communications	Brian Kean	513-745-0500x244
521	Defense Intelligence Agency	Barry Atkinson	DIA-DMS@DDN-CONUS.DDN.MIL
522	Telesystems SLW Inc.	Joe Magony	416-441-9966
523	APT Communications	David Kloper	301-831-1182
524	Delta Airlines	Jim Guy	404-715-2948

Decimal	Name		References
525	California Microwave	Kevin Braun	408-720-6520
526	Avid Technology Inc	Steve Olynyk	508-640-3328
527	Integro Advanced Computer Systems	Pascal Turbiez	+33-20-08-00-40
528	RPTI	Chris Shin	886-2-918-3006
529	Ascend Communications Inc.	Marc Hyman	510-769-6001
530	Eden Computer Systems Inc.	Louis Brando	305-591-7752
531	Kawasaki-Steel Corp	Tomoo Watanabe	nrd@info.kawasaki-steel.co.jp
532	Barclays	Malcolm Houghton	+44 202 671 212
533	B.U.G., Inc.	Isao Tateishi	tateishi@bug.co.jp
534	Exide Electronics	Brian Hammill	hamill@dolphin.exide.com
535	Superconducting Supercollider Lab.	Carl W. Kalbfleisch	cwk@irrational.ssc.gov
536	Triticom	Jim Bales	(612) 937-0772
537	Universal Instruments Corp.	Tom Dinnel	BA06791%BINGVAXA.bitnet@ CUNYVM.CUNY.EDU
538	Information Resources, Inc.	Jeff Gear	jjg@infores.com
539	Applied Innovation, Inc.	Dean Dayton	dean@aicorp.cmhnet.org
540	Crypto AG	Roland Luthi	luthi@iis.ethz.ch
541	Infinite Networks, Ltd.	Sean Harding	+44 923 710 277
542	Rabbit Software	Bill Kwan	kwan@rabbit.com
543	Apertus Technologies	Stuart Stanley	stuarts@apertus.com
544	Equinox Systems, Inc.	Monty Norwood	1-800-275-3500 x293
545	Hayes Microcomputer Products	Joe Pendergrass	jpendergrass@hayes.com
546	Empire Technologies Inc.	Cheryl Krupczak	cheryl@cc.gatech.edu
547	Glaxochem, Ltd.	Andy Wilson	0229 522-61547
548	Software Professionals, Inc	Gordon Vickers	gordon@netpartners.com
549	Agent Technology, Inc.	Ibi Dhilla	idhilla@genesis.nred.ma.us
550	Dornier GMBH	Arens Heinrech	49-7545-8 ext 9337
551	Telxon Corporation	Frank Ciotti	frankc@teleng.telxon.com
552	Entergy Corporation	Louis Cureau	504-364-7630
553	Garrett Communications Inc.	Igor Khasin	(408) 980-9752
554	Agile Networks, Inc.	Dave Donegan	ddonegan@agile.com
555	Larscom	Sameer Jayakar	415-969-7572
556	Stock Equipment	Karl Klebenow	216-543-6000
557	ITT Corporation	Kevin M. McCauley	kmm@vaxf.acdnj.itt.com
558	Universal Data Systems, Inc.	Howard Cunningham	70400.3671@compuserve.com
559	Sonix Communications, Ltd.	David Webster	+44 285 641 651
560	Paul Freeman Associates, Inc.	Pete Wilson	pwilson@world.std.com
561	John S. Barnes, Corp.	Michael Lynch	704-878-4107
562	Northern Telecom, Ltd.	Glenn Waters	613-763-3933 <gwaters@bnr.ca>
563	CAP Debris	Patrick Preuss	ppr@lfs.hamburg.cap-debris.de
564	Telco Systems NAC	Harry Hirani	Harry@telco-nac.com
565	Tosco Refining Co	Fred Sanderson	510-602-4358
566	Russell Info Sys	Atul Desai	714-362-4040
567	University of Salford	Richard Letts	R.J.Letts@salford.ac.uk
568	NetQuest Corp.	Jerry Jacobus	netquest@tigger.jvnc.net

Decimal	Name		References
569	Armon Networking Ltd.	Yigal Jacoby	yigal@armon.hellnet.org
570	IA Corporation	Didier Fort	Didier.Fort@lia.com
571	AU-System Communicaton ABT	orbjorn Ryding	8-7267572
572	GoldStar Information & Communications, Ltd.	Soo N. Kim	ksn@giconet.gsic.co.kr
573	SECTRA AB	Tommy Pedersen	tcp@sectra.se
574	ONEAC Corporation	Bill Elliot	ONEACWRE@AOL.COM
575	Tree Technologies	Michael Demjanenko	(716) 688-4640
576	GTE Government Systems	John J. Holzhauer	holzhauerj@mail.ndhm.gtegsc.com
577	Denmac Systems, Inc.	Andy Denenberg	(708) 291-7760
578	Interlink Computer Sciences, Inc.	Mike Mazurek	mfm@interlink.com
579	Bridge Information Systems, Inc.	Stephen Harvey	(314) 567-8482
580	Leeds and Northrup Australia (LNA)	Nigel Cook	nigelc@lna.oz.au
581	BHA Computer	David Hislop	rob@bha.oz.au
582	Newport Systems Solutions, Inc.	Pauline Chen	paulinec@cisco.com
583	azel Corporation	Narender Reddy Vangati	vnr@atrium.com
584	ROBOTIKER	Maribel Narganes	maribel@teletek.es
585	PeerLogic Inc.	Ratinder Ahuja	ratinder@peerlogic.com
586	Digital Transmittion Systems	Bill VerSteeg	bvs@ver.com
587	Far Point Communications	Bill VerSteeg	bvs@ver.com
588	Xircom	Bill VerSteeg	bvs@ver.com
589	Mead Data Central	Stephanie Bowman	steph@meaddata.com
590	Royal Bank of Canada	N. Lim	(416) 348-5197
591	Advantis, Inc.	Janet Brehm	813 878-4298
592	Chemical Banking Corp.	Paul McDonnell	pmcdonnl@world.std.com
593	Eagle Technology	Ted Haynes	(408) 441-4043
594	British Telecom	Ray Smyth	rsmyth@bfsec.bt.co.uk
595	Radix BV	P. Groenendaal	project2@radix.nl
596	TAINET Communication System Corp.	Joseph Chen	+886-2-6583000 (R.O.C.)
597	Comtek Services Inc.	Steve Harris	(703) 506-9556
598	Fair Issac	Steve Pasadis	apple.com!fico!sxp (415) 472-2211
599	AST Research Inc.	Bob Beard	bobb@ast.com
600	Soft*Star s.r.l. Ing.	Enrico Badella	softstar@pol88a.polito.it
601	Bancomm	Joe Fontes	jwf@bancomm.com
602	Trusted Information Systems, Inc.	James M. Galvin	galvin@tis.com
603	Harris & Jeffries, Inc.	Deepak Shahane	deepak@hjinc.com
604	Axel Technology Corp.	Henry Ngai	(714) 455-1688
605	GN Navtel, Inc.	Joe Magony	416-479-8090
606	CAP debis	Patrick Preuss	+49 40 527 28 366
607	Lachman Technology, Inc.	Steve Alexander	stevea@lachman.com
608	Galcom Networking Ltd.	Zeev Greenblatt	galnet@vax.trendline.co.il
609	BAZIS	M. van Luijt	martin@bazis.nl
610	SYNAPTEL	Eric Remond	remond@synaptel.fr
611	Investment Management Services, Inc.	J. Laurens Troost	rens@stimpys.imsi.com
612	Taiwan Telecommunication Lab	Dennis Tseng	LOUIS%TWNMOCTL.BITNET@ pucc.Princeton.EDU

Decimal	Name		References
613	Anagram Corporation	Michael Demjanenko	(716) 688-4640
614	Univel	John Nunneley	jnunnele@univel.com
615	University of California, San Diego	Arthur Bierer	abierer@ucsd.edu
616	CompuServe	Ed Isaacs, Brian Biggs	SYSADM@csi.compuserve.com
617	Telstra—OTC Australia	Peter Hanselmann	peterhan@turin.research.otc.com.au
618	Westinghouse Electric Corp.	Ananth Kupanna	ananth@access.digex.com
619	DGA Ltd.	Tom L. Willis	twillis@pintu.demon.co.uk
620	Elegant Communications Inc.	Robert Story	Robert.Story@Elegant.COM
621	Experdata	Claude Lubin	+33 1 41 28 70 00
622	Unisource Business Networks Sweden AB	Goran Sterner	gsr@tip.net
623	Molex, Inc.	Steven Joffe	molex@mcimail.com
624	Quay Financial Software	Mick Fleming	mickf@quay.ie
625	VMX Inc.	Joga Ryali	joga@vmxi.cerfnet.com
626	Hypercom, Inc.	Noor Chowdhury	(602) 548-2113
627	University of Guelph	Kent Percival	Percival@CCS.UoGuelph.CA
628	DIaLOGIKa	Juergen Jungfleisch	0 68 97 9 35-0
629	NBASE Switch Communication	Sergiu Rotenstein	75250.1477@compuserve.com
630	Anchor Datacomm B.V.	Erik Snoek	sdrierik@diamond.sara.nl
631	PACDATA	John Reed	johnr@hagar.pacdata.com
632	University of Colorado	Evi Nemeth	evi@cs.colorado.edu
633	Tricom Communications Limited	Robert Barrett	0005114429@mcimail.com
634	Santix Software GmbH	Michael Santifaller	santi%mozart@santix.guug.de
635	FastComm Communications Corp.	Bill Flanagan	70632.1446@compuserve.com
636	The Georgia Institute of Technology	Michael Mealling	michael.mealling@oit.gatech.edu
637	Alcatel Data Networks	Douglas E. Johnson	doug.e.johnson@adn.sprint.com
638	GTECH	Brian Ruptash	bar@gtech.com
639	UNOCAL Corporation	Peter Ho	ho@unocal.com
640	First Pacific Network	Randy Hamilton	408-703-2763
641	Lexmark International	Don Wright	don@lexmark.com
642	Qnix Computer	Sang Weon, Yoo	swyoo@qns.qnix.co.kr
643	Jigsaw Software Concepts (Pty) Ltd.	Willem van Biljon	wvb@itu2.sun.ac.za
644	VIR, Inc.	Mark Cotton	(215) 364-7955
645	SFA Datacomm Inc.	Don Lechthaler	lech@world.std.com
646	SEIKO Communication Systems, Inc.	Lyn T. Robertson	ltr@seikotsi.com
647	Unified Management	Andy Barnhouse	(612) 561-4944
648	RADLINX Ltd.	Ady Lifshes	ady%rndi@uunet.uu.net
649	Microplex Systems Ltd.	Henry Lee	hyl@microplex.com
650	Objecta Elektronik & Data AB	Johan Finnved	jf@objecta.se
651	Phoenix Microsystems	Bill VerSteeg	bvs@ver.com
652	Distributed Systems International, Inc.	Ron Mackey	rem@dsiinc.com
653	Evolving Systems, Inc.	Judith C. Bettinger	judy@evolving.com
654	SAT GmbH	Walter Eichelburg	100063.74@compuserve.com
655	CeLAN Technology, Inc.	Mark Liu	886-35-772780

Decimal	Name		References
656	Landmark Systems Corp.	Steve Sonnenberg	steves@socrates.umd.edu
657	Netone Systems Co., Ltd.	YongKui Shao	syk@new-news.netone.co.jp
658	Loral Data Systems	Jeff Price	jprice@cps070.lds.loral.com
659	Cellware Broadband Technology	Michael Roth	mike@cellware.de
660	Mu-Systems	Gaylord Miyata	miyata@musys.com
661	IMC Networks Corp.	Jerry Roby	(714) 724-1070
662	Octel Communications Corp.	Alan Newman	(408) 321-5182
663	RIT Technologies LTD.	Ghiora Drori	drori@dcl.hellnet.org
664	Adtran	Jeff Wells	205-971-8000
665	PowerPlay Technologies, Inc.	Ray Caruso	rayman@csn.org
666	Oki Electric Industry Co., Ltd.	Shigeru Urushibara	uru@cs1.cs.oki.co.jp
667	Specialix International	Jeremy Rolls	jeremyr@specialix.co.uk
668	INESC (Instituto de Engenharia de Sistemas e Computadores)	Pedro Ramalho Carlos	prc@inesc.pt
669	Globalnet Communications	Real Barrier	(514) 651-6164
670	Product Line Engineer SVEC Computer Corp.	Rich Huang	msumgr@enya.cc.fcu.edu.tw
671	Printer Systems Corp.	Bill Babson	bill@prsys.com
672	Contec Micro Electronics USA	David Sheih	(408) 434-6767
673	Unix Integration Services	Chris Howard	chris@uis.com
674	Dell Computer Corporation	Steven Blair	sblair@dell.com
675	Whittaker Electronic Systems	Michael McCune	mccune@cerf.net
676	QPSX Communications	David Pascoe	davidp@qpsx.oz.au
677	Loral WDl	Mike Aronson	Mike_Aronson@msgate.wdl.loral.com
678	Federal Express Corp.	Randy Hale	(901) 369-2152
679	E-COMMS Inc.	Harvey Teale	(206) 857-3399
680	Software Clearing House	Tom Caris	ca@sch.com
681	Antlow Computers LTD.	C. R. Bates	44-635-871829
682	Emcom Corp.	Mike Swartz	emcom@cerf.net
683	Extended Systems, Inc.	Al Youngwerth	alberty@tommy.extendsys.com
684	Sola Electric	Mike Paulsen	(708) 439-2800
685	Esix Systems, Inc.	Anthony Chung	esix@esix.tony.com
686	3M/MMM	Chris Amley	ccamley@mmm.com
687	Cylink Corp.	Ed Chou	ed@cylink.com
688	Znyx Advanced Systems Division, Inc.	Alan Deikman	aland@netcom.com
689	Texaco, Inc.	Jeff Lin	linj@Texaco.com
690	McCaw Cellular Communication Corp.	Tri Phan	tri.phan@mccaw.com
691	ASP Computer Product Inc.	Elise Moss	71053.1066@compuserve.com
692	HiPerformance Systems	Mike Brien	+27-11-806-1000
693	Regionales Rechenzentrum	Sibylle Schweizer	unrz54@daphne.rrze.uni-erlangen.de
694	SAP AG	Dr. Uwe Hommel	+49 62 27 34 0
695	ElectroSpace System Inc.	Dr. Joseph Cleveland	e03353@esitx.esi.org

Decimal	Name		References
696	(Unassigned)		
697	MultiPort Software	Reuben Sivan	72302.3262@compuserve.com
698	Combinet, Inc.	Samir Sawhney	samir@combinet.com
699	TSCC	Carl Wist	carlw@tscc.com
700	Teleos Communications Inc.	Bill Nayavich	wln@teleoscom.com
701	Alta Research	Jack Moyer	ian@altarsrch.com
702	Independence Blue Cross	Bill Eshbach	esh@ibx.com
703	ADACOM Station Interconnectivity LTD.	Itay Kariv	+9 72 48 99 89 9
704	MIROR Systems	Frank Kloes	+27 12 911 0003
705	Merlin Gerin	Adam Stolinski	(714) 557-1637 x249
706	Owen-Corning Fiberglas	Tom Mann	mann.td@ocf.compuserve.com
707	Talking Networks Inc.	Terry Braun	tab@lwt.mtxinu.com
708	Cubix Corporation	Rebekah Marshall	(702) 883-7611
709	Formation Inc.	Bob Millis	bobm@formail.formation.com
710	Lannair Ltd.	Pablo Brenner	pablo@lannet.com
711	LightStream Corp.	Chris Chiotasso	chris@lightstream.com
712	LANart Corp.	Doron I. Gartner	doron@lanart.com
713	University of Stellenbosch	Graham Phillips	phil@cs.sun.ac.za
714	Wyse Technology	Bill Rainey	bill@wyse.com
715	DSC Communications Corp.	Colm Bergin	cbergin@cpdsc.com
716	NetEc	Thomas Krichel	netec@uts.mcc.ac.uk
717	Breltenbach Software Engineering GmbH	Hilmar Tuneke	tuneke@namu01.gwdg.de
718	Victor Company of Japan, Limited	Atsushi Sakamoto	101176.2703@compuserve.com
719	Japan Direx Corporation	Teruo Tomiyama	+81 3 3498 5050
720	NECSY Network Control Systems S.p.A.	Piero Fiozzo	fip@necsy.it
721	ISDN Systems Corp.	Jeff Milloy	p00633@psilink.com
722	Zero-One Technologies, LTD.	Curt Chen	+88 62 56 52 32 33
723	Radix Technologies, Inc.	Steve Giles	giless@delphi.com
724	National Institute of Standards and Technology	Jim West	west@mgmt3.ncsl.nist.gov
725	Digital Technology Inc.	Chris Gianattasio	gto@lanhawk.com
726	Castelle Corp.	Waiming Mok	wmm@castelle.com
727	Presticom Inc.	Martin Dube	76270.2672@compuserve.com
728	Showa Electric Wire & Cable Co., Ltd.	Robert O'Grady	kfn@tanuki.twics.co.jp
729	SpectraGraphics	Jack Hinkle	hinkle@spectra.com
730	Connectware Inc.	Rick Downs	rxd4@acsysinc.com
731	Wind River Systems	Emily Hipp	hipp@wrs.com
732	RADWAY International Ltd.	Doron Kolton	0005367977@mcimail.com
733	System Management ARTS, Inc.	Alexander Dupuy	dupuy@smarts.com
734	Persoft, Inc.	Steven M. Entine	entine@pervax.persoft.com
735	Xnet Technology Inc.	Esther Chung	estchung@xnet-tech.com
736	Unison-Tymlabs	Dean Andrews	ada@unison.com
737	Micro-Matic Research	Patrick Lemli	73677.2373@compuserve.com
738	B.A.T.M. Advance Technologies	Nahum Killim	bcrystal@actcom.co.il

Decimal	Name		References
739	University of Copenhagen	Kim Hlglund	shotokan@diku.dk
740	Network Security Systems, Inc.	Carleton Smith	rpitt@nic.cerf.net
741	JNA Telecommunications	Sean Cody	seanc@jna.com.au
742	Encore Computer Corporation	Tony Shafer	tshafer@encore.com
743	Central Intelligent Agency	Carol Jobusch	703 242-2485
744	ISC (GB) Limited	Mike Townsend	miket@cix.compulink.co.uk
745	Digital Communication Associates	Ravi Shankar	shankarr@dca.com
746	CyberMedia Inc.	Unni Warrier	unni@cs.ucla.edu
747	Distributed Systems International, Inc.	Ron Mackey	rem@dsiinc.com
748	Peter Radig EDP-Consulting	Peter Radig	+49 69 9757 6100
749	Vicorp Interactive Systems	Phil Romine	phil@vis.com
750	Inet Inc.	Bennie Lopez	brl@inetinc.com
751	Argonne National Laboratory	Jeffrey S. Curtis	curtis@anl.gov
752	Tek Logix	Peter Palsall	905 625-4121
753	North Western University	Phil Draughon	jpd@nwu.edu
754	Astarte Fiber Networks	James Garnett	garnett@catbelly.com
755	Diederich & Associates, Inc.	Douglas Capitano	dlcapitano@delphi.com
756	Florida Power Corporation	Bob England	rengland@fpc.com
757	ASK/INGRES	Howard Dernehl	howard@ingres.com
758	Open Network Enterprise	Spada Stefano	+39 39 245-8101
759	The Home Depot	Allen Thomas	art01@homedepot.com
760	Pan Dacom Telekommunikations	Jens Andresen	+49 40 644 09 71
761	NetTek	Steve Kennedy	steve@gbnet.com
762	Karlnet Corp.	Doug Kall	kbridge@osu.edu
763	Efficient Networks, Inc.	Thirl Johnson	(214) 991-3884
764	Fiberdata	Jan Fernquist	+46 828 8383
765	Lanser	Emil Smilovici	(514) 485-7104
766	Telebit Communications A/S	Peder Chr. Norgaard	pcn@tbit.dk
767	HILAN GmbH	Markus Pestinger	markus@lahar.ka.sub.org
768	Network Computing Inc.	Fredrik Noon	fnoon@ncimail.mhs.compuserve.com
769	Walgreens Company	Denis Renaud	(708) 317-5054 (708) 818-4662
770	Internet Initiative Japan Inc.	Toshiharu Ohno	tony-o@iij.ad.jp
771	GP van Niekerk Ondernemings	Gerrit van Niekerk	gvanniek@dos-lan.cs.up.ac.za
772	Queen's University Belfast	Patrick McGleenon	p.mcgleenon@ee.queens-belfast.ac.uk
773	Securities Industry Automation Corporation	Chiu Szeto	cszeto@prism.poly.edu
774	SYNaPTICS	David Gray	david@synaptics.ie
775	Data Switch Corporation	Joe Welfeld	jwelfeld@dasw.com
776	Telindus Distribution	Karel Van den Bogaert	kava@telindus.be
777	MAXM Systems Corporation	Gary Greathouse	ggreathouse@maxm.com
778	Fraunhofer Gesellschaft	Jan Gottschick	jan.gottschick@isst.fhg.de
779	EQS Business Services	Ken Roberts	kroberts@esq.com
780	CNet Technology Inc.	Repus Hsiung	idps17@shts.seed.net.tw
781	Datentechnik GmbH	Thomas Pischinger	+43 1 50100 266
782	Network Solutions Inc.	Dave Putman	davep@netsol.com

Decimal	Name		References
783	Viaman Software	Vikram Duvvoori	info@viman.com
784	Schweizerische Bankgesellschaft Zuerich	Roland Bernet	Roland.Bernet@zh014.ubs.ubs.ch
785	University of Twente—TIOS	Aiko Pras	pras@cs.utwente.nl
786	Simplesoft Inc.	Sudhir Pendse	sudhir@netcom.com
787	Stony Brook, Inc.	Ken Packert	p01006@psilink.com
788	Unified Systems Solutions, Inc.	Steven Morgenthal	smorgenthal@attmail.com
789	Network Appliance Corporation	Varun Mehta	varun@butch.netapp.com
790	Ornet Data Communication Technologies Ltd.	Haim Kurz	haim@ornet.co.il
791	Computer Associates International	Glenn Gianino	giagl01@usildaca.cai.com
792	Multipoint Network Inc.	Michael Nguyen	mike@multipoint.com
793	NYNEX Science & Technology	Lily Lau	llau@nynexst.com
794	Commercial Link Systems	Wiljo Heinen	wiljo@freeside.cls.de
795	Adaptec Inc.	Tom Battle	tab@lwt.mtxinu.com
796	Softswitch	Charles Springer	cjs@ssw.com
797	Link Technologies, Inc.	Roy Chu	royc@wyse.com
798	IIS	Olry Rappaport	iishaifa@attmail.com
799	Mobile Solutions Inc.	Dale Shelton	dshelton@srg.srg.af.mil
800	Xylan Corp.	Burt Cyr	burt@xylan.com
801	Airtech Software Forge Limited	Callum Paterson	tsf@cix.compulink.co.uk
802	National Semiconductor	Maurice Turcotte	mturc@atlanta.nsc.com
803	Video Lottery Technologies	Angelo Lovisa	ange@awd.cdc.com
804	National Semiconductor Corp	Waychi Doo	wcd@berlioz.nsc.com
805	Applications Management Corp	Terril (Terry) Steichen	tjs@washington.ssds.com
806	Travelers Insurance Company	Eric Miner	ustrv67v@ibmmail.com
807	Taiwan International Standard Electronics Ltd.	B. J. Chen	bjchen@taisel.com.tw
808	US Patent and Trademark Office	Rick Randall	randall@uspto.gov
809	Hynet, LTD.	Amir Fuhrmann	amf@teleop.co.il
810	Aydin, Corp.	Rick Veher	(215) 657-8600
811	ADDTRON Technology Co., LTD.	Tommy Tasi	+8 86-2-4514507
812	Fannie Mae	David King	s4ujdk@fnma.com
813	MultiNET Services	Hubert Martens	martens@multinet.de
814	GECKO mbH	Holger Dopp	hdo@gecko.de
815	Memorex Telex	Mike Hill	hill@raleng.mtc.com
816	Advanced Communications Networks (ACN) SA	Antoine Boss	+41 38 247434
817	Telekurs AG	Jeremy Brookfield	bkj@iris.F2.telekurs.ch
818	Victron bv	Jack Stiekema	jack@victron.nl
819	CF6 Company	Francois Caron	+331 4696 0060
820	Walker Richer and Quinn Inc.	Rebecca Higgins	rebecca@elmer.wrq.com
821	Saturn Systems	Paul Parker	paul_parker@parker.fac.cs.cmu.edu

Decimal	Name		References
822	Mitsui Marine and Fire Insurance Co. LTD.	Kijuro Ikeda	+813 5389 8111
823	Loop Telecommunication International, Inc.	Charng-Show Li	+886 35 787 696
824	Telenex Corporation	James Krug	(609) 866-1100
825	Bus-Tech, Inc.	Charlie Zhang	chun@eecs.cory.berkley.edu
826	ATRIE	Fred B.R. Tuang	cmp@fddi3.ccl.itri.org.tw
827	Gallagher & Robertson A/S	Arild Braathen	arild@gar.no
828	Networks Northwest, Inc.	John J. Hansen	jhansen@networksnw.com
829	Conner Peripherials	Richard Boyd	rboyd@mailserver.conner.com
830	Elf Antar France	P. Noblanc	+33 1 47 44 45 46
831	Lloyd Internetworking	Glenn McGregor	glenn@lloyd.com
832	Datatec Industries, Inc.	Chris Wiener	cwiener@datatec.com
833	TAICOM	Scott Tseng	cmp@fddi3.ccl.itri.org.tw
834	Brown's Operating System Services Ltd.	Alistair Bell	alistair@browns.co.uk
835	MiLAN Technology Corp.	Gopal Hegde	gopal@milan.com
836	NetEdge Systems, Inc.	Dave Minnich	Dave_Minnich@netedge.com
837	NetFrame Systems	George Mathew	george_mathew@netframe.com
838	Xedia Corporation	Colin Kincaid	colin%madway.uucp@dmc.com
839	Pepsi	Niraj Katwala	niraj@netcom.com
840	Tricord Systems, Inc.	Mark Dillon	mdillon@tricord.mn.org
841	Proxim Inc.	Russ Reynolds	proxim@netcom.com
842	Applications Plus, Inc.	Joel Estes	joele@hp827.applus.com
843	Pacific Bell	Aijaz Asif	saasif@srv.PacBell.COM
844	Scorpio Communications	Sharon Barkai	sharon@supernet.com
845	TPS-Teleprocessing Systems	Manfred Gorr	gorr@tpscad.tps.de
846	Technology Solutions Company	Niraj Katwala	niraj@netcom.com
847	Computer Site Technologies	Tim Hayes	(805) 967-3494
848	NetPort Software	John Bartas	jbartas@sunlight.com
849	Alon Systems	Menachem Szus	70571.1350@compuserve.com
850	Tripp Lite	Lawren Markle	72170.460@compuserve.com
851	NetComm Limited	Paul Ripamonti	paulri@msmail.netcomm.pronet.com
852	Precision Systems, Inc. (PSI)	Fred Griffin	cheryl@empiretech.com
853	Objective Systems Integrators	Ed Reeder	Ed.Reeder@osi.com
854	Simpact, Inc.	Ron Tabor	rtabor@simpact.com
855	Systems Enhancement Corporation	Steve Held	71165.2156@compuserve.com
856	Information Integration, Inc.	Gina Sun	iiii@netcom.com
857	CETREL S.C.	Louis Reinard	ssc-re@cetrel.lu
858	Platinum Technology, Inc.	Theodore J. Collins III	ted.collins@vtdev.mn.org
859	Olivetti North America	Tom Purcell	tomp@mail.spk.olivetti.com
860	WILMA	Nikolaus Schaller	hns@ldv.e-technik.tu-muenchen.de
861	ILX Systems Inc.	Peter Mezey	peterm@ilx.com
862	Total Peripherals Inc.	Mark Ustik	(508) 393-1777
863	SunNetworks Consultant	John Brady	jbrady@fedeast.east.sun.com
864	Arkhon Technologies, Inc.	Joe Wang	rkhon@nic.cerf.net
865	Computer Sciences Corporation	George M. Dands	dands@sed.csc.com

Decimal	Name		References
866	Philips Communication d'Entreprise	Claude Lubin	+331412870 00
867	Katron Technologies Inc.	Robert Kao	+88 627 991 064
868	Transition Engineering Inc.	Hemant Trivedi	hemant@transition.com
869	Altos Engineering Applications, Inc.	Wes Weber or Dave Erhart	altoseng@netcom.com
870	Nicecom Ltd.	Arik Ramon	arik@nicecom.nice.com
871	Fiskars/Deltec	Carl Smith	(619) 291-2973
872	AVM GmbH	Andreas Stockmeier	stocki@avm-berlin.de
873	Comm Vision	Richard Havens	(408) 923 0301 x22
874	Institute for Information Industry	Peter Pan	peterpan@pdd.iii.org.tw
875	Legent Corporation	Gary Strohm	gstrohm@legent.com
876	Network Automation	Doug Jackson	+64 6 285 1711
877	NetTech	Marshall Sprague	marshall@nettech.com
878	Coman Data Communications Ltd.	Zvi Sasson	coman@nms.cc.huji.ac.il
879	Skattedirektoratet	Karl Olav Wroldsen	+47 2207 7162
880	Client-Server Technologies	Timo Metsaportti	timo@itf.fi
881	Societe Internationale de Telecommunications Aeronautiques	Chuck Noren	chuck.noren@es.atl.sita.int
882	Maximum Strategy Inc.	Paul Stolle	pstolle@maxstrat.com
883	Integrated Systems, Inc.	Michael Zheng	mz@isi.com
884	E-Systems	Rick Silton	rsilton@melpar.esys.com
885	Reliance Comm/Tec	Mark Scott	73422.1740@compuserve.com
886	Summa Four Inc.	Paul Nelson	(603) 625-4050
887	J & L Information Systems	Rex Jackso	(818) 709-1778
888	Forest Computer Inc.	Dave Black	dave@forest.com
889	Palindrome Corp.	Jim Gast	jgast@palindro.mhs.compuserve.com
890	ZyXEL Communications Corp.	Harry Chou	howie@csie.nctu.edu.tw
891	Network Managers (UK) Ltd,	Mark D Dooley	mark@netmgrs.co.uk
892	Sensible Office Systems Inc.	Pat Townsend	(712) 276-0034
893	Informix Software	Anthony Daniel	anthony@informix.com
894	Dynatek Communications	Howard Linton	(703) 490-7205
895	Versalynx Corp.	Dave Fisler	(619) 536-8023
896	Potomac Scheduling Communications Company	David Labovitz	del@access.digex.net
897	Sybase Inc.	Dave Meldrum	meldrum@sybase.com
898	DiviCom Inc.	Eyal Opher	eyal@divi.com
899	Datus elektronische Informationssysteme GmbH	Hubert Mertens	marcus@datus.uucp
900	Matrox Electronic Systems Limited	Marc-Andre Joyal	marc-andre.joyal@matrox.com
901	Digital Products, Inc.	Ross Dreyer	rdreyer@digprod.com
902	Scitex Corp. Ltd.	Yoav Chalfon	yoav_h@ird.scitex.com
903	RAD Vision	Oleg Pogorelik	radvis@vax.trendline.co.il
904	Tran Network Systems	Bill Hamlin	billh@revco.com
905	Scorpion Logic	Sean Harding	+09 2324 5672
906	Inotech Inc.	Eric Jacobs	(703) 641-0469

Decimal	Name		References
907	Controlled Power Co.	Yu Chin	76500,3160@compuserve.com
908	Elsag Bailey Incorporate	Derek McKearney	mckearney@bailey.com
909	J.P. Morgan	Chung Szeto	szeto_chung@jpmorgan.com
910	Clear Communications Corp.	Kurt Hall	khall@clear.com
911	General Technology Inc.	Perry Rockwell	(407) 242-2733
912	Adax Inc.	Jory Gessow	jory@adax.com
913	Mtel Technologies, Inc.	Jon Robinson	552-3355@mcimail.com
914	Underscore, Inc.	Jeff Schnitzer	jds@underscore.com
915	SerComm Corp.	Ben Lin	+8 862-577-5400
916	Baxter Healthcare Corporation	Joseph Sturonas	sturonaj@mpg.mcgawpark.baxter.com
917	Tellus Technology	Ron Cimorelli	(510) 498-8500
918	Continuous Electron Beam Accelerator Facility	Paul Banta	banta@cebaf.gov
919	Canoga Perkins	Margret Siska	(818) 718-6300
920	R.I.S Technologies	Fabrice Lacroix	+33 7884 6400
921	INFONEX Corp.	Kazuhiro Watanabe	kazu@infonex.co.jp
922	WordPerfect Corp.	Douglas Eddy	eddy@wordperfect.com
923	NRaD	Russ Carleton	roccor@netcom.com
924	Hong Kong Telecommunications Ltd.	K. S. Luk	+8 52 883 3183
925	Signature Systems	Doug Goodall	goodall@crl.com
926	Alpha Technologies LTD.	Guy Pothiboon	(604) 430-8908
927	PairGain Technologies, Inc.	Ken Huang	kenh@pairgain.com
928	Sonic Systems	Sudhakar Ravi	sudhakar@sonicsys.com
929	Steinbrecher Corp.	Kary Robertson	krobertson@delphi.com
930	Centillion Networks, Inc.	Derek Pitcher	derek@lanspd.com
931	Network Communication Corp.	Tracy Clark	ncc!central!tracyc@netcomm. attmail.com
932	Sysnet A.S.	Carstein Seeberg	case@sysnet.no
933	Telecommunication Systems Lab	Gerald Maguire	maguire@it.kth.se
934	QMI	Scott Brickner	Scott_Brickner.QMI-DEV@FIDO. qmi.mei.com
935	Phoenixtec Power Co., LTD.	An-Hsiang Tu	+8 862 646 3311
936	Hirakawa Hewtech Corp.	H. Ukaji	lde02513@niftyserve.or.jp
937	No Wires Needed B.V.	Arnoud Zwemmer	roana@cs.utwente.nl
938	Primary Access	Kerstin Lodman	lodman@priacc.com
939	FD Software AS	Dag Framstad	dag.framstad@fdsw.no
940	Grabner & Kapfer GnbR	Vinzenz Grabner	zen@wsr.ac.att
941	Nemesys Research Ltd.	Michael Dixon	mjd@nemesys.co.uk
942	Pacific Communication Sciences, Inc. (PSCI)	Yvonne Kammer	mib-contact@pcsi.com
943	Level One Communications, Inc.	Moshe Kochinski	moshek@level1.com
944	Fast Track, Inc.	Andrew H. Dimmick	adimmick@ftinc.com
945	Andersen Consulting, OM/NI Practice	Greg Tilford	p00919@psilink.com
946	Bay Technologies Pty Ltd.	Paul Simpson	pauls@baytech.com.au
947	Integrated Network Corp.	Daniel Joffe	wandan@integnet.com
948	Epoch, Inc.	David Haskell	deh@epoch.com
949	Wang Laboratories Inc.	Pete Reilley	pvr@wiis.wang.com
950	Polaroid Corp.	Sari Germanos	sari@temerity.polaroid.com

Decimal	Name		References
951	Sunrise Sierra	Gerald Olson	(510) 443-1133
952	Silcon Group	Bjarne Bonvang	+45 75 54 22 55
953	Coastcom	Peter Doleman	pdoleman@coastcom.com
954	4th DIMENSION SOFTWARE LTD.	Thomas Segev/Ely Hofner	autumn@zeus.datasrv.co.il
955	SEIKO SYSTEMS Inc.	Kiyoshi Ishida	ishi@ssi.co.jp
956	PERFORM	Jean-Hugues Robert	+33 42 27 29 32
957	TV/COM International	Jean Tellier	(619) 675-1376
958	Network Integration, Inc.	Scott C. Lemon	slemon@nii.mhs.compuserve.com
959	Sola Electric, A Unit of General Signal	Bruce Rhodes	72360,2436@compuserve.com
960	Gradient Technologies, Inc.	Geoff Charron	geoff@gradient.com
961	Tokyo Electric Co., Ltd.	A. Akiyama	+81 558 76 9606
962	Codonics, Inc.	Joe Kulig	jjk@codonics.com
963	Delft Technical University	Mark Schenk	m.schenk@ced.tudelft.nl
964	Carrier Access Corp.	Roger Koenig	tomquick@carrier.com
965	eoncorp	Barb Wilson	wilsonb@eon.com
966	Naval Undersea Warfare Center	Mark Lovelace	lovelace@mp34.nl.nuwc.navy.mil
967	AWA Limited	Mike Williams	+61 28 87 71 11
968	Distinct Corp.	Tarcisio Pedrotti	tarci@distinct.com
969	National Technical University of Athens	Theodoros Karounos	karounos@phgasos.ntua.gr
970	BGS Systems, Inc.	Amr Hafez	amr@bgs.com
971	McCaw Wireless Data Inc.	Brian Bailey	bbailey@airdata.com
972	Bekaert	Koen De Vleeschauwer	kdv@bekaert.com
973	Epic Data Inc.	Vincent Lim	vincent_lim@epic.wimsey.com
974	Prodigy Services Co.	Ed Ravin	elr@wp.prodigy.com
975	First Pacific Networks (FPN)	Randy Hamilton	randy@fpn.com
976	Xylink Ltd.	Bahman Rafatjoo	100117.665@compuserve.com
977	Relia Technologies Corp.	Fred Chen	fredc@relia1.relia.com.tw
978	Legacy Storage Systems Inc.	James Hayes	james@lss-chq.mhs.compuserve.com
979	Digicom, SPA	Claudio Biotti	+39 3312 0 0122
980	Ark Telecom	Alan DeMars	alan@arktel.com
981	National Security Agency (NSA)	(301) 688-1058 Cynthia Beighley	maedeen@romulus.ncsc.mil
982	Southwestern Bell Corporation	Brian Bearden	bb8840@swuts.sbc.com
983	Virtual Design Group, Inc.	Chip Standifer	70650.3316@compuserve.com
984	Rhone Poulenc	Olivier Pignault	+33 1348 2 4053
985	Swiss Bank Corporation	Neil Todd	toddn@gb.swissbank.com
986	ATEA N.V.	Walter van Brussel	p81710@banyan.atea.be
987	Computer Communications Specialists, Inc.	Carolyn Zimmer	cczimmer@crl.com
988	Object Quest, Inc.	Michael L. Kornegay	mlk@bir.com
989	DCL System International, Ltd.	Gady Amit	gady-a@dcl-see.co.il
990	SOLITON SYSTEMS	K.K. Masayuki Yamai	+81 33356 6091
991	U S Software	Don Dunstan	ussw@netcom.com
992	Systems Research and Applications Corporation	Todd Herr	herrt@smtplink.sra.com

Decimal	Name		References
993	University of Florida	Todd Hester	todd@circa.ufl.edu
994	Dantel, Inc.	John Litster	(209) 292-1111
995	Multi-Tech Systems, Inc.	Dale Martenson	(612) 785-3500 x519
996	Softlink Ltd.	Moshe Leibovitch	softlink@zeus.datasrv.co.il
997	ProSum	Christian Bucari	+33.1.4590.6231
998	March Systems Consultancy, Ltd.	Ross Wakelin	r.wakelin@march.co.uk
999	Hong Technology, Inc.	Walt Milnor	brent@oceania.com
1000	Internet Assigned Numbers Authority		iana@isi.edu
1001	PECO Energy Co.	Rick Rioboli	u002rdr@peco.com
1002	United Parcel Service	Steve Pollini	nrd1sjp@nrd.ups.com
1003	Storage Dimensions, Inc.	Michael Torhan	miketorh@xstor.com
1004	ITV Technologies, Inc.	Jacob Chen	itv@netcom.com
1005	TCPSI	Victor San Jose	Victor.Sanjose@sp1.y-net.es
1006	Promptus Communications, Inc.	Paul Fredette	(401) 683-6100
1007	Norman Data Defense Systems	Kristian A. Bognaes	norman@norman.no
1008	Pilot Network Services, Inc.	Rob Carrade	carrade@pilot.net
1009	Integrated Systems Solutions Corporation	Chris Cowan	cc@austin.ibm.com
1010	SISRO	Kamp Alexandre	100074.344@compuserve.com
1011	NetVantage	Kevin Bailey	speed@kaiwan.com
1012	Marconi S.p.A.	Giuseppe Grasso	gg@relay.marconi.it
1013	SURECOM	Mike S. T. Hsieh	+886.25.92232
1014	Royal Hong Kong Jockey Club	Edmond Lee	100267.3660@compuserve.com
1015	Gupta	Howard Cohen	hcohen@gupta.com
1016	Tone Software Corporation	Neil P. Harkins	(714) 991-9460
1017	Opus Telecom	Pace Willisson	pace@blitz.com
1018	Cogsys Ltd.	Ryllan Kraft	ryllan@ryllan.demon.co.uk
1019	Komatsu, Ltd.	Akifumi Katsushima	+81 463.22.84.30
1020	ROI Systems, Inc	Michael Wong	(801) 942-1752
1021	Lightning Instrumentation SA	Mike O'Dowd	odowd@lightning.ch
1022	TimeStep Corp.	Stephane Lacelle	slacelle@newbridge.com
1023	INTELSAT	Ivan Giron	i.giron@intelsat.int
1024	Network Research Corporation Japan, Ltd.	Tsukasa Ueda	100156.2712@compuserve.com
1025	Relational Development, Inc.	Steven Smith	rdi@ins.infonet.net
1026	Emerald Systems, Corp.	Robert A. Evans Jr.	(619) 673-2161 x5120
1027	Mitel, Corp.	Tom Quan	tq@software.mitel.com
1028	Software AG	Peter Cohen	sagpc@sagus.com
1029	MillenNet, Inc.	Manh Do	(510) 770-9390
1030	NK-EXA Corp.	Ken'ichi Hayami	hayami@dst.nk-exa.co.jp
1031	BMC Software	Chris Sharp	csharp@patrol.com
1032	StarFire Enterprises, Inc.	Lew Gaiter	lg@starfire.com
1033	Hybrid Networks, Inc.	Doug Muirhead	dougm@hybrid.com
1034	Quantum Software GmbH	Thomas Omerzu	omerzu@quantum.de
1035	Openvision Technologies Limited	Andrew Lockhart	alockhart@openvision.co.uk

Decimal	Name		References
1036	Healthcare Communications, Inc. (HCI)	Larry Streepy	streepy@healthcare.com
1037	SAIT Systems	Hai Dotu	+3223.7053.11
1038	SAT	Mleczko Alain	+33.1.4077.1156
1039	CompuSci Inc.,	Bob Berry	bberry@compusci.com
1040	Aim Technology	Ganesh Rajappan	ganeshr@aim.com
1041	CIESIN	Kalpesh Unadkat	kalpesh@ciesin.org
1042	Systems & Technologies International	Howard Smith	ghamex@aol.com
1043	Israeli Electric Company (IEC)	Yoram Harlev	yoram@yor.iec.co.il
1044	Phoenix Wireless Group, Inc.	Gregory M. Buchanan	buchanan@pwgi.com
1045	SWL	Bill Kight	wkightgrci.com (410) 290.7245
1046	nCUBE	Greg Thompson	gregt@ncube.com
1047	Cerner, Corp.	Dennis Avondet	(816) 221.1024 X2432
1048	Andersen Consulting	Mark Lindberg	mlindber@andersen.com
1049	Lincoln Telephone Company	Bob Morrill	root@si6000.ltec.com
1050	Acer	Jay Tao	jtao@Altos.COM
1051	Cedros	Juergen Haakert	+49.2241.9701.80
1052	AirAccess	Ido Ophir	100274.365@compuserve.com
1053	Expersoft Corporation	David Curtis	curtis@expersoft.com
1054	Eskom	Sanjay Lakhani	h00161@duvi.eskom.co.za
1055	SBE, Inc.	Vimal Vaidya	vimal@sbei.com
1056	EBS, Inc.	Emre Gundogan	baroque@ebs.com
1057	American Computer and Electronics, Corp.	Tom Abraham	tha@acec.com
1058	Syndesis Limited	Wil Macaulay	wil@syndesis.com
1059	Isis Distributed Systems, Inc.	Ken Chapman	kchapman@isis.com
1060	Priority Call Management	Greg Schumacher	gregs@world.std.com
1061	Koelsch & Altmann GmbH	Christian Schreyer	100142.154@compuserve.com
1062	WIPRO INFOTECH LTD.	Chandrashekar Kapse	kapse@wipinfo.soft.net
1063	Controlware	Uli Blatz	ublatz@cware.de
1064	Mosaic Software	W.van Biljon	willem@mosaic.co.za
1065	Canon Information Systems	Victor Villalpando	vvillalp@cisoc.canon.com
1066	AmericaOnline	Andrew R. Scholnick	andrew@aol.net
1067	Whitetree Network Technologies, Inc.	Carl Yang	cyang@whitetree.com
1068	Xetron Corp.	Dave Alverson	davea@xetron.com
1069	Target Concepts, Inc.	Bill Price	bprice@tamu.edu
1070	DMH Software	Yigal Hochberg	72144.3704@compuserve.com
1071	Innosoft International, Inc.	Jeff Allison	jeff@innosoft.com
1072	Controlware GmbH	Uli Blatz	ublatz@cware.de
1073	Telecommunications Industry Association (TIA)	Mike Youngberg	mikey@synacom.com
1074	Boole & Babbage	Rami Rubin	rami@boole.com
1075	System Engineering Support, Ltd.	Vince Taylor	+44 454.614.638

Decimal	Name		References
1076	SURFnet	Ton Verschuren	Ton.Verschuren@surfnet.nl
1077	OpenConnect Systems, Inc.	Mark Rensmeyer	mrensme@oc.com
1078	PDTS (Process Data Technology and Systems)	Martin Gutenbrunner	GUT@pdts.mhs.compuserve.com
1079	Cornet, Inc.	Nat Kumar	(703) 658-3400
1080	NetStar, Inc.	John K. Renwick	jkr@netstar.com
1081	Semaphore Communications, Corp.	Jimmy Soetarman	(408) 980-7766
1082	Casio Computer Co., Ltd.	Shouzo Ohdate	ohdate@casio.co.jp
1083	CSIR	Frikkie Strecker	fstreck@marge.mikom.csir.co.za
1084	APOGEE Communications	Olivier Caleff	caleff@apogee-com.fr
1085	Information Management Company	Michael D. Liss	mliss@imc.com
1086	Wordlink, Inc.	Mike Aleckson	(314) 878-1422
1087	PEER	Avinash S. Rao	arao@cranel.com
1088	Telstra Corp.	Michael Scollay	michaels@ind.tansu.com.au
1089	Net X, Inc.	Sridhar Kodela	techsupp@netx.unicomp.net
1090	PNC PLC	Gordon Tees	+44 716.061.200
1091	DanaSoft, Inc.	Michael Pierce	mpierce@danasoft.com
1092	Yokogawa-Hewlett-Packard	Hisao Ogane	hisao@yhp.hp.com
1093	Universities of Austria/Europe	Manfred R. Siegl	siegl@edvz.tuwien.ac.at
1094	Link Telecom, Ltd.	Michael Smith	michael@ska.com
1095	Xirion bv	Frans Schippers	frans@xirion.nl
1096	Centigram Communications, Corp.	Mike Nguyen	michael.nguyen@centigram.com
1097	Gensym Corp.	Greg Stanley	gms@gensym.com
1098	Apricot Computers, Ltd.	Paul Bostock	paulb@apricot.co.uk
1099	CANAL+	Philippe Desarzens	desarzen@canal-plus.fr
1100	Cambridge Technology Partners	Zaki Alam	zalam@ctp.com
	(Backup Person)	Peter Wong	pwong@ctp.com
1101	MoNet Systems, Inc.	Frank Jiang	fjiang@irvine.dlink.com
1102	Metricom, Inc.	Harold E. Austin	austin@metricom.com
1103	Xact, Inc	Keith Wiles	keith@iphase.com
1104	First Virtual Holdings Incorporated	Marshall T. Rose	mrose@dbc.fv.com
1105	NetCell Systems, Inc.	Frank Jiang	fjiang@irvine.dlink.com
1106	Uni-Q	Lennart Norlander	lennart.norlander@uniq.se or mib@uniq.se
1107	DISA Space Systems Development Division	William Reed	reedw@cc.ims.disa.mil
1108	INTERSOLV	Gary Greenfield	Gary_Greenfield@intersolv.com
1109	Vela Research, Inc.	Ajoy Jain	cheryl@empiretech.com
1110	Tetherless Access, Inc.	Richard Fox	kck@netcom.com
1111	Magistrat Wien, AT	Michael Gsandtner	gsa@adv.magwien.gv.at
1112	Franklin Telecom, Inc.	Mike Parkhurst	mikes@fdihq.com
1113	EDA Instruments, Inc.	Alex Chow	alexc@eda.com
1114	EFI Electronics, Corporation	Tim Bailey	efiups@ix.netcom.com
1115	GMD	Ferdinand Hommes	Ferdinand.Hommes@gmd.de
1116	Voicetek, Corp	Joe Micozzi	jam@voicetek.com

Decimal	Name		References
1117	Avanti Technology, Inc.	Steve Meyer	avanti@netcom.com
1118	ATLan LTD	Emanuel Wind	ew@actcom.co.il
1119	Lehman Brothers	Niten Ved	nved@mango.cities.lehman.com
1120	LAN-hopper Systems, Inc.	Jim Baugh	76227.307@compuserve.com
1121	Web-Systems	Cecile Mulder	web@aztec.co.za
1122	Piller GmbH	Stephan Leschke	100063.3642@compuserve.com
1123	Symbios Logic, Inc	Mark Johnson	mark.johnson@@wichitaks.hmpd.com
1124	NetSpan, Corp.	Lawrence Halcomb	214-690-8844
1125	Nielsen Media Research	Andrew R. Reese	reesear@msmail.dun.nielsen.com
1126	Sterling Software	Greg Rose	Greg_Rose@sydney.sterling.com
1127	Applied Network Technology, Inc.	Abbot Gilman	gilman@antech.com
1128	Union Pacific Railroad	Ed Hoppe	emhoppe@notes.up.com
1129	Tec Corporation	Tomoaki Suzuki	nab00570@niftyserve.or.jp
1130	Datametrics Systems, Corporation	Karl S. Friedrich	friedrich@datametrics.com
1131	Intersection Development Corporation	Michael McCrary	mikem43190@aol.com
1132	BACS Limited, GB	Andy Sewell	puck@pookhill.demon.co.uk
1133	Engage Communication	Peter Gibson	peterg@cruzio.com
1134	Fastware, S.A.	Christian Berge	+33 4748 0616
1135	LONGSHINE Electronics Corp.	C.T. Tseng	via@tpts1.seed.net.tw
1136	BOW Software Inc.	Bill Otto	bill@bowsoft.com
1137	emotion, Inc.	Jesus Ortiz	jesus_ortiz@emotion.com
1138	Rautaruukki steel factory, Information systems	Raine Haapasaari	rhaapasa@ratol.fi
1139	EMC Corp	Kevin Flanagan	flanagan@emc.com
1140	University of West England	Tom Johnson	tom-x@csd.uwe.ac.uk
1141	Com21	Randy Miyazaki	randy@com21.com
1142	Compression Tehnologies Inc.	Paul Wilson	paul@compression.com
1143	Buslogic Inc.	Janakiraman Gopalan	janaki@buslogic.com
1144	Firefox Corporation	John Severs	johns@firefox.co.uk
1145	Mercury Communications Ltd	David Renshaw	ag13@cityscape.co.uk
1146	COMPUTER PROTOCOL MALAYSIA SDN. BHD.	Ronald Khoo	ronald@cpm.com.my
1147	Institute for Information Industry	Shein-Tung Wu	hunter@netrd.net.tw
1148	Pacific Electric Wire & Cable Co. Ltd.	Cheng Chen	tony@tpts1.seed.net.tw
1149	MPR Teltech Ltd	Chris Sullivan	sullivan@mprott.ott.mpr.ca
1150	P-COM, Inc	Joe Shiran	joesh@netcom.com
1151	Anritsu Corporation	Manabu Usami	usami@accpd1.anritsu.co.jp
1152	SPYRUS	Russ Housley	housley@spyrus.com
1153	NeTpower, Inc.	Mark Davoren	markd@netpower.com
1154	Diehl ISDN GmbH	Larry Butler	lrb@diehl.de
1155	CARNet	Nevenko Bartolincic	Nevenko.Bartolincic@CARNet.hr
1156	AS-TECH	Jean Pierre Joerg	+33 6770 8926
1157	SG2 Innovation et Produits	Pascal Donnart	bcouderc@altern.com
1158	CellAccess Technology, Inc.	Steve Krichman	cati@netcom.com
1159	Bureau of Meteorology	Paul Hambleton	p.hambleton@bom.gov.au

Decimal	Name		References
1160	Hi-TECH Connections, Inc.	Kiran Goparaju	kiran@htconn.com
1161	Thames Water Utilities Limited	Derek Manning	+44 1734 591159
1162	Micropolis, Corp.	Jerry Sorcsek	jerome_sorcsek@microp.com
1163	Integrated Systems Technology	William Marshall	marshall@kingcrab.nrl.navy.mil
1164	Brite Voice Systems, Inc.	John Morrison	john.morrison@brite.com
1165	Associated Grocer	Michael Zwarts	(206) 764-7506
1166	General Instrument	Fred Gotwald	fgotwald@gi.com
1167	Stanford Telecom	Luther Edwards	ledwards@fuji.sed.stel.com
1168	ICOM Informatique	Jean-Luc Collet	100074,36@compuserve.com
1169	MPX Data Systems Inc.	Bill Hayes	bhayes@mpx.com
1170	Syntellect	Kevin Newsom	kevin@syntellect.com
1171	Perihelion Technology Ltd	Dave Stoneham	dave@polyhedra.com
1172	Shoppers Drug Mart	Aleksandar Simic	aleks@sdm.shoppersdrugmart.ca
1173	Apollo Travel Services	Judith Williams-Murphy	judyats@cscns.com
1174	Time Warner Cable, Inc.	Masuma Ahmed	mxa@cablelabs.com
1175	American Technology Labs Inc.	Laura Payton	(301) 695-1547
1176	Dow Jones & Company, Inc.	John Ruccolo	(609) 520 5505
1177	FRA	Per Hansson	Per.Hansson@fra.se
1178	Equitable Life Assurance Society	Barry Rubin	75141,1531@compuserve.com
1179	Smith Barney Inc.	James A. LaFleur	(212) 723-3919
1180	Compact Data Ltd	Stephen Ades	sa@compactdata.co.uk
1181	I.Net Communications	Stephane Appleton	+33 1607 20205
1182	YAMAHA Corporation	Tsuneyuki Koikeda	koikeda@lab3.yamaha.co.jp
1183	Illinois State University	Scott Genung	sagenung@ilstu.edu
1184	RADGuard Ltd.	Omer Karp	omer@radguard.co.il
1185	Calypso Software Systems, Inc.	Paul J. LaFrance	lafrance@calsof.com
1186	ACT Networks Inc.	Jacek Pastuszka	jpastusz@acti.com
1187	Kingston Communications	Nick Langford	+49 0127 9600016
1188	Incite	Susan M. Sauter	ssauter@intecom.com
1189	VVNET, Inc.	C. M. Heard	heard@vvnet.com
1190	Ontario Hydro	Bruce A Nuclear	robc@flute.candu.aecl.ca
1191	CS-Telecom	Bertrand Velle	bertrand.velle@csee-com.fr
1192	ICTV Inc.	Mark Tom	marktom@ictv.com
1193	CORE International Inc.	Bill Cloud	(407) 997-6033
1194	Mibs4You	David T. Perkins	dperkins@scruznet.com
1195	ITK	Jan Elliger	jan.elliger@itk.de
1196	Network Integrity, Inc.	Mark Fox	mfox@netint.com
1197	BlueLine Software, Inc.	Paul K. Moyer	moyer002@gold.tc.umn.edu
1198	Migrant Computing Services,Inc.	Gil Nardo	gil@netcom.com
1199	Linklaters & Paines	Suheil Shahryar	sshahrya@landp.co.uk
1200	EJV Partners,	L.P. Shean-Guang	Chang schang@ejv.com
1201	Software and Systems Engineering Ltd.	Adrian Colley	Adrian.Colley@sse.ie
1202	VARCOM Corporation	Prathibha Boregowda or Judy Smith	pboregowda@varcom.com or jsmith@ varcom.com
1203	Equitel	Marcelo Raseira	m.raseira.sulbbs%ttbbs@ibase.org.br
1204	The Southern Company	John Sciranko	/G = John/I = F/S = Sciranko/ O = SCS@mhs-southern.attmail.com

Decimal	Name		References
1205	Dataproducts Corporation	Ron Bergman	rbergma@dpc.com
1206	National Electrical Manufacturers Association	Michael M. McCrary	mikem43190@aol.com
1207	RISCmanagement, Inc.	Roger Hale	roger@riscman.com
1208	GVC Corporation	Timon Sloane	timon@timonWare.com
1209	timonWare Inc.	Timon Sloane	timon@timonWare.com
1210	Capital Resources Computer Corporation	Jeff Lee	jeff@capres.com
1211	Storage Technology Corporation	Dominique Ambach	Dominique_Ambach@stortek.com
1212	Tadiran Telecomunications L.T.D	Meir Barack	Meirb@Telecomm.Tadiran.co.il
1213	NCP	Reiner Walter	rwa@ncp.de
1214	Operations Control Systems (OCS)	Christine Young	cyoung@ocsinc.com
1215	The NASDAQ Stock Market Inc.	Hibbard Smith	(203) 385-4580
1216	Tiernan Communications, Inc.	Kyle Woodward	woodward@tiernan.com
1217	Goldman, Sachs Company	Steven Polinsky	polins@gsco.com
1218	Advanced Telecommunications Modules Ltd	William Stoye	wrs@atml.co.uk
1219	Phoenix Data Communications	Michel Robidoux	phoenix@cam.org
1220	Quality Consulting Services	Alan Boutelle	alanb@quality.com
1221	MILAN	Deh-Min Wu	wu@fokus.gmd.de
1222	Instrumental Inc.	Henry Newman	hsn@instrumental.com
1223	Yellow Technology Services Inc.	Martin Kline	(913)344-5341
1224	Mier Communications Inc.	Edwin E. Mier	ed@mier.com
1225	Cable Services Group Inc.	Jack Zhi	j.zhi@gonix.gonix.com
1226	Forte Networks Inc.	Mark Copley	mhc@fortenet.com
1227	American Management Systems, Inc.	Robert Lindsay	robert_lindsay@mail.amsinc.com
1228	Choice Hotels Intl.	Robert Peters	robert@sunnet.chotel.com
1229	SEH Computertechnik Gm	Rainer Ellerbrake	re@sehgmbh.bi.eunet.de
1230	McAFee Associates Inc.	Perry Smith	pcs@cc.mcafee.com
1231	Network Intelligent Inc.	Bob Bessin	(415) 494-6473
1232	Luxcom Technologies, Inc.	Tony Szanto	(631) 825-3788
1233	ITRON Inc.	Roger Cole	rogersc@itron-ca.com
1234	Linkage Software Inc.	Brian Kress	briank@linkage.com
1235	Spardat AG	Wolfgang Mader	mader@telecom.at
1236	VeriFone Inc.	Richard Cowan	richard_c@verifone.com
1237	Revco D.S., Inc.	Paul Winkeler	paulw@revco.com
1238	HRB Systems, Inc.	Craig R. Watkins	crw@icf.hrb.com
1239	Litton Fibercom	Mark Robison	robison@fibercom.com
1240	XCD, Incorporated	Lee Aydelotte	lca@xcd.com
1241	ProsjektLeveranser AS	Rolf Frydenberg	rolff@kinfix.no
1242	Halcyon Inc.	Mark Notten	mnotten@swi.com
1243	SBB	Michel Buetschi	michel.buetschi@sbb.ch
1244	LeuTek	W. Kruck	(0711) 790067
1245	Zeitnet, Inc	Mario Garakani	mario.garakani@zeitnet.com

Decimal	Name		References
1246	Visual Networks, Inc.	Tom Nisbet	nisbet@po.mctec.com
1247	Coronet Systems Corporation	Scott Kaplan	scott@coronet.com
1248	SEIKO EPSON	Nagahashi Toshinori	nagahasi@hd.epson.co.jp
1249	DnH Technologies	Aleksandar Simic	aasimic@mobility.com
1250	Deluxe Data	Mike Clemens	mclemens@execpc.com
1251	Michael A. Okulski Inc.	Mike Okulski	mike@okulski.com
1252	Saber Software Corporation	David Jackson	(214) 361-8086
1253	Mission Systems, Inc.	Mark Lo Chiano	p00231@psilink.com
1254	Siemens Plessey Electronics Systems	Jim Hunt	jrh@roke.co.uk
1255	Applied Communications Inc,	Al Doney	/s = doneya/o = apcom/p = apcom.oma/ admd = telemail/c = us/@sprint.com
1256	Transaction Technology, Inc.	Bill Naylor	naylor@tti.com
1257	HST Ltd	Ricardo Moraes Akaki	ricardo.araki@mpcbbs.ax.apc.org
1258	Michigan Technological University	Onwuka Uchendu	ouchendu@mtu.edu
1259	Next Level Communications	James J. Song	jsong@nlc.com
1260	Instinet Corp.	John Funchion	funchion@instinet.com
1261	Analog & Digital Systems Ltd.	Brijesh Patel	jay@ads.axcess.net.in
1262	Ansaldo Trasporti SpA	Giovanni Sorrentino	mibadm@ansaldo.it
1263	ECCI	Scott Platenberg	scottp@ecci.com
1264	Imatek Corporation	Charlie Slater	cslater@imatek.com
1265	PTT Telecom bv	Heine Maring	marin002@telecom.ptt.nl
1266	Data Race, Inc.	Lee Ziegenhals	lcz@datarace.com
1267	Network Safety Corporation	Les Biffle	les@anasazi.com
1268	Application des Techniques Nouvelles en Electronique	Michel Ricart	mricart@dialup.francenet.fr
1269	MFS Communications Company	Steve Feldman	feldman@mfsdatanet.com
1270	Information Services Division	Phil Draughon	jpd@is.rpslmc.edu
1271	Ciena Corporation	Wes Jones	wjones@ciena.com
1272	Ascom Nexion	Bill Anderson	anderson@maelstrom.timeplex.com
1273	Standard Networks, Inc	Tony Perri	tony@stdnet.com
1274	Scientific Research Corporation	James F. Durkin	jdurkin@scires.com
1275	micado SoftwareConsult GmbH	Markus Michels	Markus_Michels.MICADO@notes. compuserve.com
1276	Concert Management Services, Inc.	Jim McWalters	CONCERT/RSMPO02/mcwaltj% Concert_-_Reston_1@mcimail.com
1277	University of Delaware	Adarsh Sethi	sethi@cis.udel.edu
1278	Bias Consultancy Ltd.	Marc Wilkinson	marc@bias.com
1279	Micromuse PLC.	Martin Butterworth	mb@marx.micromuse.co.uk
1280	Translink Systems	Richard Fleming	richard@finboro.demon.co.uk
1281	PI-NET	Kirk Trafficante	pinet@netcom.com
1282	Amber Wave Systems	Bruce Kling	bkling@amberwave.com
1283	Superior Electronics Group Inc.	Bob Feather	seggroup@packet.net
1284	Network Telemetrics Inc	Jonathan Youngman	jyoungman@telemtrx.com

Decimal	Name		References
1285	BSW-Data	P.P. Stander	philip@bsw.co.za
1286	ECI Telecom Ltd.	Yuval Ben-Haim	yuval@ecitele.com
1287	BroadVision	Chuck Price	cprice@broadvision.com
1288	ALFA, Inc.	Jau-yang Chen	cjy@alfa.com.tw
1289	TELEFONICA SISTEMAS, S.A.	Luis Colorado	luis@grea1.ts.es
1290	Image Sciences, Inc.	Al Marmora	ajm@sail.iac.net
1291	MITSUBISHI ELECTRIC INFORMATION NETWORK Corporation (MIND)	CHIKAO IMAMICHI	imamichi@mind.melco.co.jp
1292	Central Flow Management Unit	Anh Le-Phuoc	lep@cfmu.eurocontrol.be
1293	Woods Hole Oceanographic Institution	Andrew R. Maffei	amaffei@whoi.edu
1294	Raptor Systems, Inc.	Alan Kirby	akirby@raptor.com
1295	TeleLink Technologies Inc.	Dean Neumann	dneum@telelink.com
1296	First Virtual Corporation	K.D. Bindra	kd@fvc.com
1297	Network Services Group	Graham King	ukking@aol.com
1298	SilCom Manufacturing Technology Inc.	Brian Munshaw	brian.munshaw@canrem.com
1299	NETSOFT Inc.	Tim Su	paullee@cameonet.cameo.com.tw
1300	Fidelity Investments	Ray Capistran	ray.capistran@fmr.com
1301	Telrad Telecommunications	Eli Greenberg	greenberg@moon.elex.co.il
1302	Arcada Software, Inc.	Bryan Booth	bbooth@arcada.com
1303	LeeMah DataCom Security Corporation	Cedric Hui	chui@cs.umb.edu
1304	SecureWare, Inc.	Luis P Caamano	lpc@sware.com
1305	USAir, Inc	Loren Cain	loren@usair.com
1306	Jet Propulsion Laboratory	Paul Springer	pls@jpl.nasa.gov
1307	ABIT Co	Matjaz Vrecko	vrecko@abit.co.jp
1308	Dataplex Pty. Ltd.	Warwick Freeman	wef@dataplex.com.au
1309	Creative Interaction Technologies, Inc.	Dave Neal	daven@ashwin.com
1310	Network Defenders, Inc.	Bill Myerson	75054.3710@compuserve.com
1311	Optus Communications	Peter Cearns	Peter_Cearns@yes.optus.com.au
1312	Klos Technologies, Inc.	Patrick Klos	klos@klos.com
1313	ACOTEC	Martin Streller	mst@acotec.de
1314	Datacomm Management Sciences Inc.	Dennis Vane	70372.2235@compuserve.com
1315	MG SOFT Co.	Andrej Duh	andrej@fiz.uni-lj.si
1316	Plessey Tellumat SA	Eddie Theart	etheart@plessey.co.za
1317	PaineWebber, Inc.	Sean Coates	coates@pwj.com
1318	DATASYS Ltd.	Michael Kodet	kodet@syscae.cz
1319	QVC Inc.	John W. Mehl	John_Mehl@QVC.Com
1320	IPL Systems	Kevin Fitzgerald	kdf@bu.edu
1321	Pacific Micro Data, Inc.	Larry Sternaman	mloomis@ix.netcom.com
1322	DeskNet Systems, Inc	Ajay Joseph	ajay@desknet.com
1323	TC Technologies	Murray Cockerell	murrayc@tctech.com.au
1324	Racotek, Inc.	Baruch Jamilly	(612) 832-9800
1325	CelsiusTech AB	Leif Amnefelt	leam@celsiustech.se
1326	Xing Technology Corp.	Jon Walker	jwalker@xingtech.com
1327	dZine n.v.	Dirk Ghekiere	100273,1157@compuserve.com
1328	Electronic merchant Services, Inc.	James B. Moore	JBM@SCEMS.COM

Decimal	Name		References
1329	Linmor Information Systems Management, Inc.	Thomas Winkler	thomas.winkler@linmor.com
1330	ABL Canada Inc.	Guy Cyr	cyrguy@abl.ca
1331	University of Coimbra	Fernando P. L. Boavida Fernandes	boavida@mercurio.uc.pt
1332	Iskratel, Ltd., Telecommunications Systems	Ante Juros	juros@iskratel.si
1333	ISA Co. Ltd.	Pan Lit WONG	plwong@hk.super.net
1334	CONNECT, Inc.	Reid Ligon	rligon@connectrf.com
1335	Digital Video	Tom Georges	tom.georges@antec.com
1336	InterVoice, Inc.	Tom Hall	thall@intervoice.com
1337	Liveware Tecnologia a Servico a Ltda	Fabio Minoru Tanada	tanada@lvw.ftpt.br
1338	Precept Software, Inc.	Karl Auerbach	karl@precept.com
1339	Heroix Corporation	Sameer J. Apte	sja@sja.heroix.com
1340	Holland House B.V.	Johan Harmsen	johan@holhouse.nl
1341	Dedalus Engenharia S/C Ltda	Philippe de M. Sevestre	dedalus.engenharia@dialdata.com.br
1342	GEC ALSTHOM I.T.	Terry McCracken	terrym@nsg.com.au
1343	Deutsches Elektronen-Synchrotron (DESY) Hamburg	Hans Kammerlocher	kammerlocher@desy.de
1344	Switchview Inc.	Jack Kesselman	jack@switchview.com
1345	Dacoll Ltd	Dan McDougall	dan@stonelaw.demon.co.uk
1346	NetCorp Inc.	Jean-Lou Dupont	dupont@netcorp.qc.ca
1347	KYOCERA Corporation	Shinji Mochizuki	SUPERVISOR@KYOCERA.CCMAIL.COMPUSERVE.COM
1348	The Longaberger Company	George Haller	75452.376@compuserve.com
1349	ILEX	J. Dominique GUILLEMET	dodo@ilex.remcomp.fr
1350	Conservation Through Innovation, Limited	Doug Hibberd	dhibberd@cti-ltd.com
1351	Software Technologies Corporation	Pete Wenzel	pete@stc.com
1352	Multex Systems, Inc.	Alex Rosin	alexr@multexsys.com
1353	Gambit Communications, Inc.	Uwe Zimmermann	gambit@gti.com
1354	Central Data Corporation	Jeff Randall	randall@cd.com
1355	CompuCom Systems, Inc.	Timothy J. Perna	tperna@compucom.com
1356	Generex Systems GMBH	F. Blettenberger	100334.1263@compuserve.com
1357	Periphonics Corporation	John S. Muller	john@peri.com
1358	Freddie Mac	John Rosner	john_rosner@freddiemac.com
1359	Digital Equipment bv	Henk van Steeg	henk.van.steeg@uto.mts.dec.com
1360	PhoneLink plc	Nick James	Nickj@Phonelink.com
1361	Voice-Tel Enterprises, Inc.	Jay Parekh	vnet@ix.netcom.com
1362	AUDILOG	Laurent EYRAUD	eyraud@audilog.fr
1363	SanRex Corporation	Carey O'Donnell	sanrex@aol.com
1364	Chloride	Jean Phillippe Gallon	33-1-60-82-04-04
1365	GA Systems Ltd	Garth Eaglesfield	geaglesfield@gasystems.co.uk
1366	Microdyne Corporation	Ron Delaney	delaneyr@mcdy.com
1367	Boston College	Eileen Shepard	eileen@bc.edu
1368	France Telecom	Vincent PRUNET	Vincent.Prunet@sophia.cnet.fr
1369	Stonesoft Corp	Jukka Maki-Kullas	juke@stone.fi
1370	A. G. Edwards & Sons, Inc.	Mike Benoist	benoisme@hqnmon1.agedwards.com

Decimal	Name		References
1371	Attachmate Corp.	Brian L. Henry	brianhe@atm.com
1372	LSI Logic	Gary Bridgewater	gjb@lsil.com
1373	interWAVE Communications, Inc.	Bruce Nelson	bruce@iwv.com
1374	mdl-Consult	Marc De Loore	marcd@mdl.be
1375	Bunyip Information Systems Inc.	Patrik Faltstrom	paf@bunyip.com
1376	Nashoba Networks Inc	Rich Curran	rcurran@nashoba.com
1377	Comedia Information AB	Rickard Schoultz	schoultz@comedia.se
1378	Harvey Mudd College	Mike Erlinger	mike@cs.hmc.edu
1379	First National Bank of Chicago	Mark J. Conroy	mark.conroy@fnb.sprint.com
1380	Department of National Defence (Canada)	Larry Bonin	burke@alex.disem.dnd.ca
1381	CBM Technologies, Inc.	George Grenley	grenley@aol.com
1382	InterProc Inc.	Frank W. Hansen	fhansen@noghri.cycare.com
1383	Glenayre R&D Inc.	Joseph Tosey	jtosey@glenayre.com
1384	Telenet GmbH Kommunikationssysteme	Mr. H. Uebelacker	uebelacker@muc.telenet.de
1385	Softlab GmbH	Martin Keller	kem@softlab.de
1386	Storage Computer Corporation	William R. Funk, III	funk@world.std.com
1387	Nine Tiles Computer Systems Ltd	John Grant	jgrant@k-net.co.uk
1388	Network People International	Dominic Fisk	dominic@nwpeople.demon.co.uk
1389	Simple Network Magic Corporation	Daris A Nevil	danevil@tddcae99.tddeng00.fnts.com
1390	Stallion Technologies Pty Ltd	Christopher Biggs	chris@stallion.oz.au
1391	Loan System	Yann Guernion	100135.426@compuserve.com
1392	DLR—Deutsche Forschungsanstalt fuer Luft- und Raumfahrt e.V.	Mr. Klaus Bernhardt	klaus.bernhardt@dlr.de
1393	ICRA, Inc.	Michael R. Wade	MWADE@ibm.com
1394	Probita	Steve Johnson	johnson@probita.com
1395	NEXOR Ltd	Colin Robbins	c.robbins@nexor.co.uk
1396	American Internation Facsimile Products	Tom Denny	denny@aifp.com
1397	Tellabs	Barry Glicklich	barry@tellabs.com
1398	DATAX	Casier Fred	100142.2571@compuserve.com
1399	IntelliSys Corporation	Pauline Sha	76600.114@compuserve.com
1400	Sandia National Laboratories	David P. Duggan	dduggan@sandia.gov
1401	Synerdyne Corp.	Dan Burns	310-453-0404
1402	UNICOM Electric, Inc.	Christopher Lin	jlo@interserv.com
1403	Central Design Systems Inc.	Bala Parthasarathy	bala@cdsi.com
1404	The Silk Road Group, Ltd.	Tim Bass	bass@silkroad.com
1405	Positive Computing Concepts	Russel Duncan	100026.1001@compuserve.com
1406	First Data Resources	Bobbi Durbin	bdurbin@marlton.1dc.com
1407	INETCO Systems Limited	Paul A. Girone	paul_girone@inetco.com
1408	NTT Mobile Communications Network Inc.	Hideaki Nishio	nishio@trans.nttdocomo.co.jp

Decimal	Name		References
1409	Target Stores	Tim Hadden	tim_hadden@target.e-mail.com
1410	Advanced Peripherals Technologies, Inc.	Yoshio Kurishita	kurishi@mb.tokyo.infoweb.or.jp
1411	Funk Software, Inc.	Cimarron Boozer	cboozer@funk.com
1412	DunsGate, a Dun and Bradstreet Company	David Willen	WILLENDC@acm.org
1413	AFP	C. MONGARDIEN	mykeeper@afp.com
1414	Comsat RSI Precision Controls Division	Travis Dye	Travisd@dfw.net
1415	Williams Energy Services Company	Matt Culver	mculver@wev.twc.com
1416	ASP Technologies, Inc.	Phil Hutchinson	VantageASP@aol.com
1417	Philips Communication Systems	Alfred Homsma	ens_homsma@nlthcl.decnet.philips.nl
1418	Dataprobe Inc.	David Weiss	dweiss@dataprobe.com
1419	ASTROCOM Corp.	DONALD A. LUCAS	612-378-7800
1420	CSTI(Communication Systems Technology, Inc.)	Ronald P.Ward	rward@csti-md.com
1421	Sprint	Makdo Abuelbassal	Majdi.M.Abuelbassal@txivg.sprint.com
1422	Syntax	Joseph A. Dudar	joe@syntax.com
1423	LIGHT-INFOCON	Mr. Santos Farias	Katyusco@cgsoft.softex.br
1424	Performance Technology, Inc.	Lewis Donzis	lew@perftech.com
1425	CXR Telecom	Nick Charles	7511.1047@CompuServe.com
1426	Amir Technology Labs	Derek Palma	dpalma@atlabs.com
1427	ISOCOR	Marcel DePaolis	marcel@isocor.com
1428	Array Technology Corportion	Mark Schnorbeger	postmaster@arraytech.com
1429	Scientific-Atlanta, Inc.	Jeff Davison	jdavison@melb.sciatl.com
1430	GammaTech, Inc.	Benny N. Ormson	ormson@ionet.net
1431	Telkom SA	Mike Brien	mikebr@hipsys.co.za
1432	CIREL SYSTEMES	Isabelle REGLEY	100142.443@compuserve.com
1433	Redflex Limited Australia	Eric Phan	epyl@mulga.cs.mu.oz.au
1434	Hermes—Enterprise Messaging LTD	Shaul Marcus	shaul@hermes.co.il
1435	Acacia Networks Inc.	Steve DesRochers	sdesrochers@acacianet.com
1436	NATIONAL AUSTRALIA BANK Ltd.	Mr. Lindsay Hall	lindsay@nabaus.com.au
1437	SineTec Technology Co., Ltd.	Louis Fu	louis@rd.sinetec.com.tw
1438	Badger Technology Inc.	Zane Morris	zane@badger.badger.com
1439	Arizona State University	Hosoon Ku	Hosoon.Ku@asu.edu
1440	Xionics Document Technologies, Inc.	Robert McComiskie	rmccomiskie@xionics.com
1441	Southern Information System Inc.	Dr. Ruey-der Lou	idps74@shts.seed.net.tw
1442	Nebula Consultants Inc.	Peter Schmelcher	nebula@mindlink.bc.ca
1443	SITRE, SA	PEDRO C. RODRIGUEZ	sitre@gapd.id.es
1444	Paradigm Technology Ltd	Roland Heymanns	roland@paradigm.co.nz
1445	Telub AB	Morgan Svensson	morgan.svensson@telub.se
1446	Communications Network Services, Virginia Tech	Dhawal Tyagi	tyagi@vt.edu
1447	Martis Oy	Seppo Hirviniemi	Seppo.Hirviniemi@martis.fi
1448	ISKRA TRANSMISSION	Lado Morela	Lado.Morela@guest.arnes.si
1449	QUALCOMM Incorporated	Frank Quick	fquick@qualcomm.com

Decimal	Name		References
1450	Netscape Communications Corp.	George Dong	gdong@netscape.com
1451	BellSouth Wireless, Inc.	Chris Hamilton	hamilton.chris@bwi.bls.com
1452	NUKO Information Systems, Inc.	Rajesh Raman	nuko@netcom.com
1453	IPC Information Systems, Inc.	Kenneth Lockhart	lockhark@ipc.com
1454	Estudios y Proyectos de Telecomunicacion, S.A.	Bruno Alonso Plaza	100746.3074@compuserve.com
1455	Winstar Wireless	Bob Hannan	bhannan@winstar.com
1456	Terayon Corp.	Amir Fuhrmann	amir@terayon.com
1457	Harris Computer Systems Corporation	David Rhein	David.Rhein@mail.hcsc.com
1458	Silicon Systems, Inc.	Todd Martin	todd.martin@tus.ssi1.COM
1459	Jupiter Technology, Inc.	Bill Kwan	billk@jti.com
1460	Delphi Internet Services	Diego Cassinera	diego@newscorp.com
1461	Kesmai Corporation	Diego Cassinera	diego@newscorp.com
1462	Compact Devices, Inc.	John Bartas	jbartas@devices.com
1463	OPTIQUEST	ERIK WILLEY	optiques@wdc.net
1464	Loral Defense Systems-Eagan	Marvin Kubischta	mkubisch@eag.unisysgsg.com
1465	OnRamp Technologies	Carl W. Kalbfleisch	cwk@onramp.net
1466	Mark Wahl	Mark Wahl	M.Wahl@isode.com
1467	Loran International Technologies, Inc.	David Schenkel	schenkel@loran.com
1468	S & S International PLC	Paul Gartside	pgartside@sands.uk.com
1469	Atlantech Technologies Ltd.	Robin A Hill	robinh@atlantec.demon.co.uk
1470	IN-SNEC	Patrick Lamourette	fauquet@calvanet.calvacom.fr
1471	Melita International Corporation	Bob Scott	rescott@melita.com
1472	Sharp Laboratories of America	Randy Turner	turnerr@sharpsla.com
1473	Groupe Decan	Nicolas lacouture	(33)78-64-31-00
1474	Spectronics Micro Systems Limited	David Griffiths	davidg@spectronics.co.uk
1475	pc-plus COMPUTING GmbH	Alois Lohner	alois@pc-plus.de
1476	Microframe, Inc.	Renee A. Morris	71242.1633@compuserve.com
1477	Telegate Global Access Technology Ltd.	Amir Wassermann	daveg@zeus.datasrv.co.il
1478	Merrill Lynch & Co., Inc.	Robert F. Marano	rmarano@ml.com
1479	JCPenney Co., Inc.	Edward Cox	cox@jcpenney.com
1480	The Torrington Company	Robert Harwood	harwood@hydra.torrington.com
1481	GS-ProActive	Giovanni Sciavicco	+3580 2787021
1482	BARCO Communication Systems	Jan Colpaert	jan.colpaert@unicall.be
1483	vortex Computersysteme GmbH	Vitus Jensen	vitus@vortex.de
1484	DataFusion Systems (Pty) Ltd	Mr. H Dijkman	dijkman@stb.dfs.co.za
1485	Allen & Overy	Peter J Williams	williamp@AllenOvery.com
1486	Atlantic Systems Group	Roy Nicholl	Roy.Nicholl@ASG.unb.ca
1487	Kongsberg Informasjonskontroll AS	Paal Hoff	ph@inko.no
1488	ELTECO a.s.	Ing. Miroslav Jergus	elteco@uvt.utc.sk

Decimal	Name		References
1489	Schlumberger Limited	David Sims	dpsims@slb.com
1490	CNI Communications Network International GmbH	Dr.Michael Bauer	Michael.Bauer@cni.net
1491	M&C Systems, Inc.	Seth A. Levy	mcsys@ix.netcom.com
1492	OM Systems International (OMSI)	Mats Andersson	mats.andersson@om.se
1493	DAVIC (Digital Audio-Visual Council)	Richard Lau	cll@nyquist.bellcore.com
1494	ISM GmbH	Bernd Richte	brichter@ism.mhs.compuserve.com
1495	E.F. Johnson Co.	Dan Bown	dbown@efjohnson.com
1496	Baranof Software, Inc.	Ben Littauer	littauer@baranof.com
1497	University of Texas Houston	William A. Weems	wweems@oac.hsc.uth.tmc.edu
1498	Ukiah Software Solutions/EDS/HDS	Tim Landers	tlanders@hds.eds.com
1499	STERIA	Christian Jamin	c.jamin@X400.steria.fr
1500	ATI Australia Pty Limited	Peter Choquenot	pchoq@@jolt.mpx.com.au
1501	The Aerospace Corporation	Michael Erlinger	erlinger@aero.org
1502	Orckit Communications Ltd.	Nimrod Ben-Natan	nimrod@orckit.co.il
1503	Tertio Limited	James Ho	jamesho@tertio.co.uk
1504	COMSOFT GmbH	Jurgen Rufleth	rufleth@comsoft.de
1505	Innovative Software	Jay Whitney	jay.whitney@innovative.com
1506	Technologic, Inc.	Perry Flinn	perry@tlogic.com
1507	Vertex Data Science Limited	Norman Fern	norman_fern@htstamp.demon.co.uk
1508	ESIGETEL	Nader Soukouti	soukouti@esigetel.fr
1509	Illinois Business Training Center	Weixiong Ho	wxho@nastg.gsfc.nasa.gov
1510	Arris Networks, Inc.	Eric Peterson	epeterson@arrisnet.com
1511	TeamQuest Corporation	Jon Hill	jdh@teamquest.com
1512	Sentient Networks	Jeffrey Price	jprice@sentientnet.com
1513	Skyrr	Elin Gautadottir	elin@skyrr.is
1514	Tecnologia y Gestion de la Innovacion	Manuel Lopez-Martin	mlm@tgi.es
1515	Connector GmbH	Matthias Reinwarth	Matthias.Reinwarth@connector.de
1516	Kaspia Systems, Inc.	Jeff Yarnell	jeffya@kaspia.com
1517	SmithKline Beecham	Campbell White	0181-975-3030
1518	NetCentric Corp.	D. O'Sullivan	d@server.net
1519	ATecoM GmbH	Michael Joost	joost@atecom.de
1520	Citibank Canada	Mike Rothwell	416-941-6007
1521	MMS (Matra Marconi Space)	PLANCHOU Fabrice	planchou@mms.matra-espace.fr
1522	Intermedia Communications of Florida, Inc	Michael Haertjens	mikeh@prog1.intermedia.com
1523	School of Computer Science, University Science of Malaysia	Mr. Sureswaran Ramadass	sures@cs.usm.my
1524	University of Limerick	Mr. Brian Adley	brian.adley@ul.ie
1525	ACTANE	Jean Vincent	actane@pacwan.mm-soft.fr
1526	Collaborative Information Technology Research Institute(CITRI)	Nam Hong Cheng	hong@catt.citri.edu.au

Decimal	Name		References
1527	Intermedium A/S	Peter Shorty	intermed@inet.uni-c.dk
1528	ANS CO + RE Systems, Inc.	Dennis Shiao	shiao@ans.net
1529	UUNET Technologies, Inc.	Louis A. Mamakos	louie@uu.net
1530	Securicor Telesciences	Hovig Heghinian	h.heghinian@telesciences.com
1531	QSC Audio Products	Ron Neely	RON_NEELY@qscaudio.com
1532	Australian Department of Employment, Education and Training	Peter McMahon	peter_mcmahon@vnet.ibm.com
1533	Network Media Communications Ltd.	Martin Butterworth	mb@netmc.com
1534	Sodalia	Giovanni Cortese	cortese@sodalia.sodalia.it
1535	Innovative Concepts, Inc.	Andy Feldstein	andy@innocon.com
1536	Japan Computer Industry Inc.	Yuji Sasaki	kyagi@po.iijnet.or.jp
1537	Telogy Networks, Inc.	Oren D. Eisner	oeisner@telogy.com
1538	Merck & Company, Inc.	Timothy Chamberlin	tim_chamberlin@merck.com
1539	GeoTel Communications Corporation	Jerry Stern	jerrys@geotel.com
1540	Sun Alliance (UK)	Peter Lancaster	+ 44 1403 234437
1541	AG Communication Systems	Pratima Shah	shahp@agcs.com
1542	Pivotal Networking, Inc.	Francis Huang	pivotal@netcom.com
1543	TSI TelSys Inc.	Jay Costenbader	mib-info@tsi-telsys.com
1544	Harmonic Systems Incorporated	Timon Sloane	timon@timonware.com
1545	ASTRONET Corporation	Chester Brummett	cbrummet@astronet.mea.com
1546	Frontec	Erik Steinholtz	Erik.Steinholtz@sth.frontec.se
1547	NetVision	Anne Gowdy	gowdy@ix.netcom.com
1548	FlowPoint Corporation	Philippe Roger	roger@flowpoint.com
1549	TRON B.V. Datacommunication	Peter Kuiper	peter@tron.nl
1550	Nuera Communication Inc.	Kuogee Hsieh	kgh@pcsi.cirrus.com
1551	Radnet Ltd.	Osnat Cogan	radnet@radmail.rad.co.il
1552	Oce Nederland BV	H. Ketelings	standards@oce.nl
1553	Air France	Chantal NEU	neuch@airfrance.fr
1554	Communications & Power Engineering, Inc.	Ken Dayton	kd@compwr.com
1555	Charter Systems	Michael Williams	mwilliams@charter.com
1556	Performance Technologies, Inc.	Rayomnd C. Ward	rcw@pt.com
1557	Paragon Networks International	Joseph Welfeld	jwelfeld@howl.com
1558	Skog-Data AS	Haakon Beitnes	hb@skogdata.no
1559	mitec a/s	Arne-J = F8rgen Auberg	mitec@sn.no
1560	THOMSON-CSF/ Departement Reseaux d'Entreprise	Eric BOUCHER	eric.boucher@rcc.thomson.fr
1561	Ipsilon Networks, Inc.	Joe Wei	jwei@ipsilon.com
1562	Kingston Technology Corporation	Barry Man	barry_man@kingston.com
1563	Harmonic Lightwaves	Abi Shilon	abi@harmonic.co.il
1564	InterActive Digital Solutions	Rajesh Raman	rraman@sgi.com
1565	Coactive Aesthetics, Inc.	Dan Hennage	dan@coactive.com

Decimal	Name		References
1566	Tech Data Corporation	Michael Brave	mbrave@techdata.com
1567	Z-Com	Huang Hsuang Wang	hhwang@center.zcom.com.tw
1568	COTEP	Didier VECTEN	100331.1626@COMPUSERVE.COM
1569	Raytheon Company	Robert Emmett	emmett@ed.ray.com
1570	Telesend Inc.	Craig Sharper	csharper@telesend.com
1571	NCC	Nathan Guedalia	natig@ncc.co.il
1572	Forte Software, Inc.	Geoff Puterbaugh	geoff@forte.com
1573	Secure Computing Corporation	Raymond Lu	rlu@sctc.com
1574	BEZEQ	Tom Lorber	ltomy@dialup.netvision.net.il
1575	Technical University of Braunschweig Detlef	J. Schmidt	djs@tu-bs.de
1576	Stac Inc.	Laurence Church	lchurch@stac.com
1577	StarNet Communications	Christopher Shepherd	belgo@winternet.com
1578	Universidade do Minho	Dr Vasco Freitas	vf@uminho.pt
1579	Department of Computer Science, University of Liverpool	Dave Shield	D.T.Shield@csc.liv.ac.uk
1580	Tekram Technology, Ltd.	Mr. Joseph Kuo	jkuo@tekram.com.tw
1581	RATP	Pierre MARTIN	pma@ratp.fr
1582	Rainbow Diamond Limited	Frank O'Dwyer	fod@brd.ie
1583	Magellan Communications, Inc	Paul Stone	paul@milestone.com
1584	Bay Networks Incorporated	Steven P. Onishi	sonishi@BayNetworks.com
1585	Quantitative Data Systems (QDS)	Joe R. Lindsay Jr.	JLindsay@QDS.COM
1586	ESYS Limited	Mark Gamble	mgamble@esys.co.uk
1587	Switched Network Technologies (SNT)	Bob Breitenstein	bstein@sntc.com
1588	Brocade Communications Systems, Inc.	Mr. Kha-Sin Teow	khasin@brocadesys.com
1589	Computer Resources International A/S (CRI)	Ole Rieck Sorensen	ors@cri.dk
1590	LuchtVerkeersBeveiliging	D. Helmer	sss@lvb.nl
1591	PST	Yakov Roitman	yakovr@pst.co.il
1592	XactLabs Corporation	Bill Carroll	billc@xactlabs.com
1593	NetPro Computing, Inc.	Gil Kirkpatrick	gilk@primenet.com
1594	TELESYNC	Fred R Stearns	fred@crl.com
1595	BOSCH Telecom	Holger Lenz	lenz@FRTIMC.DECNET.BOSCH.DE
1596	INS GmbH	Andreas Frackowiak	af@ins.de
1597	Distributed Processing Technology	Joseph A. Ballard	ballard@dpt.com
1598	Tivoli Systems Inc.	Greg Kattawar	greg.kattawar@tivoli.com
1599	Network Management Technologies	Mark Hammett	mhammett@nmt.com.au
1600	SIRTI	Mr. Angelo ZUCCHETI	rossima@mbox.vol.it
1601	TASKE Technology Inc.	Dennis Johns	dennis@taske.com
1602	CANON Inc.	Masatoshi Otani	otani@cptd.canon.co.jp
1603	Systems and Synchronous, Inc.	Dominic D. Ricci	ddr@ssinc.com
1604	XFER International	Fred Champlain	fred@xfer.com
1605	Scandpower A/S	Bjorn Brevig	Bjorn.Brevig@halden.scandpower.no
1606	Consultancy & Projects Group srl	Monica Lausi	monica@cpg.it

Decimal	Name		References
1607	STS Technologies, Inc.	Scott Chaney	ststech@icon-stl.net
1608	Mylex Corporation	Vytla Chandramouli	mouli@mylex.com
1609	CRYPTOCard Corporation	Greg Carter	gregc@cryptocard.com
1610	LXE, Inc.	Don Hall	dhh2347@lxe.com
1611	BDM International, Inc.	John P. Crouse	jcrouse@bdm.com
1612	GE Spacenet Services Inc.	Alan Schneider	spacenet.aschneid@capital.ge.com
1613	Datanet GmbH	Mr. Juraj Synak	juraj@datanet.co.at
1614	Opcom, Inc.	Walter D. Ballew	(405) 733-1011
1615	Mlink Internet Inc.	Patrick Bernier	pat@4p.com
1616	Netro Corporation	Isaac Oren	isaaco@netro-corp.com
1617	Net Partners Inc.	Deepak Khosla	dkhosla@npartners.com
1618	Peek Traffic—Transyt Corp.	Robert De Roche	Robert2161@aol.com
1619	Comverse Information Systems	Carlo San Andres	Carlo_San_Andres@Comverse.com
1620	Data Comm for Business, Inc.	John McCain	jmccain@dcbnet.com
1621	CYBEC Pty. Ltd.	Jia Dong HUANG	100240.3004@compuserve.com
1622	Mitsui Knowledge Industry Co.,Ltd.	Daisuke Kumada	kumada@pc.mki.co.jp
1623	NORDX/CDT Inc	Matthieu Lachance	Mathieu_Lachance@nt.com
1624	Blockade Systems Corp.	Konstantin Iavid	kj@interlog.com
1625	Nixu Oy	Pekka Nikander	Pekka.Nikander@nixu.fi
1626	Australian Software Innovations (Services) Pty. Ltd.	Andrew Campbell	andrew@asi.oz.au
1627	Omicron Telesystems Inc.	Martin Gadbois	mgadb@ibm.net
1628	DEMON Internet Ltd.	William John Hulley	enterprise-mib@demon.net
1629	PB Farradyne, Inc.	Alan J. Ungar	UngarA@farradyne.com
1630	Telos Corporation	Sharon Sutherlin	sharon.sutherlin@telos.com
1631	Manage Information Technologies	Kim N. Le	72124.2250@compuserve.com
1632	Harlow Butler Broking Services Ltd.	Kevin McCarthy	+ 44 171 407 5555 x 5246
1633	Eurologic Systems Ltd	Brian Meagher	eurologic@attmail.com
1634	Telco Research Corporation	Bizhan Ghavami	ghavami@telcores.com
1635	Mercedes-Benz AG	Dr. Martin Bosch	Martin.Bosch@Sifi.Mercedes-Benz.com
1636	HOB electronic GmbH	Ernst Petermann	Petermann@hob.de
1637	NOAA	Ken Sragg	ksragg@sao.noaa.gov
1638	Cornerstone Software	Jennifer Parent	JenniferParent@corsof.com
1639	Wink Communications	Dave Brubeck	dave.brubeck@wink.com
1640	Thomson Electronic Information Resources (TEIR)	John Roberts	jroberts@teir.com
1641	HITT Holland Institute of Traffic Technology B.V.	C. van den Doel	vddoel@hitt.nl
1642	KPMG	Richard Ellis	richard.ellis@kpmg.co.uk
1643	Loral Federal Systems	Mike Gulden	mgulden@lfs.loral.com
1644	S.I.A.—Societa Interbancaria per l'Automazione	Fiorenzo Claus	claus@sia.it
1645	United States Cellular Corp.	Curtis Kowalski	kowalski@uscc.com
1646	AMPER DATOS S.A.	Angel Miguel Herrero	34-1-8040909
1647	Carelcomp Forest Oy	Rauno Hujanen	Rauno.Hujanen@im.ccfo.carel.fi

Decimal	Name		References
1648	Open Environment Australia	Geoff Bullen	gbullen@jarrah.com.au
1649	Integrated Telecom Technology, Inc.	Suresh Rangachar	suresh@igt.com
1650	Langner Gesellschaft fuer Datentechnik mbH	Heiko Bobzin	hb@langner.com
1651	Wayne State University	Mark Murphy	mark@opus.pass.wayne.edu
1652	SICC (SsangYong Information & Communications Corp.)	Mi-Young, Lee	traum@toody.sicc.co.kr
1653	THOMSON—CSF	De Zaeytydt	33 1 69 33 00 47
1654	Teleconnect Dresden GmbH	Matthias Lange	mlange@teleconnect.de
1655	Panorama Software Inc.	Bill Brookshire	bbrooksh@pansoft.com
1656	CompuNet Systemhaus GmbH	Heiko Vogeler	hvo@compunet.de
1657	JAPAN TELECOM CO., Ltd.	Seiji Kuroda	kuroda@japan-telecom.co.jp
1658	TechForce Corporation	Mark Dauscher	mdauscher@techforce.com
1659	Granite Systems Inc.	Michael Fine	mfine@arp.com
1660	Bit Incorporated	Tom Alexander	talex@bitinc.com
1661	Companhia de Informatica do Parana—Celepar	Armando Rech Filho	armando@lepus.celepar.br
1662	Rockwell International Corporation	Ted Sickles	tgsickles@corp.rockwell.com
1663	Ancor Communications	Bently H. Preece	benp@ancor.com
1664	Royal Institute of Technology, Sweden (KTH)	Rickard Schoultz	staff@kth.se
1665	SUNET, Swedish University Network	Rickard Schoultz	staff@sunet.se
1666	Sage Instruments, Inc.	Jack Craig	jackc@sageinst.com
1667	Candle Corporation	Dannis Yang	DannisYang@msn.com
1668	CSO GmbH	Andreas Kientopp	100334.274@compuserve.com
1669	M3i Systems Inc.	Louis St-Pierre	lstpier@m3isystems.qc.ca
1670	CREDINTRANS	Pascal BON	BON@credintrans.fr
1671	BIT Communications	Mr Alistair Swales	bitcomm@cix.compulink.co.uk
1672	Pierce & Associates	Fred Pierce	fred@sccsi.com
1673	Real Time Strategies Inc.	Jay Moskowitz	jmosk@email.rts-inc.com
1674	R.I.C. Electronics	Andrew Philip	102135.1051@compuserve.com
1675	Amoco Corporation	Tim Martin	tlmartin@amoco.com
1676	Qualix Group, Inc.	Takeshi Suganuma	tk@qualix.com
1677	Sahara Networks, Inc.	Thomas K Johnson	johnson@saharanet.com
1678	Hyundai Electronics Industries Co., Ltd.	Ha-Young OH	hyoh@super5.hyundai.co.kr
1679	RICH, Inc.	Yuri Salkinder	yuri.salkinder@richinc.com
1680	Amati Communications Corp.	Mr. Gail Cone	gpc@amati.com
1681	P.H.U. RysTECH	Rafal Fagas	apc@silter.silesia.ternet.pl
1682	Data Labs Inc.	Raul Montalvo	raul.montalvo@datalabsinc.com
1683	Occidental Petroleum Services, Inc.	Glenn A. Blakley	glenn_blakley@oxy.com
1684	Rijnhaave Internet Services	Thierry van Herwijnen	t.vanherwijnen@rijnhaave.net

Decimal	Name		References
1685	Lynx Real-Time Systems, Inc.	Ganesan Vivekanandan	ganesan@lynx.com
1686	Pontis Consulting	Mr A. R. Price	priceto@ukh.pontis.com
1687	SofTouch Systems, Inc.	Kody Mason	405-947-8080
1688	Sonda S.A.	Hermann von Borries	h_vborries@sonda.cl
1689	McCormick Nunes Company	Charles Craft	chuckc@mn.com
1690	Ume E5 Universitet	Roland Hedberg	Roland.Hedberg@umdac.umu.se
1691	NetiQ Corporation	Ching-Fa Hwang	cfh@netiq.com
1692	Starlight Networks	Jim Nelson	jimn@starlight.com
1693	Informacion Selectiva S.A. de C.V. (Infosel)	Francisco Javier Reyna Castillo	freyna@infosel.com.mx
1694	HCL Technologies Limited	Murugan R.	murugan@hclt.com
1695	Maryville Data Systems, Inc.	Samuel T. Denton, III	sam.denton@maryville.com
1696	EtherCom Corp	Nafis Ahmad	nafis@ethercom.com
1697	MultiCom Software	Ari Hallikainen	ari.hallikainen@multicom.fi
1698	BEA Systems Ltd.	Garth Eaglesfield	geaglesfield@beasys.co.uk
1699	Advanced Technology Ltd.	Yuda Sidi	atlsidi@inter.net.il
1700	Mobil Oil	Tony Dworznicki	afdworzn@dal.mobil.com
1701	Technical Software	Colin Haxton	Colin.Haxton@wang.co.nz
1702	Netsys International (Pty) Ltd	Wayne Botha	wayne@inetsys.alt.za
1703	Titan Information Systems Corp.	Edgar St.Pierre	edgar@titan.com
1704	Cogent Data Technologies	Wade Andrews	wadea@cogentdata.com
1705	Reliasoft Corporation	Chao-Li Tarng	chaoli@reliasoft.com
1706	Midland Business Systems, Inc.	Bryan Letcher	bletcher@ccmailgw.str.com
1707	Optimal Networks	John Graham-Cumming	jgc@optimal.com
1708	Gresham Computing plc	Tony Churchill	tchurchill@gresham.co.uk
1709	Science Applications International Corporation (SAIC)	Martin Hobson	martin@mail.apd.saic.com
1710	Acclaim Communications	Pratima Janakir	janakir@acclaiminc.com
1711	BISS Limited	David Bird	dbird@biss.co.uk
1712	Caravelle Inc.	Ron Stone	rstone@caravelle.com
1713	Diamond Lane Communications Corporation	Bill Hong	hong@dlcc.com
1714	Infortrend Technology, Inc.	Michael Schnapp	michael@infortrend.com.tw
1715	Orda-B N.V.	Steef Van Braband	steef.van.braband@ordab.com
1716	Ariel Corporation	Allan Chu	allan.chu@ariel.com
1717	Datalex Communications Ltd.	David Tracey	d_tracey@datalex.ie
1718	Server Technology Inc.	Brian P. Auclair	brian@servertech.com
1719	Unimax Systems Corporation	Bill Sparks	bsparks@unimax.com
1720	DeTeMobil GmbH	Olaf Geschinske	Olaf.Geschinske@ms.DeTeMobil.de
1721	INFONOVA GmbH	Ing. Alois Hofbauer	alois.hofbauer@infonova.telecom.at
1722	Kudelski SA	Eric Chaubert	chaubert@nagra-kudelski.ch
1723	Pronet GmbH	Juergen Littwin	jl@pronet.de

Decimal	Name		References
1724	Westell, Inc.	Rodger D. Higgins	rhiggins@westell.com
1725	Nupon Computing, Inc.	Tim Lee	tim@nupon.com
1726	CIANET Ind e Com Ltda (CIANET Inc.)	Norberto Dias	cianet@bbsoptions.com.br
1727	Aumtech of Virginia (amteva)	Deepak Patil	dpatil@amteva.com
1728	CheongJo data communication, Inc.	HyeonJae Choi	cyber@cdi.cheongjo.co.kr
1729	Genesys Telecommunications Laboratories Inc. (Genesys Labs.)	Igor Neyman	igor@genesyslab.com
1730	Progress Software	Anil Vayaliparambath	anilv@bedford.progress.com
1731	ERICSSON FIBER ACCESS	George Lin	gglin@rides.raynet.com
1732	Open Access Pty Ltd	Guy Elliott	guy@oa.com.au
1733	Sterling Commerce	Dale Moberg	dale_moberg@stercomm.com
1734	Predictive Systems Inc.	Adam Steckelman	asteckelman@predictive.com
1735	Architel Systems Corporation	Natalie Chew	n.chew@architel.com
1736	US West !nterAct	Jeff Konz	jkonz@uswest.net
1737	Eclipse Technologies Inc.	Alex Holland	alexh@eclipse-technologies.com
1738	Navy	Ryan Huynh	huynhr@manta.nosc.mil
1739	Bindi Technologies, Pty Ltd	Tim Potter	bindi@ozemail.com.au
1740	Hallmark Cards Inc.	Kevin Leonard	kleonard@hallmark.com
1741	Object Design, Inc.	George M. Feinberg	gmf@odi.com
1742	Vision Systems	Gregory Frascadore	gaf@vsys.com
1743	Zenith Data Systems (ZDS)	Daniel G. Peters	dg.peters@zds.com
1744	Gobi Corp.	Kenneth Hart	khart@cmp.com
1745	Universitat de Barcelona	Ricard de Mingo	ricardo@ub.es
1746	Institute for Simulation and Training (IST)	Seng Tan	stan@ist.ucf.edu
1747	US Agency for International Development	Ken Roko	kroko@usaid.gov
1748	Tut Systems, Inc.	Mark Miller	markm@tutsys.com
1749	AnswerZ Pty Ltd (Australia)	Bernie Ryan	bernier@answerz.com.au
1750	H.Bollmann Manufacturers Ltd (HBM)	Klaus Bollmann	mallen@hbmuk.com
1751	Lucent Technologies	Richard Bantel	richard.bantel@lucent.com
1752	phase2 networks Inc.	Jeffrey Pickering	p01152@psilink.com
1753	Unify Corporation	Bill Bonney	sa@unify.com
1754	Gadzoox Microsystems Inc.	Kim Banker	408-399-4877
1755	Network One, Inc.	David Eison	deison@faxjet.com
1756	MuLogic b.v.	Jos H.J. Beck	jos_beck@euronet.nl
1757	Optical Microwave Networks, Inc.	Joe McCrate	elaw@omnisj.com
1758	SITEL, Ltd.	Boris Jurkovic	jurkovic@iskratel.si
1759	Cerg Finance	Philippe BONNEAU	101605.1403@compuserve.com

Decimal	Name		References
1760	American Internet Corporation	Brad Parker	brad@american.com
1761	PLUSKOM GmbH	Norbert Zerrer	zerrer@ibm.net
1762	Dept. of Communications, Graz University of Technology Dipl.-Ing.	Thomas Leitner	tom@finwds01.tu-graz.ac.at
1763	MindSpring Enterprises, Inc.	Allen R. Thomas	athomas@mindspring.com
1764	Db-Tech, Inc.	Raj K. Salgam	rsalgam@attmail.com
1765	Apex Voice Communications, Inc.	Osvaldo Gold	osvaldo@apexvoice.com
1766	National DataComm Corporation	Ms. Changhua Chiang	101400.242@compuserve.com
1767	Telenor Conax AS	Aasmund Skomedal	Asmund.Skomedal@oslo.conax.telenor.no
1768	Patton Electronics Company	William B. Clery III	benson@ari.net
1769	The Fulgent Group Ltd.	Mark Flewelling	mflewell@mail.transdata.ca
1770	BroadBand Technologies, Inc.	Keith R. Schomburg	krs@bbt.com
1771	Myricom, Inc.	Bob Felderman	feldy@myri.com
1772	DecisionOne	Doug Green	douggr@world.std.com
1773	Tandberg Television	Olav Nybo	saxebol@sn.no
1774	AUDITEC SA	Pascal CHEVALIER	pascalch@dialup.remcomp.fr
1775	PC Magic	Tommy Cheng	TommyCheng@compuserve.com
1776	Philips Electronics NV	Bob Rossiter	Robert.M.Rossiter@nl.cis.philips.com
1777	ORIGIN	Bob Rossiter	Robert.M.Rossiter@nl.cis.philips.com
1778	CSG Systems	Gordon Saltz	gsaltz@probe.net
1779	Alphameric Technologies Ltd	Mr Tim Raby	100034.1074@compuserve.com
1780	NCR Austria	Michael Ostendorf	Michael.Ostendorf@Austria.NCR.COM
1781	ChuckK, Inc.	Chuck Koleczek	chuckk@well.net
1782	PowerTV, Inc.	David Ma	dma@powertv.com
1783	Active Software, Inc.	Steve Jankowski	steve@activesw.com
1784	Enron Capitol & Trade Resources	Steven R. Lovett	slovett@ect.enron.com
1785	ORBCOMM	Todd Hara	thara@orbcomm.net
1786	Jw direct shop	pavel deng	ivb00285@192.72.158.10
1787	B.E.T.A.	Brian Mcgovern	mcgovern@spoon.beta.com
1788	Healtheon	Marco Framba	framba@hscape.com
1789	Integralis Ltd.	Andy Harris	Andy.Harris@Integralis.co.uk
1790	Folio Corporation	Eric Isom	eisom@folio.com
1791	ECTF	Joe Micozzi	508-250-7953 jam@voicetek.com
1792	WebPlanet	Ray Taft	Ray_Taft@webplanet.com
1793	nStor Corporation	Bret Jones	Bret.Jones@4dmg.net
1794	Deutsche Bahn AG	peter ritzert	peter.ritzert@bku.db.de
1795	Paradyne	Kevin Baughman	klb@eng.paradyne.com
1796	Nastel Technologies, Inc.	Krish Shetty	nastel@nyc.pipeline.com
1797	Metaphase Technology, Inc.	Michael Engbrecht	Michael.R.Engbrecht@cdc.com
1798	Zweigart & Sawitzki	Mr. Andreas Georgii	100316.2050@compuserve.com
1799	PIXEL	Mauro Ieva	mieva@mbox.vol.it
1800	WaveAccess Inc.	Yoram Feldman	yoram@waveaccess.com
1801	The SABRE Group	Richard Buentello	richb@fastlane.net
1802	Far Point Systems	Kody Mason	405-570-4068

To request an assignment of an enterprise number send the complete company name, address, fax number, and phone number and the contact person's complete name, address, phone number, and e-mail mailbox in an e-mail message to <iana-mib@isi.edu>.

Chapter

19

Internet Services

Internet Access Methods

Due to the expense, a full-time, dedicated, 56-kbps to 1.544-Mbps or faster connection to the Internet is currently a rare and expensive commitment. If you (or your organization's networks) have that access, your system will be running TCP/IP and you will be able to connect directly to other systems and Internet resources.

The Integrated Services Digital Network (ISDN) that integrates voice and data into a single digital telephone network is up and running in many areas. The cable TV access alternative is out of the starting gate. If you live in one of a few major metropolitan areas or a selected test market, you could have that full-time Ethernet-style access in your home at 3 Mbps or faster. You would then run a TCP/IP package on your system and have direct access to other Internet connected systems.

The newest player may come through a TV service provider or a local telephone company or even an unlikely alliance of the two. Asymmetrical Digital Subscriber Line (ADSL) will offer access to the Internet at a slower, interactive dial-up speed (like 33.6 kbps or the recently announced 56 kbps) while returning digital data at much higher rates like T-1 (1.544 Mbps) or a multi-megabit rate.

Since the high-speed return rate will work via the local cable TV provider or one of the new, digital satellite providers, you may soon have your choice. That would enable a short request for a web page to go out at our human typing speed and the web page return at the higher rate for the higher volume of computer-driven data.

If you, like most of us, are not in one of those select groups, your access to the network of networks (Internet) is through a dial-up modem connection. While your modem takes care of the hardware to connect to one of the hundreds of IAPs, the available software presents many other choices.

You can use your choice of modem software packages to connect to the access provider's UNIX system and use that system's TCP/IP capability. We call this access a *shell account* because you use a command shell to access the Internet. The main alternatives are PPP or SLIP.

PPP and SLIP Access

Serial Line Internet Protocol and Point-to-Point Protocol are network access or data link layer protocols for WANs, just as Ethernet and token ring are for LANs. Both operate over dial-up serial lines and modems (although PPP will also run over ISDN). With connections between only two machines, we do not need the source and destination hardware addresses. This is where the similarity between the two ends.

As you may remember from our discussion in Chap. 3, SLIP supports only asynchronous communications and IP. PPP adds support for synchronous communications and can handle multiple protocols. SLIP lacks PPP's error detection. SLIP typically has an MTU of 1006 bytes while PPP supports an MTU of 1500 bytes.

Both protocols work much the same for TCP/IP Internet access. They offer network or individual access. For network access, the organization's gateway (router) must support the same protocol as the ISP's gateway. For individual use, your SLIP or PPP software must be compatible with the access provider's software.

In both situations, your system(s) will run TCP/IP software. You have direct access to Internet resources by passing your TCP/IP traffic through the ISP's access point. The main advantage of PPP and SLIP over community access television (CATV), i.e., cable, and full-time dedicated circuits is the lower cost, while having the same direct services.

There is also the added security of controlling the time that you make the connection to the Internet. The disadvantages are slower speeds and the lack of a local, full-time server that others can access from the Internet.

Archie

Archie is not Archie Andrews of Riverdale High School and comic strip fame. It is the result of missing a keystroke when typing in "archive." The Archie service helps Internet users find the host, directory path, and file or directory name of interest for doing an anonymous FTP retrieval.

As the list in Fig. 19.1 shows, Archie operates through a number of servers. Each server holds the same information: a database of over 100 gigabytes of pointers to keep track of more than 2 million files and directories available through anonymous FTP from more than 1500 different hosts. Students at the School of Computer Science, McGill University, in Montreal, Canada, originally developed Archie as a project. From there, it became a matter of pride that Archie servers collect anonymous FTP resources from their part of the Internet and share them with the other Archie servers around the rest of the world.

archie.au	139.130.4.6	Australia
archie.edvz.uni-linz.ac.at	140.78.3.8	Austria
archie.univie.ac.at	131.130.1.23	Austria
archie.uqam.ca	132.208.250.10	Canada
archie.funet.fi	128.214.6.102	Finland
archie.th-darmstadt.de	130.83.22.60	Germany
archie.cs.huji.ac.il	132.65.20.254	Israel
archie.unipi.it	131.114.21.10	Italy
archie.kuis.kyoto-u.ac.jp	130.54.20.1	Japan
archie.wide.ad.jp	133.4.3.6	Japan
archie.nz	130.195.9.4.	New Zealand
archie.kr	128.134.1.1	South Korea
archie.sogang.ac.kr	163.239.1.11	South Korea
archie.rediris.es	130.206.1.2	Spain
archie.luth.se	130.240.18.4	Sweden
archie.switch.ch	130.59.1.40	Switzerland
archie.ncu.edu.tw	140.115.19.24	Taiwan
archie.doc.ic.ac.uk	146.169.11.3	United Kingdom
archie.sura.net	128.167.254.195	USA (Md.)
archie.unl.edu	129.93.1.14	USA (Neb.)
archie.rutgers.edu	128.6.18.15	USA (N.J.)
archie.ans.net	147.225.1.10	USA (N.Y.)
archie.internic.net	198.49.45.10	USA (N.Y.)

Figure 19.1 Archie servers.

Archie has done a great job during its time, but technology has outgrown it over the past few years. While Archie continues to work in a limited way, most of the Internet searches happen through the World Wide Web and the user's web browser.

Gopher

The Gopher service, developed in April 1991 at the University of Minnesota, was the closest thing to a one-service-fits-all-abilities function available to every user on the Internet. To use it you would make selections from a menu like the one in Fig. 19.2. Some of those menu items (noted with a slash /) are to access submenus; others access files. Still others access programs to search Gopherspace (the several thousand, interconnected Gopher servers around the world) for a file or topic contained in a file. Like Archie, however, the World Wide Web has overtaken the Gopher technology. While both Gopher and Archie did a great job at their time, they are fading fast in the bright headlights of the Web-based search engines.

World Wide Web

The most popular reason to use the Internet and the most user-friendly way to navigate the Internet has more names than any other resource. You may

```
>gopher

Press ? for Help, q to Quit                              Retrieving Directory../

                  Internet Gopher Information Client 2.0 pl5

                    Root gopher server: gopher.netcom.com

 1. About this GOPHER server.
 2. Information about NETCOM/
 3. Internet information/
 4. Jughead-Search High-Level Gopher Menus via Washington &  Lee <?>
 5. Search Gopherspace using Veronica/
 6. Other Internet Gopher Servers (via U.C. Santa Cruz)/
 7. Weather (via U. Minnesota)/
 8. Worldwide Directory Services (via Notre Dame)/
 9. Interesting items/
10. ATTENTION NETCOM users.

Press item number: ___                                   Retrieving Directory..\
Press ? for Help, q to Quit
```

Figure 19.2 Gopher menu.

call it WWW or the World Wide Web or W3 or, as many do, the Web. Either way, the Web is an attempt to organize an ever-expanding collection of information into an easily usable, linked format.

Developed at CERN (Europe's high-energy physics lab in Switzerland) and released in August 1991, it took time for the National Center for Supercomputing Applications at the University of Illinois to make a key to use it. Mosaic, the legendary graphical Web browser, hit the streets in February 1993.

The key to the Web is hypertext, which is data that contains links to other data. Hypertext has two parts: the anchor and the address. The anchor is the text or graphic hot spot that the user clicks to jump to that page of information. The address, a uniform resource locator, points to the document that the browser will load when the user clicks the anchor.

All Web clients display pages that an author has constructed with Hypertext Markup Language (HTML). While HTML is not a highly advanced desktop publishing language, it offers far more presentation styles than simple ASCII characters on a screen. The HTML specification is constantly being updated to add new features.

Web browsers

Speaking of the Web, we need to talk about popular Web browsers. Rather than try to include all of those on the market today, let's look at some of the best known: Mosaic, Netscape Navigator, Microsoft Internet Explorer, and Lynx.

Mosaic was the first graphical browser for MS Windows and is now also available for Macintosh and X-Windows. While it is free, it has no tech support. With its widespread use, NCSA has been diligent about fixing user-found bugs. Mosaic works well over high-speed connections (56 kbps and above), but is not as smooth over slower modem speeds.

One author of NCSA Mosaic (Marc Andreesen) started the Mosaic Communications Corp. (MCC) that developed Netscape. MCC changed its name to Netscape Communications Corp. on November 14, 1994. Netscape Navigator is a graphical browser that is available for a very low cost. Netscape supports Mac, Windows, and X-Windows. It is the leading choice for Web browser software.

While they joined the browser competition late, Microsoft has closely followed the lead of Netscape with its Internet Explorer browser. In fact, its latest version has many of the same functions that Netscape Navigator supports. Since both Netscape and Microsoft are constantly enhancing their products frequently, we will not even attempt to compare the two. Instead, check your favorite computer magazine.

Lynx was developed at the University of Kansas for shell account access. It is usually the fastest Web browser for modems because it runs on VMS and UNIX systems that have high-speed Internet access. The drawbacks are that Lynx has limited forms support and does not display graphics.

URL

The Uniform Resource Locator is a standard scheme for compactly identifying any document on any compatible Web server anywhere in the world. It contains six component parts, of which three are required and three are optional.

The required parts of the URL are

- The service type tells the browser how to contact the server for the requested data. In Fig. 19.3 it is Hypertext Transport Protocol (HTTP). Alternates include: Gopher, WAIS, ftp, Telnet, mailto, netnews, and file.
- The system name is the Fully Qualified Domain Name (FQDN) of the server. In Fig. 19.3 that is system.org.dom.

http://system.org.dom:999/dirA/dir1/file.html?gold

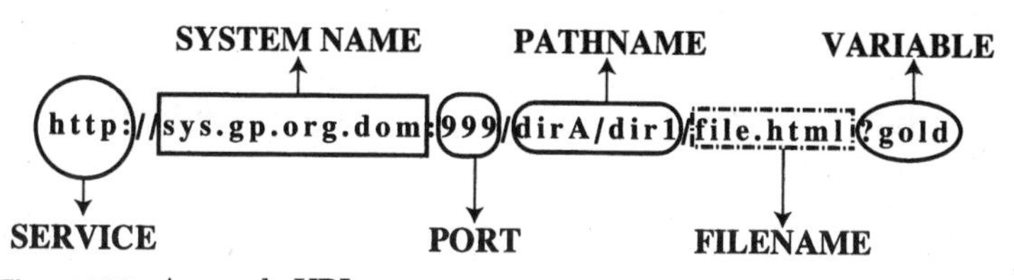

Figure 19.3 A sample URL.

- The path name is the directory path for the file. In Fig. 19.3 that is /dirA/dir1/ (some servers allow shortening it to an alias).

The optional parts of the URL are:

- The port tells the client which port to use if it does not use the default port (80). In Fig. 19.3, that is 999.
- If the URL lists no file name, it will provide a listing of the last directory the user selected. The URL in Fig. 19.3 sets that to file.html.
- If the URL is a request to search a database, the URL sets that query text to follow a ? or a # in the URL. The URL in Fig. 19.3 chose a question mark and the variable *gold*.

At that point, in a browser and requesting a URL, we invoke HTTP to carry the request.

The Hypertext Transfer Protocol

Web browsers (clients) and Web servers communicate using HTTP (RFC 1945), which is a simple protocol with very little overhead. In fact, an HTTP connection has only four stages, as follows:

1. Open the connection. Each request from the client opens a new TCP connection to the server at the URL identified location and port number (the default is port 80).
2. Request the service. That request contains HTTP request headers which define the method for the transaction while providing information about the client's capabilities. Those methods include
   - GET: Retrieve the designated URL
   - HEAD: Retrieve the HTTP header information from the identified URL
   - POST: Send the data to the specified URL
   - PUT: Place the POSTed data in the designated URL, replacing the existing data
   - DELETE: Delete the identified URL resource (limited implementation)
   - LINK: Link an existing object with another object (not implemented)
   - UNLINK: Remove the inserted link information (not implemented)
3. Respond. The server sends a message consisting of response headers which describe the status of the transaction, the type of data being sent, and the data itself.
4. Close the connection. Once the requested document has been transmitted to the client, the connection is closed. The server does not keep any information on that transaction.

Most TCP/IP-based systems can let the browser open more than one connection at a time, transferring multiple items of text and graphics simultaneously.

Code	Meaning	Code	Meaning
200	Request successful	401	Missing authorization field information
201	POST or PUT successful	402	No valid Chargeto field
202	Accepted for processing; results unknown	403	Forbidden resource requested
203	GET or HEAD request OK; partial info returned	404	Cannot find URL
204	Request fulfilled, no new info returned	405	Disallowed access method
		406	Resource found, client cannot accept type
300	Multiple resource sites; select preferred site	410	Resource no longer available
301	Permanent move; see new URL (provided)		
302	Data found at another site	500	Internal server error
304	Filter-requested resource does not pass filter	501	Server does not support method
		502	Invalid response from secondary server
400	Syntax error	503	Server busy
		504	Timeout from secondary server

Figure 19.4 HTTP status codes.

Figure 19.4 identifies the numbers and describes the meanings of the different HTTP status codes that show up in the response stage of an HTTP transaction. As we saw with FTP and SMTP, the lower the numbers, the more positive the response. The 200 through 299 codes identify successful transactions. The codes in the 300s redirect transaction. The codes between 400 and 599 report errors.

There are many types of data that can be used in a Web document. Web browsers and servers negotiate data types using MIME header information (see Chap. 15). Using this negotiation, the browser can activate additional "helper" applications to display nonnative data formats (e.g., MPEG videos, TIFF graphics) to the user.

One of the last remaining roadblocks to full commercialization of the Internet is the ability to process financial transactions. Work is being done to enhance the security of Web services with the development of Secure HTTP. With Secure HTTP, a user could affix digital signatures that are binding and auditable and could transmit sensitive information, such as credit card numbers, that would be encrypted to prevent others from misusing the information.

Java

If you have been on vacation and far from the Internet for the last year, you may not have heard of Java. It has been hyped as the tool that will change the Internet and the greatest thing since the World Wide Web.

Saying Java is a computer language is like saying Porsche is a car. While true, there is a lot more to it than that. Java is an object-oriented language patterned on C++. The difference is that it is a more simple language than C++ and much safer. The safety comes from the strict enforcement (and automation) of features such as memory allocation and variable typing.

Java is designed for writing smaller applications or "applets" that can run on any computer processor. It has special features that make this work in a way that is safe for the processor and the system running the applet. This works well because the three Java components work so well together. In short,

- The Java language keeps the process simple and safe.
- The Java compiler produces processor-independent byte code. While doing that, it encrypts a checksum into the file to prevent tampering with the code.
- The Java runtime system is the part that runs the program on the client system. It also has checks built into it to prevent tampering and to protect the resources on the client system.

We are now on the edge of the possibilities of what Java can do. As you browse the Web, you will see animated objects done with Java as well as user-support processes. As corny as it may sound, the sky is the limit, and even that may be a vast understatement.

Common Gateway Interface

Web browsers can directly access many types of information services, but not every type. What if you owned a bookstore and wanted your customers to be able to check to see if the book they wanted was currently in stock? Are you going to rewrite your Web page every time a book is received or sold? Wouldn't it be better to have your Web server access your inventory database for the information?

The Common Gateway Interface (CGI) provides a mechanism where information can be passed from the Web server, possibly from user input in a form, and passed to an external application. That application reads the information passed to it, acts on that information, and passes the results, in HTML format, back to the Web server. The Web server, in turn, transmits the HTML document to the client that requested the information.

Applications of the gateway interface are limited only by your imagination. Automakers provide a virtual showroom that allows you to configure and price your own car. Paging companies allow you to send e-mail to a customer's alphanumeric pager. One author even allows you to talk to his cat using a voice synthesis unit connected to his Web server.

Suggested Reading List

The 29 FYIs from ds.internic.net:

Building Internet Firewalls by D. Brent Chapman and Elizabeth D. Zwicky
Computer Crime by David Icove, Karl Seger, and William VonStorch
Computer Networks and Internets by Douglas E. Comer
Computer Security Basics by Deborah Russell and G. T. Gangemi, Sr.
DNS and BIND by Paul Albitz and Cricket Liu
Doing Business on the Internet by Mary J. Cronin
Emerging Communications Technologies by Uyless Black
Firewalls and Internet Security by Wm. R. Cheswick and Steven M. Bellovin
Hands-on Internet by David Sachs and Henry Stair
Interconnections, Bridges and Routers by Radia Perlman
Internet for Dummies by John R. Levine and Carol Baroudi
Internetworking by Mark Miller
Internetworking with TCP/IP (Vol. 1) by Douglas E. Comer (3/e)
IPNG and the TCP/IP Protocols by Stephen A. Thomas
IPNG Internet Protocol Next Generation, ed. by Scott Bradner and Allison Mankin
Managing Internetworks with SNMP by Mark Miller
Netiquette by Virginia Shea
Network Administration by Craig Hunt
Network Management Standards by Uyless Black
Network Management: A Practical Prospective by Allan Leinwand and Karen Fang Conroy (2/e)
Networking Personal Computers with TCP/IP by Craig Hunt
Practical Unix and Internet Security by Simson Garfinkel and Gene Spafford
Routing in the Internet by Christian Huitema
sendmail by Bryan Costales, with Eric Allman and Neil Rickert
SNMP, SNMPv2 and RMON by William Stallings (2/e)
TCP/IP and Related Protocols by Uyless Black
TCP/IP Architecture, Protocols, and Implementation by Dr. Sidnie Feit (2/e)
TCP/IP Clearly Explained by Pete Loshin (2/e)
TCP/IP Illustrated (Vol. 1) by W. Richard Stevens
TCP/IP Network Administration by Craig Hunt
TCP/IP: Running a Successful Network by Kevin Washburn and Jim Evans
The Internet Companion by Tracy LaQuey with Jeanne C. Ryer
The Internet Complete Reference by Harley Hahn and Rick Stout
The Simple Book by Marshall Rose (2/e)
The Whole Internet by Ed Krol (2/e)
Total SNMP by Sean Harnedy

About the CD

The enclosed CD was created by Computer and Communication Services, Inc. of Rockledge, Fla., which can be contacted at (407) 639-8100. The LANWatch demonstration software on the CD is included through the kind cooperation of FTP Software, Inc. of North Andover, Mass., reachable at (508) 685-4000. The shareware subnet program is provided by The Empowerment Group of Hickory, N.C.; its telephone number is (704) 431-4400.

The CD contains four directories: FYI, RFC, LWDEMO, and WIN95. The FYI directory contains the 29 FYIs referenced in Chap. 1.

The RFC directory contains all the RFCs that were available at DS.InterNIC.Net on the day the CD was created.

The LWDEMO directory contains FTP Software's LANWatch demo files and some that were added to give you more hands-on experience with protocol analysis and TCP/IP. The best place to start is by reading the demo.txt file. It gives you some good pointers and steps you through some standard protocol analysis functions.

Since LANWatch is an MS-DOS application, you will need to go to Windows 95's MS-DOS prompt. Once there, type DEMO-LW to load the LANWatch demo. You can load the selected dump file (named with a .dmp extension) by typing the letter i and then specifying the dump file you want, such as TELNET.DMP, PING.DMP, or PICT.DMP.

The additional LAB(1 through 9).bat files will automatically load the demo and the correct file to give you nine different traffic captures for review with LANWatch. That way, from a live production network, you can see (in both hex and decoded) the Ethernet header, an ARP message, an IP header, a TCP header, a UDP header, and full application sessions such as FTP.

In each of the capture files, select the desired packet by moving the cursor up or down and then pressing Enter twice to display the hex, ASCII, and decoded versions in a single window. Please note that the LANWatch decoding does not match the header fields in exact sequence.

You will probably find the LWDEMO directory most useful as reinforcement for the text and as a double check on your understanding of the fields in the header layouts.

The WIN95 directory contains a single SUBNET directory that includes all you need to let your system calculate subnet values for you. Start the program by running SETUP. As I am sure you will find it hugely helpful, please remember to contact The Empowerment Group for the current shareware fee and the latest version of the software.

Index

Abstract Syntax Notation One (ASN.1), 303–304
Acceptable Use Policy (AUP), 8
Address Lifetime Expectations (ALE) group, 127
Address Resolution Protocol (ARP), 19, 79–91, 146
  cache, 80–81, 87–91
  exchange, 88–90
  hardware length, 84
  hardware type, 83
  operation, 85
  parameters, 83
  protocol length, 85
  reply, 81, 86, 89–90
  request, 81–82, 85–88
  source hardware address, 85
  source protocol address, 86–87
  target hardware address, 87
  target protocol address, 87
Advanced Networks and Services (ANS), 9
Advanced Research Projects Agency network (ARPAnet), 3, 6
ALE (*see* Address Lifetime Expectations group)
Alias, 73
American Standard Code for Information Interchange (ASCII), 253–254, 257, 277, 295, 304, 358
Anonymous FTP, 13, 307
Anycast address, 142
API (*see* Application Program Interface)
AppleTalk, 302
Application message, 22
Application Program Interface (API), 152
Archie, 8, 356–357
ARP (*see* Address Resolution Protocol)
ASCII (*see* American Standard Code for Information Interchange)
Assigned Numbers RFC, 13, 208, 275
Asymmetrical Digital Subscriber Line (ADSL), 355
Autonomous system, 177–178, 188
Bastion host, 243
BGP (*see* Border Gateway Protocol)
Big endian, 304
Binary, 295
Boggs, David, 1
Bolt, Beranek and Newman (BBN), 7
Boot Protocol (BootP), 20–21, 154–155, 279, 282–284
  parameters, 289–290
  request, 282–283, 291
  response, 283–284, 292
Border Gateway Protocol (BGP), 179–180, 188–192
  attribute, 190
  message, 191–192
  routing information base, 190–191
Bridge, 173–175
Broadcast, 81

Carrier Sense, Multiple Access with Collision Detection (CSMA/CD), 29
CATNIP (*see* Common Architecture for the Internet Protocol)
Central processing unit (CPU), 203, 259
Cerf, Vinton, 6, 9
CERT (*see* Computer Emergency Response Team)
CGI (*see* Common Gateway Interface)
Classless Interdomain Routing (CIDR), 127, 132, 189–190
Client, 24–25
CLNP (*see* Connectionless Network Protocol)
CMIP (*see* Common Management Information Protocol)
CMOT (*see* Common Management Information Protocol over TCP/IP)
COAST (*see* Computer Operations, Audit and Security Technology)
Commercial Internet Exchange (CIX), 8
Common Architecture for the Internet Protocol (CATNIP), 128–131
Common Gateway Interface (CGI), 362

Common Management Information Protocol (CMIP), 301, 305
Common Management Information Protocol over TCP/IP (CMOT), 301, 305
Community access television (CATV), 356
Computer Emergency Response Team (CERT), 8, 244
Computer Operations, Audit and Security Technology (COAST), 245
Congestion, 215–216
Connectionless Network Protocol (CLNP), 128, 184
Convergence, slow vs. fast, 180–181

Data Encryption Standard (DES), 241
Datagram Delivery Protocol (DDP), 302
Datagram storms, 180
DDP (*see* Datagram Delivery Protocol)
Decimal, 81
Defense Advanced Research Projects Agency (DARPA), 7, 107
Defense Communications Agency (DCA), 7
Defense Data Network (DDN), 7
Defense Information Systems Agency (DISA), 7
Department of Defense (DoD), 7
DES (*see* Data Encryption Standard)
Desktop Management Interface (DMI), 309
Desktop Management Task Force (DMTF), 308–309
DHCP (*see* Dynamic Host Configuration Protocol)
Direct routing, 174–176
Diskless workstation, 279
DMI (*see* Desktop Management Interface)
DMTF (*see* Desktop Management Task Force)
Domain Name System (DNS), 20–21, 73–79
  .ac, 77
  .co, 77
  .us, 77
  case, 74
  fully qualified domain name (FQDN), 74
  label, 74–77
  resource records, 78–79, 98–102
  server, 74, 77–79
  top level domain (TLD), 75–76
  tree, 74–77
  zone, 74–75, 77–79
Dotted decimal notation, 86
DS-1, 8
DS.InterNIC.Net, 76–77
Dynamic Host Configuration Protocol (DHCP), 142, 279, 284–292
  address acquisition, 285
  address release, 285–286
  address renewal, 286
  bound state, 285
  discover message, 285
  initialize state, 285
  message format, 287
  NACK message, 286
  options, 287–288
  PACK message, 285
  parameters, 289–290
  REBIND state, 286–287
  RENEW state, 286
  REQUEST message, 285
  select state, 285

EBCDIC (*see* Extended Binary Coded Decimal Interchange Code)
EGP (*see* Exterior Gateway Protocol)
Ether type, 84, 280
Ethernet, 1–2, 85, 87, 109–110, 173–174, 182, 302
  broadcast address, 27–29
  cyclic redundancy check, 31
  DEC Intel Xerox (DIX), 27
  driver, 31, 87
  frame, 29
  giant packet, 31
  header, 29–31, 82, 84, 106
  multicast address, 27–29
  padding, 87–88
  protocol type, 29–31, 47–55
  runt packet, 31
  unicast address, 27–29
  vendor address component, 28, 35–46, 85–86
Ethernet II address, 27
European Laboratory for Particle Physics Research in Switzerland (CERN), 9, 358
Extended Binary Coded Decimal Interchange Code (EBCDIC), 304
Exterior gateway, 179
Exterior Gateway Protocol (EGP), 7, 179–180, 188–192, 303

Fiber Distributed Data Interface (FDDI), 110

File Transfer Protocol (FTP), 6, 20–21, 247–266
commands, 248–267
access control commands, 261–262
CWD, 261
DELE, 265
MKD, 265
NLST, 253–254, 257, 266
PASS, 251–252, 261
PORT, 252–253, 256, 262
PWD, 252, 265
QUIT, 262
RETR, 256–257, 264
RMD, 265
service commands, 263–267
STOR, 264
transfer parameter commands, 262 –263
TYPE, 263
USER, 251, 261
control session, 249–250, 254–255
control session end, 260
data session, 254, 256
data session end, 255–256, 258–259
file request, 256–257
file transfer, 255–256
line mode, 250
login, 250–251
multiple packets, 257–258
multiple sessions, 247
network virtual terminal (NVT), 248
password, 251–252
response codes, 248–249
sample session, 249–260
service ready, 250
session status, 257
TCP, 247, 249–251, 254, 259–260
Firewall, 242–243
FIRST (*see* Forum of Incident Response and Security Teams)
Flame, 9
For Your Information (FYI) RFCs, 15–17
Forum of Incident Response and Security Teams (FIRST), 243–244
FQDN (*see* Fully qualified domain name)
Fragmentation, 22
Frequently asked questions (FAQs), 15–17
FTP (*see* File Transfer Protocol)
Fully qualified domain name (FQDN), 74, 359

Gateway, 174–185, 200–201
Gateway-to-Gateway Protocol (GGP), 178–179
Global address bit, 28, 33
Gopher, 8, 357
Government Open Systems Interconnect Profile (GOSIP), 8, 9
Graphical user interface (GUI), 308

Hardware address, 87–90
Header, 22
Hexadecimal, 27, 81
Host-to-host layer, 19, 20
Hosts file, 73–74
HTML (*see* Hypertext Markup Language)
HTTP (*see* Hypertext Transfer Protocol)
Hypertext Markup Language (HTML), 358, 362
Hypertext Transfer Protocol (HTTP), 20–21, 360–361

IAB (*see* Internet Architecture Board)
IANA (*see* Internet Assigned Numbers Authority)
ICMP (*see* Internet Control Message Protocol)
IDEA (*see* International Data Encryption Algorithm)
IEEE (*see* Institute of Electrical and Electronic Engineers)
IETF (*see* Internet Engineering Task Force)
IGRP (*see* Interior Gateway Routing Protocol)
Indirect routing, 174–176
Institute of Electrical and Electronic Engineers (IEEE), 28–34, 109, 173
802.3, 29–32, 109, 173
802.5, 32–34
Integrated Services Digital Network (ISDN), 355–356
Interface, 73
Interior gateway protocols, 179–188
Interior Gateway Routing Protocol (IGRP), 179–180
International Data Encryption Algorithm (IDEA), 241
International Standards Organization (ISO), 3, 76–77, 92–98, 304
Internet, 7, 9–10, 355
access methods, 355
Internet Architecture Board (IAB), 7–9, 78
Internet Assigned Numbers Authority (IANA), 30, 132, 151, 307
Internet Configuration Control Board (ICCB), 7

Internet Control Message Protocol (ICMP), 19, 109, 146, 193–205
code, 194–195
destination unreachable, 111–112, 197–198
destination unreachable codes, 198
diagnostic, 193–194
echo request / response, 195–197
examples, 204–205
failed IP data, 197–202
failed IP header, 197–202
ID number, 195–196
parameter problem, 109, 202
pointer, 202
redirect, 199–200
sequence number, 195–196
source quench, 198–199
subnet mask request / response, 204
time exceeded, 201
timestamp request / response, 202–204
type, 194–195
variation reporting, 193–194
Internet Engineering Task Force (IETF), 8, 10, 128, 305
Internet Explorer, 358–359
Internet layer, 22–23
Internet Packet Exchange (IPX), 302
Internet Protocol (IP):
connectionless, 105
datagram, 22
diagnostics, 105
fault tolerance, 105
fragmentation, 105, 110–113
functions, 105–106
minimum acceptable data size, 113
next generation (IPng), 106, 127–148, 189
proposal comparisons, 131
quality of service, 105, 107–108
routing, 106–107, 112, 120, 122–124
self-healing, 105
Version 4, 105–125, 127, 132, 135–136, 138, 140–141
header, 105–125
checksum, 115
copy-through-gate, 121
datagram ID number, 109–110
fragment area, 110–112
length, 107
options, 120–125
precedence, 107–108
protocol field, 115–119
route-based options, 122–124
samples, 125
source IP address, 115, 120
target IP address, 120
Internet Protocol, Version 4, header (*Cont.*):
time-to-live (TTL), 113–114
timestamp options, 124–125
total IP length, 109
type-of-service (TOS), 107–109
version, 106
Version 6, 128, 132–148, 184
address:
anycast, 142
format, 142
IPv4-compatible, 143–144
IPv4-mapped, 143–144
loopback, 143
multicast, 142
prefixes, 143
provider-based unicast, 144–145
registry, 144
subnet-router, 143
unicast, 142
unspecified, 143
ether type, 133
extension headers, 137–141
features, 132–133
functions, 145–146
fragment, 140–141
header, 133–141
flow label, 134
hop limit, 135
next header, 135
payload length, 134
priority, 133–134
source address, 135
target address, 135
version, 133
options:
hop-by-hop, 137–138
jumbo payload, 138
routing, 138–140
transition plan, 146–147
version comparison, 135–136
versions, 128–129
Internet Research Task Force (IRTF), 8, 10
Internet service provider (ISP), 8, 9, 60, 198
Internet services, 355–362
Internet Society, 9
Internetworking Working Group (INWG), 6
InterNIC, 11, 127
IP (*see* Internet Protocol)
IP address, 57–71, 73–75, 83, 87–90
broadcast, 65
Class A, 57–59, 65–66
Class B, 57, 59–61, 65–70
Class C, 57, 59–60, 62, 64–65, 67–69
Class D, 57, 60

IP address (*Cont.*):
first byte value, 58–60
host, 58, 61, 63–68
interface, 58
locally administered bits, 58–59
loop-back, 58
multicast, 60
network, 57–61, 63–68
subnet, 58, 61, 63–68
IP communications logic process, 60–71
network, 60–61, 68–70
physical address, 70–71, 79–80
subnet, 68–71
IPng (*see* Internet Protocol, next generation)
IPX (*see* Internet Packet Exchange)
ISP (*see* Internet service provider)

Java, 361–362

Kahn, Bob, 6
Kerberos, 241

Little endian, 304
Lock-step acknowledgement, 295
Logical link control (LLC), 32
Lynx, 358–359

MAC (*see* Media access control; Message authentication code)
Mail, 295
Management Information Base:
version 1 (MIB I), 305
version 2 (MIB II), 303
Maximum transmission unit (MTU), 25, 30–31, 33, 113, 212, 356
Media access control (MAC), 28, 33, 71, 79, 83, 141, 145–146, 173
Merit Network, Inc., 8
Message authentication code (MAC), 240
Metcalfe, Bob, 1, 6
MIB I (*see* Management Information Base version 1)
MIB II (*see* Management Information Base version 2)
MILnet, 7
MIME (*see* Multipurpose Internet Mail Extension)
Mosaic, 9, 358–359
MTU (*see* Maximum transmission unit)
Multicast bit, 27–28, 33
Multipurpose Internet Mail Extension (MIME), 274–277
encoding, 277
headers, 275–277

NAL (*see* Network access layer)
Name resolver, 74
National Center for Supercomputing Applications (NCSA), 9, 358
National Computer Security Center, 239
National Institute for Standards and Technology (NIST), 9
National Research Education Network (NREN), 8
National Science Foundation (NSF), 8
NCSA (*see* National Center for Supercomputing Applications)
Netascii, 295
Netscape Navigator, 358–359
Netware, 302
Network access layer (NAL), 23, 110, 136, 142, 174, 207
Network Information Center (NIC), 127
Network interface card, 28–29, 279–280
Network interface layer (NIL), 23–24, 142, 174, 207
(*See also* Network access layer)
Network management, 16
Network management station (NMS), 86
Network security, 235–245
Network talk (ntalk), 154–155
NIC (*see* Network Information Center)
NIL (*see* Network interface layer)
NMS (*see* Network management station)
NSCS (*see* National Computer Security Center)

Object identifier (OID), 303–304
Octet, 27
OID (*see* Object identifier)
Open architecture, 2
Open Shortest Path First (OSPF), 179–180, 185–188
alternate paths, 186
area, 185
area border router, 185
authentication, 186
boundary router, 185
example, 187–188
link state, 185–186
multiaccess area, 185
multicast, 185–186
variable subnet masking, 186–188

Open Systems Interconnect (OSI), 3–6, 8, 9, 19
  Application Layer, 6, 19
  Data Link Layer, 4–5, 19
  Network Layer, 5, 19
  Physical Layer, 4, 19
  Presentation Layer, 6, 19
  Session Layer, 6, 19
  Transport Layer, 5–6, 19
OSPF (*see* Open Shortest Path First)

Packet, 29–31, 33
Path MTU discovery, 113
PGP (*see* Pretty Good Privacy)
Physical address, 79
Ping, 195–197
Point-to-Point Protocol (PPP), 35, 356
Poison reverse, 181
Port, 151–171
  random, 151–152
  registered, 151–152, 164–171
  well-known, 151–163
Pretty Good Privacy (PGP), 8, 241
Process layer, 19, 20
Protocol analyzer, 27, 81–82
Protocol suite, 1–6
Proxy server, 242–243

Random port, 208, 293–294
RARP (*see* Reverse Address Resolution Protocol)
Request for comments (RFC), 6, 11–17
Retransmit, 215–216, 296
Reverse Address Resolution Protocol (RARP), 32, 279–282
  failure, 281–282
  reply, 281
  request, 280
  server, 279–282
RFC 760, 85
RFC 791, 85
RFC 951, 290
RFC 952, 74
RFC 959, 248, 260–266
RFC 1028, 301
RFC 1035, 74
RFC 1065, 301
RFC 1066, 301
RFC 1098, 301
RFC 1122, 242
RFC 1123, 74
RFC 1155, 301
RFC 1156, 301
RFC 1157, 301, 308
RFC 1158, 301
RFC 1191, 113
RFC 1213, 301
RFC 1214, 301
RFC 1323, 109, 212
RFC 1341, 274
RFC 1349, 109
RFC 1441, 308
RFC 1452, 308
RFC 1521, 275
RFC 1522, 275
RFC 1531, 290
RFC 1533, 290
RFC 1543, 11
RFC 1752, 131
RFC 1901, 308
RFC 1905, 308
RFC 1908, 308
RFC 1920, 179, 305
RFC 1933, 146–147
RFC 1945, 360
RIP (*see* Routing Information Protocol)
Rivest, Shamir, Adleman (RSA) encryption, 241
ROAD (*see* Routing and addressing group)
Router, 61–62, 67, 70, 107–108, 113–114, 134, 173–181, 185–187, 199–201
Routing:
  automatic, 178–192
  manual, 178
Routing and addressing group (ROAD), 127
Routing Information Protocol (RIP), 20–21, 154–155, 179–185
  active mode, 180
  command, 182–183
  cost, 184
  distance, 184
  family, 183
  group address, 183–184
  hop count 180, 184
  message, 181–185
  metric, 184
  passive mode, 180
  route loops, 180–181
  routed, 180
  vector distance routing, 180
  version 1, 181–184
  version 2, 184–185
Routing table, 176–178, 189–190
  default, 178
  extended, 177–178
  fall through, 177
  local, 176–177

RS.InterNIC.Net, 57–61, 189
RSA (*see* Rivest, Shamir, Adleman encryption)

Security Administrator for Analyzing Networks (SATAN), 245
Security procedures, 237–245
  access control, 239
  authentication, 239
  data integrity, 240
  digital signature standard (DSS), 240
  encryption, 241
  network scanning, 245
  passwords, 239
  routing control, 241–242
  security servers, 242–243
  smart cards, 239
  traffic padding, 242
Security threats, 235–237
  authenticity attacks, 235–237
  availability, 235–236
  bacteria, 237
  confidentiality, 235–236
  crackers, 235
  dumpster diving, 236
  hackers, 235–236, 238, 242, 245
  human nature, 236
  impersonation, 237
  integrity risks, 235–236
  leakage, 237
  logic bomb, 237
  modeling, 237
  protocol analyzer, 236
  salami slicing, 237
  scavenging, 237
  simulation, 237
  trap door, 237
  trojan horse program, 236
  virus, 237
  worm, 237
Sequencing, 4
Serial Line Interface Protocol (SLIP), 34–35, 356
Server, 24–25
SGMP (*see* Simple Gateway Management Protocol)
Shell account, 355–356
Silly window syndrome, 213
Simple Gateway Management Protocol (SGMP), 301
Simple Internet Protocol Plus (SIPP), 128–132
Simple Mail Transfer Protocol (SMTP), 20–21, 24, 267–277
  @, 267
  commands, 268–269
  DATA, 269, 272–273
  e-mail addresses, 267–268
  HELO, 269, 271
  mail data, 272
  MAIL FROM, 269, 272
  mail origin, 271–272
  mail recipient, 272
  mail server, 268
  message end, 273
  QUIT, 269, 274
  RCPT TO, 269, 272
  recommended commands, 269
  required commands, 269
  response codes, 269–270
  sample session, 270–274
  session end, 274
  session start, 270–271
Simple Network Management Protocol (SNMP), 20–21, 154–155, 301–354
  agent, 302–303
  groups, 305–306
  management information base (MIB), 302–307
  model, 301–302
  network management application (NMA), 307
  network management station (NMS), 302–303, 305
  private enterprise numbers, 310–353
  private MIBs, 307
  protocol data unit (PDU), 308
  version 1 (SNMPv1), 308
  version 2 (SNMPv2), 308
SIPP (*see* Simple Internet Protocol Plus)
SMI (*see* Structure of management information)
SNMP (*see* Simple Network Management Protocol)
Socket, 152–153
Sockets Secure (SOCKS) server, 242–243
Split horizon, 181
Standard, 9–15
Stanford Research International (SRI), 6, 11
Structure of management information (SMI), 302–304
Subnet, 57–70, 174, 177
  analogy, 61
  calculations, 63–65
  decision events, 62–63
  interfaces lost, 68–70

Subnet (*Cont.*):
mask, 63–70
mask binary, 63–70
mask decimal, 63–70
plan, 67–68
reason, 61
Subnetwork Access Protocol (SNAP), 32
Supernet mask, 189
Supernetting, 189–190
Sustained traffic peaks, 173
Systems Network Architecture (SNA), 2–3

TCP (*see* Transmission Control Protocol)
TCP and UDP over Big Addresses (TUBA), 128–132
Telnet, 6, 20–21, 219–233
3270, 232–233
binary, 233
client, 219
echo, 226
Interpret as Command (IAC), 220
keyboard buffer, 227
login, 226–228
multibyte subnegotiation field, 224–225
network virtual terminal (NVT), 219–220
option negotiations, 220–221
options, 221
password, 228–229
sample session, 222–232
security, 227, 230
server, 219
session end, 231
suppress go ahead, 222–224
terminal type, 223–225
TFTP (*see* Trivial File Transfer Protocol)
TLD (*see* Domain Name System, top-level domain)
Token ring, 23, 32–34
access control byte, 32
bit order, 33
frame check sequence (FCS), 33
frame control byte, 32–33
frame status byte, 33
TOS (*see* Internet Protocol, Version 4, header, type-of-service)
TP/IX (*see* Common Architecture for the Internet Protocol)
Transmission Control Protocol (TCP), 20–21, 149, 188, 207–218, 270, 274
acknowledgment sequence number, 209
buffer, 215
checksum, 211
final data sent, 216

Transmission Control Protocol (*Cont*):
handshake, 208, 214–215
step 1, 214
step 2, 214–215
step 3, 215
header, 207–217
header length, 209–210
maximum segment size, 212–213
maximum segment size default, 212
options, 211–212
pseudo-header, 211
reliable transport, 207
reset session, 216–217
sample session, 217–218
segment, 21–22
sender window size, 210–211
session flags, 210
source port, 207–208
source sequence number, 208–209
target port, 207–208
urgent data size, 211
virtual connection, 214
Trivial File Transfer Protocol (TFTP), 20–21, 153–155, 293–299
acknowledgment, 295
challenges, 296–297
data, 295
data block, 295, 298–299
errors, 296
operation codes, 294
ports, 293–294
read request, 294–295, 297–298
request layouts, 294–295
sample sessions, 297–299
write request, 294–295, 298–299
TTL (*see* Internet Protocol, Version 4, header, time-to-live)
TUBA (*see* TCP and UDP over Big Addresses)

UDP (*see* User Datagram Protocol)
unicast, 89
Uniform Resource Locator (URL), 241, 243–245, 290, 309, 359–360
Unix, 7
Unix-to-Unix Copy Protocol (UUCP), 7
URL (*see* Uniform Resource Locator)
Usenet, 7
User Datagram Protocol (UDP), 21, 149–155, 182, 197–198, 293–294, 303
applications, 153–154
checksum, 150–151
datagram, 21, 149–150

User Datagram Protocol (*Cont.*):
header, 150–151
retransmission, 149
sample exchanges, 153–154

Vector distance, 180, 188
Very-high-speed Backbone Network Service (vBNS), 9
Virtual IP networking, 288

Well-known port, 208
Wide Area Information Systems (WAIS), 8
Window size, 251, 254–255, 258
Workgroups, 62
World Wide Web (WWW), 9, 357–359

Xerox Network Systems (XNS), 1–2

ABOUT THE AUTHOR

Paul Simoneau is president of NeuroLink, Ltd., an international consulting, research, and education company that services such clients as Cisco, AT&T, Chase Manhattan, US West, the U.S. Armed Forces, Hewlett-Packard, and Sprint. He is also a senior instructor and course director for American Research Group, the systems and communications training company that pioneered the course on which this book/CD-ROM package is based, and a principal of Libra Square Ltd., an Internet marketing company. A graduate of the State University of New York at Albany and Webster University, Mr. Simoneau is based in Walkersville, Maryland.

SOFTWARE AND INFORMATION LICENSE

The software and information on this diskette (collectively referred to as the "Product") are the property of The McGraw-Hill Companies, Inc. ("McGraw-Hill") and are protected by both United States copyright law and international copyright treaty provision. You must treat this Product just like a book, except that you may copy it into a computer to be used and you may make archival copies of the Products for the sole purpose of backing up our software and protecting your investment from loss.

By saying "just like a book," McGraw-Hill means, for example, that the Product may be used by any number of people and may be freely moved from one computer location to another, so long as there is no possibility of the Product (or any part of the Product) being used at one location or on one computer while it is being used at another. Just as a book cannot be read by two different people in two different places at the same time, neither can the Product be used by two different people in two different places at the same time (unless, of course, McGraw-Hill's rights are being violated).

McGraw-Hill reserves the right to alter or modify the contents of the Product at any time.

This agreement is effective until terminated. The Agreement will terminate automatically without notice if you fail to comply with any provisions of this Agreement. In the event of termination by reason of your breach, you will destroy or erase all copies of the Product installed on any computer system or made for backup purposes and shall expunge the Product from your data storage facilities.

LIMITED WARRANTY

McGraw-Hill warrants the physical diskette(s) enclosed herein to be free of defects in materials and workmanship for a period of sixty days from the purchase date. If McGraw-Hill receives written notification within the warranty period of defects in materials or workmanship, and such notification is determined by McGraw-Hill to be correct, McGraw-Hill will replace the defective diskette(s). Send request to:

Customer Service
McGraw-Hill
Gahanna Industrial Park
860 Taylor Station Road
Blacklick, OH 43004-9615

The entire and exclusive liability and remedy for breach of this Limited Warranty shall be limited to replacement of defective diskette(s) and shall not include or extend to any claim for or right to cover any other damages, including but not limited to, loss of profit, data, or use of the software, or special, incidental, or consequential damages or other similar claims, even if McGraw-Hill has been specifically advised as to the possibility of such damages. In no event will McGraw-Hill's liability for any damages to you or any other person ever exceed the lower of suggested list price or actual price paid for the license to use the Product, regardless of any form of the claim.

THE McGRAW-HILL COMPANIES, INC. SPECIFICALLY DISCLAIMS ALL OTHER WARRANTIES, EXPRESS OR IMPLIED, INCLUDING BUT NOT LIMITED TO, ANY IMPLIED WARRANTY OF MERCHANTABILITY OR FITNESS FOR A PARTICULAR PURPOSE. Specifically, McGraw-Hill makes no representation or warranty that the Product is fit for any particular purpose and any implied warranty of merchantability is limited to the sixty day duration of the Limited Warranty covering the physical diskette(s) only (and not the software or in-formation) and is otherwise expressly and specifically disclaimed.

This Limited Warranty gives you specific legal rights; you may have others which may vary from state to state. Some states do not allow the exclusion of incidental or consequential damages, or the limitation on how long an implied warranty lasts, so some of the above may not apply to you.

This Agreement constitutes the entire agreement between the parties relating to use of the Product. The terms of any purchase order shall have no effect on the terms of this Agreement. Failure of McGraw-Hill to insist at any time on strict compliance with this Agreement shall not constitute a waiver of any rights under this Agreement. This Agreement shall be construed and governed in accordance with the laws of New York. If any provision of this Agreement is held to be contrary to law, that provision will be enforced to the maximum extent permissible and the remaining provisions will remain in force and effect.